Perspectives in Neural C

Springer
*London
Berlin
Heidelberg
New York
Barcelona
Hong Kong
Milan
Paris
Singapore
Tokyo*

Also in this series:

Dimitris C. Dracopoulos
Evolutionary Learning Algorithms for Neural Adaptive Control
3-540-76161-6

Gustavo Deco and Dragan Obradovic
An Information-Theoretic Approach to Neural Computing
0-387-94666-7

Thomas Lindblad and Jason M. Kinser
Image Processing using Pulse-Coupled Neural Networks
3-540-76264-7

Maria Marinaro and Roberto Tagliaferri (Eds)
Neural Nets - WIRN VIETRI-98
1-85233-051-1

Amanda J.C. Sharkey (Ed.)
Combining Artificial Neural Nets
1-85233-004-X

Dirk Husmeier
Neural Networks for Conditional Probability Estimation
1-85233-095-3

Achilleas Zapranis and Apostolos-Paul Refenes
Principles of Neural Model Identification, Selection and Adequacy
1-85233-139-9

Mark Girolami
Self-Organising Neural Networks
1-85233-066-X

Walter J. Freeman
Neurodynamics: An Exploration in Mesoscopic Brain Dynamics
1-85233-616-1

Paulo J.G.Lisboa, Emmanuel C. Ifeachor and Piotr S. Szczepaniak (Eds)
Artificial Neural Networks in Biomedicine
1-85233-005-8

H. Malmgren, M Borga and L. Niklasson (Eds)
Artificial Neural Networks in Medicine and Biology
1-85233-289-1

Mark Girolami (Ed.)
Advances in Independent Component Analysis
1-85233-263-8

Robert M. French and Jacques P. Sougné (Eds)

Connectionist Models of Learning, Development and Evolution

Proceedings of the Sixth Neural Computation and Psychology Workshop, Liège, Belgium, 16-18 September 2000

 Springer

Robert M. French, PhD
Jacques P. Sougné, PhD
Quantitative Psychology and Cognitive Science, Department of Psychology (B32),
University of Liège, 4000 Liège, Belgium

Series Editor
J.G. Taylor, BA, BSc, MA, PhD, FInstP
Centre for Neural Networks, Department of Mathematics, King's College,
Strand, London WC2R 2LS, UK

ISSN 1431-6854
ISBN 1-85233-354-5 Springer-Verlag London Berlin Heidelberg

British Library Cataloguing in Publication Data
Connectionist models of learning, development and evolution
 : proceedings of the Sixth Neural Computation and
Psychology Workshop, Liege, Belgium, 16-18 September 2000.
 - (Perspectives in neural computing)
1.Neural networks (Neurobiology) 2.Connectionism
I.French, Robert M. (Robert Matthew), 1951- (II.Sougne,
Jacques P. III.Neural Computation and Psychology Workshop
(6th : 2000 : Liege, Belgium)
612.8
ISBN 1852333545

Library of Congress Cataloging-in-Publication Data
Neural Computation and Psychology Workshop (6th : 2000 : Liège, Belgium)
    Connectionist models of learning, development and evolution: proceedings of the Sixth
    Neural Computation and Psychology Workshop, Liège, Belgium, 16-18 September 2000
    /editors, Robert M. French, Jacques P. Sougné
        p. cm.
    Includes bibliographical references and index.
    ISBN 1-85233-354-5 (alk. paper)
        1. Neural networks (Neurobiology)--Congresses. 2. Connectionism--Congresses. 3.
    Cognitive neuroscience--Congresses. I. French, Robert M. II. Sougné, Jacques P. III.
    Title
    QP363.3 .N395 2000
    612.8--dc21   00-069843

Apart from any fair dealing for the purposes of research or private study, or criticism or review, as
permitted under the Copyright, Designs and Patents Act 1988, this publication may only be reproduced,
stored or transmitted, in any form or by any means, with the prior permission in writing of the publishers,
or in the case of reprographic reproduction in accordance with the terms of licences issued by the
Copyright Licensing Agency. Enquiries concerning reproduction outside those terms should be sent to the
publishers.

© Springer-Verlag London Limited 2001
Printed in Great Britain

The use of registered names, trademarks etc. in this publication does not imply, even in the absence of a
specific statement, that such names are exempt from the relevant laws and regulations and therefore free
for general use.

The publisher makes no representation, express or implied, with regard to the accuracy of the information
contained in this book and cannot accept any legal responsibility or liability for any errors or omissions
that may be made.

Typesetting: Camera ready by contributors
Printed and bound at the Athenæum Press Ltd., Gateshead, Tyne and Wear
34/3830-543210 Printed on acid-free paper SPIN 10774570

Preface

This is the Proceedings of the Sixth Neural Computation and Psychology Workshop, a gathering of neural net modelers explicitly concerned with various aspects of cognition. This was the first year that the Workshop was held on the Continent and, in spite of two days of torrential rain, the meeting lived up to our most optimistic expectations. It drew participants from Belgium, Brazil, Britain, Denmark, France, Germany, Netherlands, Hungary, Italy, Switzerland, and the United States. The overarching theme of the Workshop this year was Evolution, Learning and Development, writ large. The single thread that runs through almost all of these papers is connectionist modeling in some form or another. This Proceedings is organized in six main sections which are: the neural basis of cognition, development and category learning, implicit learning, social cognition, evolution and, finally, semantics.

Connectionist modeling continues to occupy an ever more important role in psychology. It provides insights into the underlying mechanisms of cognitive processes as diverse as word and grammar learning, dyslexia, category recognition, and amnesia, face recognition, to name just a very few cognitive areas to which neural network models have been applied. As neural models begin to approach the complexity of networks of real neurons, the explanations they provide and, importantly, the predictions they will make will become better and better. The papers in the present Proceedings represent yet another small step in that direction.

The first section, the Neural Basis of Cognition, begins with a paper by Monaghan and Shillcock in which the authors posit a separation of coarse and fine coding in the two different hemispheres of the brain. They use an RBF model to implement this separation of coarse and fine coding areas. They show that damaging one or the other of these areas can explain certain types of visuospatial neglect. French, Ans, and Rousset discuss "dual-network" models of memory, which simulate the hippocampal-neocortical separation in the brain that allows us to overcome the problem of catastrophic interference that plagues single-network architectures. The authors point out that, while there are many advantages to this type of architecture, there are also a number of shortcomings that must still be overcome. Sougné presents a learning algorithm for a distributed model of "synfire" chains (i.e., chains of neurons that exhibit extraordinary temporal firing accuracy, as precise as 1 ms). De Mazière and Van Hulle present an elegant new method for analyzing neuro-imaging data based on results from chaos theory and compare this method with a standard principal components analysis. Shapiro, Wearden and Barone develop a model of time learning in which the surprising fact of a scalar relation between the time interval learned and the variance in response time is explained using a Poisson-like model. Bryson and Stein discuss modularity in the brain and its use in various approaches to artificial intelligence and the impact this could have on future connectionist models. Diniz-Filho and Ludermir also deal with the issue of modularity and present a neurally motivated, modular network that can learn the relation between conditioned and unconditioned stimuli. Lőrincz, Szatmáry, Szirtes, and Takács use a temporal independent component analysis to differentiate the neural firing histograms associated with new stimuli from previously learned stimuli. Raijmakers and Molenaar show that certain transitions in development correspond to bifurcation points in output activity diagrams obtained by modifying various parameters of an ART network.

The next section involves Development and Category Learning and begins with a paper by Hartley in which she develops a connectionist model of categorization of animals versus plants based on the size of feature-space for the members of each of these two categories. Page develops a connectionist model that focuses on the appropriate dimensions among many irrelevant dimensions required for a particular categorization task. Schlesinger and Parisi model an infant's learning to reach based on coordinating proprioceptive, tactile and visual information. Contrary to our intuition that visual information is the key to this task, they show that, in fact, tactile information is the most important. Westermann develops a neural network algorithm that dynamically adapts its architecture during the learning task. These networks obtain the same U-shaped learning curve that children show when learning the past tense. Cooper and Glasspool present an interactive activation model of learning Gibsonian affordances and action sequences. Bartos and Le Voi present a configural-cue model of category learning which shows similar learning difficulty curves as people when learning certain types of categories. Teuscher and Sanchez show that Turing was one of the first persons to consider building machines from neuron-like elements and discuss how these Turing "connectionist" machines can be used for pattern classification. Done, Gale and Frank discuss a Kohonen-network which is given bitmap images of various objects and is capable of reproducing human-like category-specific deficits based the novel notion of how "crowded" each of the images is.

This is followed by a section on Implicit Learning and kicks off with an article by Tillman, Bharucha and Bigand who use a self-organizing map to simulate implicit learning of hierarchical categorization of tone, chord and key information. Timmermans and Cleeremans present a simple recurrent network (SRN) model of implicit learning of biconditional grammars and show that this task can be learned without requiring explicit rule abstraction. Visser, Raijmakers and Molenaar also consider SRNs and develop a method based on hidden Markov models for analyzing the hidden layers of these networks.

Social Cognition is the theme of the next section. The first article by Labiouse and French develops an auto-encoder model of person perception and stereotype formation. The article by Van Overwalle and Timmermans presents a localist connectionist model of causal reasoning and shows how this model fits human social reasoning data.

Next we move to the theme of Evolution, beginning with an article by Bullinaria concerning a leaky-integrator connectionist model of learning, evolution and maturation in a control-system environment. Central to this model is the so-called "Baldwin effect" whereby individual learning has affects, over many generations, on the evolution of the configuration of the networks. Hüning presents "catalytic networks" which are based on an analogy with the dynamics of chemical molecular evolution. Related to Bullanaria's work on the Baldwin Effect,

Ferdinando, Calbretta and Parisi compare neural networks that evolve by means of a genetic algorithm alone to others that evolve in a similar manner but also learn during their existence. Dickins and Levy discuss the importance of a dynamical systems approach to cognitive modeling.

The final section concerns Semantics. Levy and Bullinaria develop a simple framework based on word frequencies that allows them to reduce the dimensionality of high-dimensional co-occurrence vectors designed to capture various lexical properties of words based on their proximity to other words in large text corpora. Wichert presents an associationist model of problem-solving that involves use of the what-where representations. Postma, Roebroeck and Lacroix compare different types of semantic networks and show that networks with a "small-world" connection scheme tend to out-perform fully-connected, randomly connected and regularly connected networks. The final article by Lowe defends the extremely counter-intuitive result that in a co-occurrence studies like those of Levy and Bullinaria, no more than two dimensions are needed to capture high-dimensional structure.

Finally, the editors would like to extend special thanks to Emmanuelle Rouy, without whose organizing skills this conference simply would not have happened. We would also like to acknowledge our debt to the Belgian National Research Foundation (FNRS) and the University of Liège for having provided funding for this Workshop. We would also like acknowledge the contribution of a IUPA grant P4/19 from the Belgian government and HPRN-CT-1999-00065 from the European Commission.

Robert M. French and Jacques P. Sougné
Liège, November 2000

Contents

SECTION I: Neural Basis of Cognition 1

1. Applying Neuroanatomical Distinctions to Connectionist Cognitive Modelling
Padraic Monaghan & Richard Shillcock 3

2. Pseudopatterns and Dual-Network Memory Models: Advantages and Shortcomings
Robert M. French, Bernard Ans, & Stéphane Rousset 13

3. A Learning Algorithm for Synfire Chains
Jacques Sougné 23

4. Towards a Spatio-Temporal Analysis Tool for fMRI Data: An Application to Depth-from-Motion Processing in Humans
P.A. De Mazière and M.M. Van Hulle 33

5. A Simple Model Exhibiting Scalar Timing
J. L. Shapiro, John Wearden, and Rossano Barone 43

6. Modularity and Specialized Learning in the Organization of Behaviour
Joanna Bryson and Lynn Andrea Stein 53

7. Modeling Modulatory Aspects in Association Processes
Jairo Diniz-Filho & Teresa Bernarda Ludermir 63

8. Recognition of Novelty Made Easy: Constraints of Channel Capacity on Generative Networks
A. L'órincz, B. Szatmáry, G. Szirtes, and B. Takács 73

9. A Biologically Plausible Maturation of an ART Network
Maartje E. J. Raijmakers & Peter C. M. Molenaar 83

SECTION II: Development and Category Learning 93

10. Developing Knowledge about Living Things: A Connectionist Investigation
Samantha J. Hartley 95

11. Paying Attention to Relevant Dimensions: A Localist Approach
Mike Page 105

12. Coordinating Multiple Sensory Modalities While Learning to Reach
Matthew Schlesinger & Domenico Parisi 113

13. Modelling Cognitive Development with Constructivist Neural Networks
Gert Westermann 123

14. Learning Action Affordances and Action Schemas
Richard Cooper & David Glasspool 133

15. A Three-Layer Configural Cue Model of Category Learning Rates
Paul Bartos and Martin Le Voi 143

16. A Revival of Turing's Forgotten Connectionist Ideas: Exploring Unorganized Machines
Christof Teuscher and Eduardo Sanchez 153

17. Visual Crowding and Category-Specific Deficits: A Neural Network Model
John Done, Tim M. Gale, Ray J. Frank 163

SECTION III: Implicit Learning 173

18. Implicit Learning of Regularities in Western Tonal Music by Self-Organization
Barbara Tillmann, Jamshed J. Bharucha & Emmanuel Bigand 175

19. Rules vs. Statistics in Implicit Learning of Biconditional Grammars
Bert Timmermans & Axel Cleeremans 185

20. Hidden Markov Model Interpretations of Neural Networks
Ingmar Visser, Maartje E. J. Raijmakers & Peter C. M. Molenaar 197

SECTION IV: Models of Social Cognition 207

21. A Connectionist Model of Person Perception and Stereotype Formation
Christophe L. Labiouse & Robert M. French 209

22. Learning about an Absent Cause: Discounting and Augmentation of Positively and Independently Related Causes
Frank Van Overwalle & Bert Timmermans 219

SECTION V: Evolution 229

23. Exploring the Baldwin Effect in Evolving Adaptable Control Systems
John A. Bullinaria 231

24. Borrowing Dynamics from Evolution: Association using Catalytic Network Models
Harald Hüning 243

25. Evolving Modular Architectures for Neural Networks
Andrea Di Ferdinando, Raffaele Calabretta & Domenico Parisi 253

26. Evolution, Development and Learning - A Nested Hierarchy?
T. E. Dickins & J. P. Levy 263

SECTION VI: Semantics 271

27. Learning Lexical Properties from Word Usage Patterns: Which Context Words Should be Used?
Joseph P. Levy & John A. Bullinaria 273

28. Associative Computation and Associative Prediction
Andrzej Wichert 283

29. The Development of Small-World Semantic Networks
Eric Postma, Alard Roebroeck, and Joyca Lacroix 293

30. What is the Dimensionality of Human Semantic Space?
Will Lowe 303

Author Index 313

Subject Index 315

List of Contributors

Bernard Ans, Laboratoire de Psychologie Expérimentale, Université de Grenoble, France
E-mail: Bernard.Ans@upmf-grenoble.fr,
http://www.upmf-grenoble.fr/LPE/Personnel/CNRS/Bernard_Ans/bernard_ans.html

Rossano Barone, Department of Psychology, Manchester University, Manchester M13 9PL, U.K.

Paul Bartos, Psychology Discipline, The Open University, Walton Hall, Milton Keynes, MK7 6AA, UK. E-mail: p.d.bartos@open.ac.uk,
http://www.open.ac.uk/socialsciences/staff/pbartos/

Jamshed J. Bharucha, Dartmouth College, Hanover, USA.
E-mail: Jamshed.J.Bharucha@dartmouth.edu,
http://www.dartmouth.edu/artsci/psych/faculty/bharucha.html

Emmanuel Bigand, Université de Bourgogne, Dijon, France.
E-mail: Emmanuel.Bigand@u-bourgogne.fr,
http://www.u-bourgogne.fr/Recherche/90603.html

Joanna Bryson, Artificial Intelligence Laboratory, Massachusetts Institute of Technology, 545, Technology Square, Cambridge MA 02139, USA. E-mail: joanna@ai.mit.edu,
http://www.ai.mit.edu/people/joanna/joanna.html

John A. Bullinaria, Department of Psychology, University of Reading, UK.
E-mail: j.bullinaria@reading.ac.uk,
http://www.rdg.ac.uk/psych/contact/directory/john_bullinaria.htm

Raffaele Calabretta, Institute of Psychology, National Research Council, Rome, Italy.
E-mail: rcalabretta@ip.rm.cnr.it, http://gral.ip.rm.cnr.it/rcalabretta/

Axel Cleeremans, Cognitive Science Research Unit, Université Libre de Bruxelles, Av. FD Roosevelt, 50, B1050 Bruxelles, Belgium. E-mail axcleer@ulb.ac.be,
http://srsc.ulb.ac.be/axcWWW

Richard Cooper, School of Psychology, Birkbeck College, University of London, Malet St., London, WC1E 7HX, UK. E-mail: R.Cooper@psyc.bbk.ac.uk,
http://www.psyc.bbk.ac.uk/staff/rc.html

Patrick A. De Mazière, Laboratory of Neurophysiology, Faculty of Medicine, Katholieke Universiteit Leuven, Herestraat, 49, B-3000 Leuven, Belgium. E-mail: patrick@neuro.kuleuven.ac.be , http://www.kuleuven.ac.be/facdep/medicine/dep_neu/neufys/

Tom. E. Dickins, Department of Psychology, London Guildhall University, London E1 7NT, UK. E-mail: dickins@lgu.ac.uk, http://www.lgu.ac.uk/psychology/dickins/home.htm

Andrea Di Ferdinando, Institute of Psychology, National Research Council, Rome, Italy. E-mail: andread@ip.rm.cnr.it, http://gral.ip.rm.cnr.it/andrea/

Jairo Diniz-Filho, Departamento de Fisiologia e Farmacologia, Universidade Federal do Ceará, Fortaleza CE, Brazil. E-mail: jdf@cin.ufpe.br.

John Done, Department of Psychology, University of Hertfordshire, UK. E-mail: d.j.done@herts.ac.uk.

Ray J. Frank, Department of Computer Science, University of Hertfordshire, UK. E-mail: comqrjf@herts.ac.uk.

Robert M. French, Cognitive Science Unit, Department of Psychology, Université de Liège, Sart Tilman B32, 4000 Liège, Belgium. E-mail: rfrench@ulg.ac.be. http://www.ulg.ac.be/cogsci/rfrench.html

Tim M. Gale, Department of Psychology, University of Hertfordshire, UK, & QEII Hospital Welwyn Garden City, UK. E-mail: T.Gale@herts.ac.uk, http://www.psy.herts.ac.uk/pub/T.Gale/hmpage.html

David Glasspool, Advanced Computation Laboratory, Imperial Cancer Research Fund, Lincoln's Inn Fields, London, WC2A 3PX, UK. E-mail: dg@acl.icnet.uk, http://www.acl.icnet.uk/lab/aclglasspool.html

Samantha J. Hartley, Dept. of Psychology, University of Liverpool, Liverpool, UK. E-mail: shartley@liverpool.ac.uk, http://www.shef.ac.uk/~abrg/sam/index.html

Harald Hüning, Electrical Engineering, Imperial College, London SW7 2BT, UK. E-mail: h.huening@ic.ac.uk, http://www.ee.ic.ac.uk/research/neural/students/harry.html

Christophe L. Labiouse, Cognitive Science Unit, Department of Psychology, Université de Liège, Sart Tilman B32, 4000 Liège, Belgium. E-mail: clabiouse@ulg.ac.be. http://www.ulg.ac.be/cogsci/clabiouse/intro.html

Joyca Lacroix, IKAT/III, Universiteit Maastricht, The Netherlands.

Martin Le Voi, Psychology Discipline, The Open University, Walton Hall, Milton Keynes, MK7 6AA, UK. E-mail: m.e.levoi@open.ac.uk, http://www.open.ac.uk/socialsciences/staff/melevoi/

Joseph P. Levy, School of Social Sciences, University of Greenwich, London SE9 2UG, UK. E-mail: j.p.levy@gre.ac.uk, http://www.gre.ac.uk/~lj31/

András Lőrincz, Department of Information Systems, Eötvös Loránd University, Budapest, Hungary. E-mail: alorincz@inf.elte.hu, http://www.inf.elte.hu/~lorincz/

Will Lowe, Center for Cognitive Studies, Tufts University, Medford, MA 02155, U.S.A. E-mail: wlowe02@tufts.edu, http://www.tufts.edu/~wlowe02/

Teresa Bernarda Ludermir, Centro de Informática, Universidade Federal de Pernambuco, Recife PE, Brazil. E-mail: tbl@di.ufpe.br, http://www.di.ufpe.br/~tbl/

Padraic Monaghan, Institute for Adaptive and Neural Computation, Division of Informatics, University of Edinburgh, 2 Buccleuch Place, Edinburgh EH8 9LW.
E-mail: Padraic.Monaghan@ed.ac.uk, http://www.hcrc.ed.ac.uk/Site/MONAGHPA.html

Peter C. M. Molenaar, Dept. of Psychology, University of Amsterdam, Roetersstraat 15, 1018 WB Amsterdam, The Netherlands. E-mail: op_molenaar@macmail.psy.uva.nl, http://macnet007.psy.uva.nl/Users/Molenaar/home.html

Mike Page, MRC Cognition and Brain Sciences Unit, 15, Chaucer Road, Cambridge, CB2 2EF, UK. E-mail: mike.page@mrc-cbu.cam.ac.uk, http://www.mrc-cbu.cam.ac.uk/People/Mike.Page.html

Domenico Parisi, Institute of Psychology, National Research Council, Rome, Italy.
E-mail: parisi@www.ip.rm.cnr.it, http://gral.ip.rm.cnr.it/parisi/

Eric Postma, IKAT/III, Universiteit Maastricht, The Netherlands.
E-mail: postma@cs.unimaas.nl, http://www.cs.unimaas.nl/~postma/

Maartje E. J. Raijmakers, Dept. of Psychology, University of Amsterdam, Roetersstraat 15, 1018 WB Amsterdam, The Netherlands. E-mail: op_raijmakers@macmail.psy.uva.nl, http://macnet007.psy.uva.nl/Users/Raijmakers/home.html

Alard Roebroeck, IKAT/III, Universiteit Maastricht, The Netherlands.

Stéphane Rousset, Laboratoire de Psychologie Expérimentale, Université de Grenoble, France. E-mail: stephane.rousset@upmf-grenoble.fr, http://www.upmf-grenoble.fr/LPE/Personnel/Enseignants/St_phane_Rousset/st_phane_rousset.html

Eduardo Sanchez, Logic Systems Laboratory, Swiss Federal Institute of Technology, CH – 1015 Lausanne, Switzerland. E-mail: Eduardo.Sanchez@epfl.ch, http://lslwww.epfl.ch/pages/staff/sanchez/home.html

Matthew Schlesinger, Psychology Department, Southern Illinois University at Carbondale, Illinois, USA. E-mail: matthews@siu.edu, http://www-anw.cs.umass.edu/~matthew/academic.htm

Jon. L. Shapiro, Department of Computer Science, Manchester University, Manchester, M13 9PL, UK. E-mail: jls@cs.man.ac.uk, http://www.cs.man.ac.uk/ai/jls/jls.html

Richard Shillcock, Institute for Adaptive and Neural Computation, Division of Informatics, Department of Psychology, University of Edinburgh, 2 Buccleuch Place, Edinburgh, EH8 9LW, E-mail: R.Shillcock@ed.ac.uk, http://www.hcrc.ed.ac.uk/Site/SHILLCRI.html

Jacques Sougné, Cognitive Science Unit, Department of Psychology, Université de Liège, Sart Tilman B32, 4000 Liège, Belgium. E-mail: J.Sougne@ulg.ac.be, http://www.ulg.ac.be/cogsci/jsougne/

Lynn Andrea Stein, Artificial Intelligence Laboratory, Massachusetts Institute of Technology, 545, Technology Square, Cambridge MA 02139, USA, also Computers and Cognition Group, Franklin W. Olin College of Engineering, 1735, Great Plain Avenue, Needham, MA 02492, USA. E-mail: las@ai.mit.edu, las@olin.edu, http://www.ai.mit.edu/people/las/las.html

Botond Szatmáry, Department of Information Systems, Eötvös Loránd University, Budapest, Hungary. E-mail: botond@inf.elte.hu

Gábor Szirtes, Department of Information Systems, Eötvös Loránd University, Budapest, Hungary. E-mail: guminyul@inf.elte.hu

Bálint Takács, Department of Information Systems, Eötvös Loránd University, Budapest, Hungary. E-mail: deim@inf.elte.hu

Christof Teuscher, Logic Systems Laboratory, Swiss Federal Institute of Technology, CH – 1015 Lausanne, Switzerland. E-mail: Christof.Teuscher@epfl.ch, http://lslwww.epfl.ch/pages/staff/teuscher/home.html

Barbara Tillmann, Dartmouth College, Hanover, USA.
E-mail: Barbara.Tillmann@dartmouth.edu

Bert Timmermans, Department of Psychology, Vrije Universiteit Brussel, Bruxelles, Belgium. E-mail: Bert.Timmermans@vub.ac.be, http://www.vub.ac.be/PESP/Timmermans.html

Marc M. Van Hulle, Laboratory of Neurophysiology, Faculty of Medicine, Katholieke Universiteit Leuven, Herestraat, 49, B-3000 Leuven, Belgium. E-mail: marc@neuro.kuleuven.ac.be, http://www.kuleuven.ac.be/facdep/medicine/dep_neu/neufys/

Frank Van Overwalle, Department of Psychology, Vrije Universiteit Brussel, Bruxelles, Belgium. E-mail: Frank.VanOverwalle@vub.ac.be, http://www.vub.ac.be/PESP/VanOverwalle.html

Ingmar Visser, Department of Psychology, University of Amsterdam, Roetersstraat 15, 1018 WB, Amsterdam, The Netherlands. E-mail: op_visser@macmail.psy.uva.nl, http://macnet007.psy.uva.nl/Users/Ingmar/

John Wearden, Department of Psychology, Manchester University, Manchester M13 9PL, UK. E-mail: wearden@psy.man.ac.uk, http://www.psy.man.ac.uk/Staff/JWeard/index.htm

Gert Westermann, Sony, Computer Science Laboratory, 6 rue Amyot, 75005 Paris, France.
E-mail: gert@csl.sony.fr, http://www.cogsci.ed.ac.uk/~gert/

Andrzej Wichert, Department of Neural Information Processing, University of Ulm, D-89069 Ulm, Germany. E-mail: wichert@neuro.informatik.uni-ulm.de, http://www.informatik.uni-ulm.de/ni/mitarbeiter/AWichert.html

Neural Basis of Cognition

Applying Neuroanatomical Distinctions to Connectionist Cognitive Modelling

Padraic Monaghan & Richard Shillcock

Abstract

We show examples of how low-level, qualitative neuroanatomical hemispheric differences can give rise to detailed psychologically realistic behaviour. We illustrate this claim with examples from the cognitive modelling of unilateral visuospatial neglect and neglect dyslexia. Our models are based on two principles: the division of information processing between the two hemispheres; and the implementation of a coarse-/fine-coding distinction between the hemispheres. Line-bisection is a standard test for visuospatial neglect, in which the patient is required to mark the centre of a straight line; in neglect caused by right-hemisphere (RH) damage, bisection points tend to be displaced to the right of centre. Models of neglect have been focussed on dysfunctions following RH damage; however, patients with comparable damage to the left hemisphere (LH) typically present with a different pattern of deficits: (a) damage to the LH less often results in visuospatial neglect, or results in smaller, more variable displacements in line-bisection; (b) when damage to the LH results in neglect, recovery is generally quicker; and (c) damage to the LH often results in neglect dyslexia without more general visuospatial neglect. These asymmetries present a challenge to current models of neglect. Our models provide a principled and parsimonious account.

1. Introduction

Connectionist modelling has been a fruitful approach to understanding normal and impaired cognition, as demonstrated by Plaut, McClelland, Seidenberg and Patterson's (1996) modelling of normal and impaired word recognition [13]. Insofar as distributed representations are a critical aspect of such models, they instantiate an important anatomical fact about the brain: the brain itself seems to rely on distributed representations. However, such models often go no further in representing anatomical reality; thus, for instance, Plaut et al. study an abstract conception of the problem of mapping the orthographic forms of words onto the corresponding phonological forms. On the one hand, this relatively functionalist approach can reveal a great deal about the nature of the particular problem that the brain is tackling: for instance, we might discover emergent effects resulting from the distribution of segments in words (the beginnings of words are more informative). On the other hand, this approach runs the risk of underestimating architectural and other constraints on the range of possible solutions, and missing parsimonious explanations of particular behaviours. We will present below a series of examples of cases in which aspects of normal and impaired cognition may be modelled by the principled exploration of very basic anatomical constraints on model architecture. One of the goals of cognitive neuropsychologists is to ground cognition in particular structures in the brain. In this paper we will argue that connectionist models of spatial processing and visual word recognition can benefit by embodying two anatomical facts: the division of the brain into two hemispheres, and the tendency of the right hemisphere (RH) to employ coarser coding than the left hemisphere (LH).

The division of the brain into two halves is the clearest fact about its anatomy, but the processing implications of this division are far from clear. Several researchers have developed models that reflect this division. Cohen, Romero, Farah and Servan-Schreiber [4] have modelled opponent attentional processors. Jacobs and Kosslyn [6] have explored categorical and coordinate processing of spatial input. Reggia, Goodall and Shkuro [15] have studied lateralisation in reading and letter recognition tasks. Elsewhere we have presented a variety of split models of neglect and of visual lexical processing [10,11,18,19], all of which contain a "vertical" division representing the divided normal brain. In modelling neglect, this approach captures Kinsbourne's [7] proposal that each hemisphere contains a quasi-autonomous attentional processor. In modelling visual word recognition, such models capture the fact that the human fovea is precisely vertically divided, meaning that the two parts of a fixated word are initially projected solely to the contralateral hemisphere.

Models which instantiate the division of the brain into two hemispheres are good candidates for explaining the data we discuss below – a related set of observed behaviours concerning the reporting and judging of horizontally presented words and lines by individuals presenting with various aspects of visuo-spatial neglect and neglect dyslexia. The data concern graded effects of both a quantitative and a qualitative nature, which interact with factors such as position in the visual field and the laterality of the damage to the patient's brain. We will claim that the behaviours emerge naturally from interactions within a split architecture embodying simple hemispheric asymmetries, and do not necessitate the addition of extra mechanisms or modules.

2. Left Neglect and Right Neglect

Visuo-spatial neglect is the frequent result of strokes in many parts of the brain, but almost invariably ensues after damage to the parietal lobule. (For an extensive review, see [5]). In unilateral visual neglect, the patient typically fails to be consciously aware of stimuli or parts of stimuli that are situated contralesionally. Thus, right parietal damage might cause the left part of a horizontal line to be ignored ("left neglect") when the patient is trying to mark the perceived midpoint of the line in the line bisection task, causing the perceived midpoint to be displaced to the right. The line bisection task [17] is a widely used tool in the exploration of neglect behaviour, yielding both quantitative and qualitative data, often of a puzzling kind. For instance, the right displacement of the midpoint in left neglect becomes smaller and smaller as shorter lines are used in the task, until a "cross-over" occurs and the midpoint becomes displaced to the *left* for the shortest lines. (For a review of accounts of the cross-over phenomenon, and an existing account based on a split architecture, see [10]). Neglect can also result from comparable damage to the LH, but the neglect is typically less severe, if it occurs at all, and spontaneous remission happens more quickly [20]. In addition, there are a number of puzzling differences between left and right neglect with respect to tasks like line bisection: Mennemeier et al. [9] have shown that LH damage produces greater qualitative variability in performance, with line bisection displacements to the left or to the right of centre, but with greater overall accuracy.

Kinsbourne [7] presented the first account of neglect based on contralaterally oriented opponent attentional processors in each hemisphere of the brain: damage to one hemisphere results in a bias of attention towards the ipsilesional hemifield. If the attentional biases are equal and opposite then they cancel each other out. A number of researchers have subsequently theorised about attention in these terms, and have introduced the idea that there are differences between the attentional processing that occurs in the two hemispheres.

3. Hemispheric Asymmetries

The division into hemispheres does not just create issues of coordination for the brain, it creates opportunities for employing different kinds of processing in the two halves. In a monolithic brain it might be more difficult to allow different versions of qualitatively similar processing to occur without interference. The issue of positive and negative emotional processing and associated approach and withdrawal behaviour is a clear example; there would seem to be clear advantages to consigning these to different hemispheres. There are longstanding claims concerning different processing styles in the two halves of the brain: the RH has been described as carrying out holistic, synthetic, global processing, in contrast to the LH's predisposition for local, analytic, focal processing. It would be possible to model such differences by building radically different processors in the two hemispheres, or by adding different modules to each hemisphere. However, a more parsimonious route should be explored first. A basic hemispheric difference has been proposed whereby the RH tends towards coarser coding, compared with the LH. Coarse-coding in this context refers to the granularity of representations with respect to the units of the representational substrate. In neural terms coarse-coding may be seen as receptive field size. Coarse-coding does not necessarily imply anything about the fineness of the distinctions which can be made; a coarse-coded set of representations can still make fine distinctions by trading on small overlaps of large receptive fields. It has been suggested that the RH has a tendency to receive projections from neurons with larger receptive fields [2]. If a coarse-coding difference between the hemispheres can be shown to explain behavioural effects, then it would constitute a particularly parsimonious, low-level explanation; no high-level processing differences need be specified.

Halligan and Marshall [5] have suggested that such a neural difference might underlie the claimed local/global difference in hemispheric processing. Below we summarise research in which we produce a computational implementation of these suggestions. Pouget and Sejnowski [14] have suggested that neurons in the parietal cortices compute radial basis functions (RBFs) on visual input. We propose that the receptive fields of these RBFs are wider in the RH than in the LH. Our models incorporate contralateral gradients and fine/coarse-coding in order to explain asymmetries in neglect following brain damage.

4. The Line Bisection Model

Figure 1 shows a split model of attentional processing. A stimulus is presented across the nodes of the input layer and the task of the network is to recreate it accurately at the output. The centre of the receptive fields of the RBF units are positioned evenly across the input space, and the RBF are all fully connected to the nodes of the output layer. The left hidden layer RBF units have smaller receptive field sizes than units in the right hidden layer. In the brain, each hemisphere's attentional processor is contralaterally oriented: at the neural level, although most parietal neurons have bilateral receptive fields, the contralateral bias is evident from the fact that the peak response from a neuron tends to be from stimulation in the contralateral receptive field [1]. We have instantiated this contralateral bias as a gradient imposed on the number of RBF units centred at a particular point in the input. Thus, in Figure 1, the gradient on the left side of the model means that the right side of the stimulus is represented by the most units, and the more leftwards parts of the stimulus are represented by progressively fewer units. The gradient on the right side of the model is in the opposite direction. The model is feedforward and connections between the

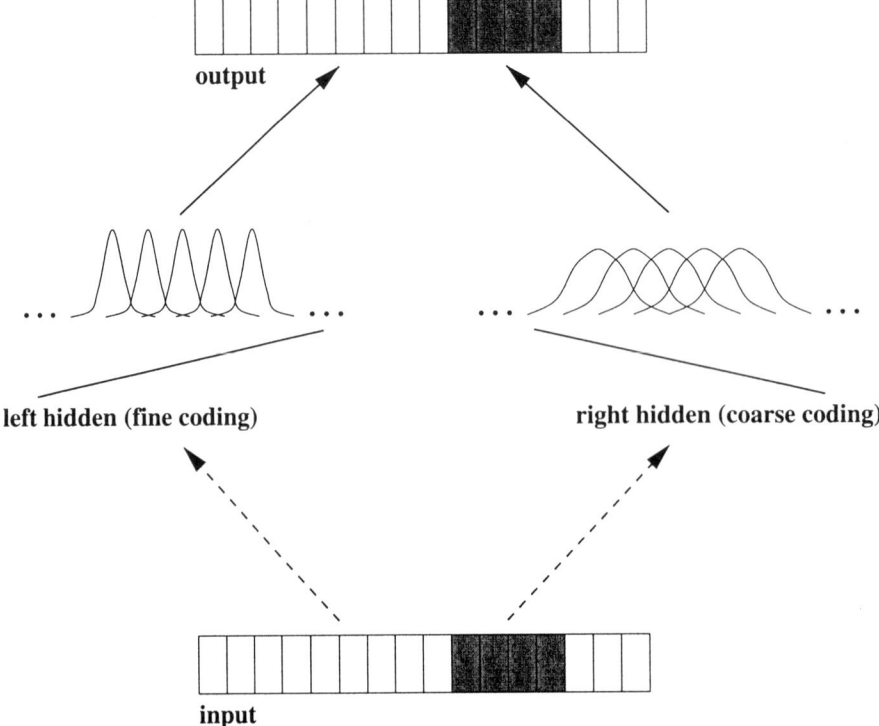

Figure 1: The model of opponent attentional processors, showing LH and RH receptive fields as fine- and coarse-coded radial basis functions, respectively. The activation gradients in the left and right halves of the model are shown beneath the representation of the radial basis function units (see text for details). The stimulus shown is a line of length 4, displaced to the right.

hidden layers and the output layer are trained according to the backpropagation learning algorithm The network is trained to divide the labour between the two sides of the model so that they complement each other and accurately recreate the input stimulus. (For further details see Monaghan and Shillcock (*forthcoming*) [11]).

5. Behaviour of the Damaged Model

Fine (LH) lesion

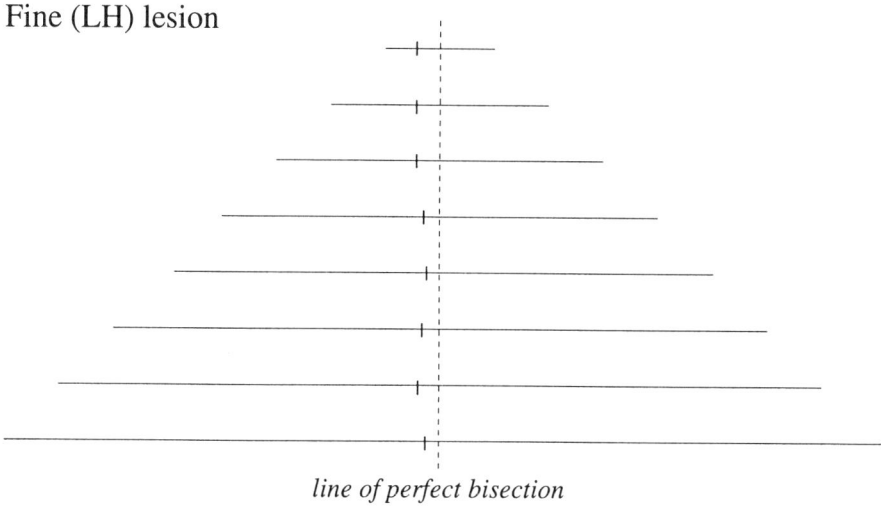

Coarse (RH) lesion

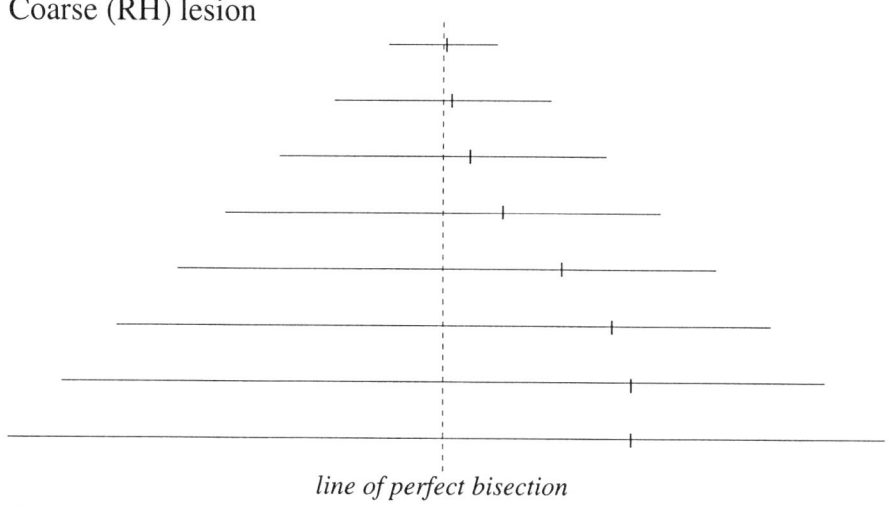

Figure 2: The line-bisection performance of the attentional model after 50% lesioning of the connections from the left and right hidden layers, respectively.

Unilateral parietal damage was simulated by reducing the strength of all connections ("lesioning") from one or other hidden layer to the output layer. Line bisection performance was judged from the centre of the activity produced at the output layer. Figure 2 summarises the behaviour of the model for central presentations of the line in the line-bisection task, with 50% lesioning to its fine-coded left half (its "LH") (top of Figure 2) and with 50% lesioning to the coarse-coded right half (its "RH") (bottom of Figure 2). Lesioning such a model produces patterns of behaviour in the line-bisection task that resemble the human behaviour reported by Mennemeier et al. [9]. Thus, in the LH-damaged model, the bisections are overall more accurate than in the RH-damaged model; indeed, it is easy to see how less extreme LH lesions might not produce large enough effects to allow neglect to be diagnosed on the basis of tests such as line-bisection. Mennemeier et al. report that LH-damaged human subjects produced a variety of both left and right displacements, in contrast to the RH-damaged subjects who overwhelmingly produced right displacements. The LH-damaged model's displacements are all to the left of the midpoint, but the addition of performance-related noise on top of this pattern (to resemble impaired human behaviour more closely) would produce the required pattern. (N.B. The trained and lesioned network's behaviour is deterministic and always produces the same output for the same input.)

Figures 3 and 4 provide fuller summaries of the behaviour of the model when it is presented with stimulus lines across all possible input positions (N.B. Figure 2 shows the data for central presentations only). Figure 3 shows the model's performance when the coarse-coded RH is lesioned by 50%. A response was judged to be correct when the centre of the line was within a quarter of a unit of the objective centre. The graph clearly shows a cross-over effect: in the longer lines the displacements are exclusively to the right, but they occur less frequently as the lines

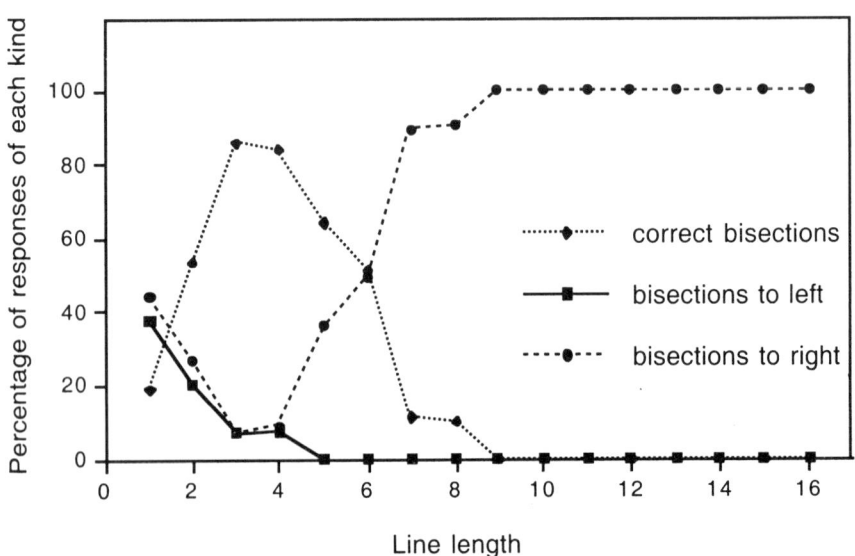

Figure 3: The performance of the model with a 50% lesion to the coarse-coded right half of the model.

themselves become smaller until, when the smallest lines are reached, they are matched in frequency by displacements to the left. Figure 4 shows the model's performance when the fine-coded LH was lesioned by 50%. As in the human data, both left and right displacements occur across a wide range of line lengths.

This demonstration of the cross-over effect emerging as a result of the effects of a fine- *versus* coarse-coding difference between the two hemispheres is a novel explanation of this challenging effect.

Finally, the requirement that there be faster spontaneous remission of LH damage compared with RH damage emerges quite naturally from the coarse- *versus* fine-coded distinction. The retraining of lesioned coarse-coded representations proceeds more slowly than fine-coded ones because of the more widespread interference that occurs in the former as new information is superimposed on existing information.

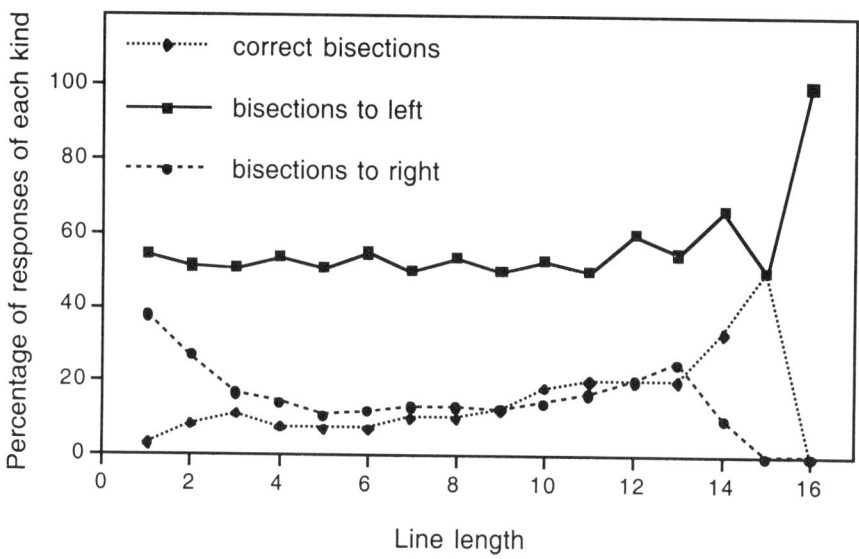

Figure 4: The performance of the model with a 50% lesion to the fine-coded left half of the model.

1. Modelling Neglect Dyslexia as a Categorical Task

Unilateral visual neglect is often accompanied by neglect dyslexia, in which the patient may ignore or misreport letters from one or other end of the word. In left neglect caused by a RH lesion, the patient might neglect the left end of the word, reporting *sunshine* as *shine*, for instance. Words have a horizontal spatial extent, just like the lines in the line-bisection task and might be expected to behave similarly, to some extent, in neglect. This prediction is verified, in that a cross-over phenomenon also exists in neglect dyslexia [3]. In the current paper we present only a simplified version of a model of neglect dyslexia, based on the previous model of attentional processing. We postpone the complexities which attend the composition of realistic lexical stimuli by using "words" in which a continuous extent of the input is occupied by one of two "letters". Figure 5 shows how each input position may contain a "letter" in which one or other node is activated. The task of the model is to recreate the input stimulus accurately at the output. Thus, the network has to process

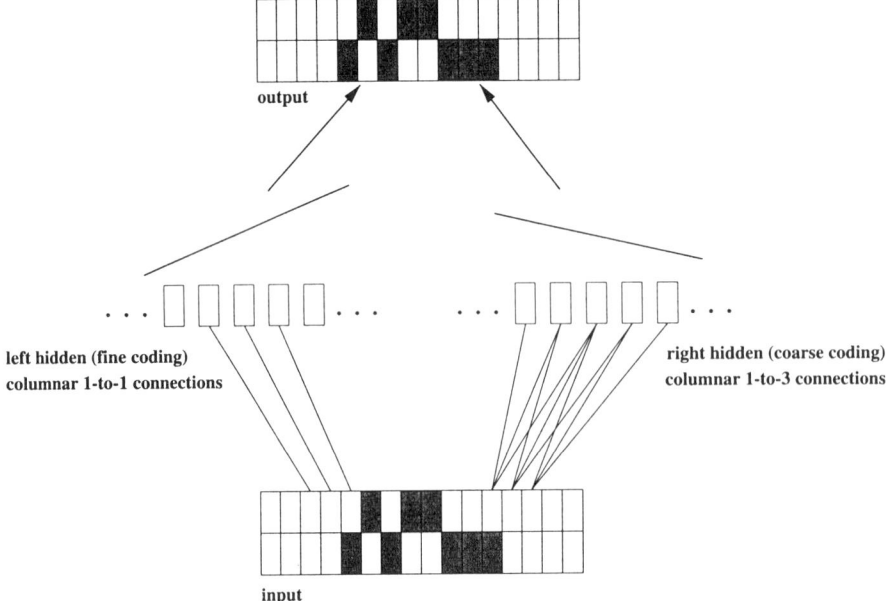

Figure 5: The model of neglect dyslexia. Each input position contains one of two "letters". The stimulus shown is a "word" of eight letters, in which letters 1, 3, 6, 7 and 8 are identical, as are 2, 4 and 5.

both the horizontal extent of the word and the categorical content of each letter position. The primary goal of this simulation is to capture the reported tendency towards the dissociation of neglect dyslexia from more general visuo-spatial neglect. Several researchers [8,12,16] report that RH lesions are likely to produce unilateral visual neglect for a range of stimulus materials, but may not produce neglect dyslexia. In contrast, LH damage is more likely to produce neglect dyslexia in the absence of general unilateral visual neglect. The dissociation is not complete, but is striking enough to suggest that different processing is responsible at some stage for the two phenomena. We hypothesised that the fine- *versus* coarse-coding distinction, with its differential implications for the representation of categorical detail *versus* simple extent, might be sufficient to produce this dissociation.

When the trained model was subjected to a set of unilateral lesions at 10%, 50%, 90% and 100%, the desired dissociation did indeed occur. Lesions to the fine-coded left half tended to produce more "neglect dyslexia", in terms of relatively poor performance in the representation of words, than did lesions to the coarse-coded right half. In contrast lesions to the coarse-coded right side tended to impair the representation of horizontal extent more than did lesions to the left side, as measured by the centre of activation at the output. Thus, within the limits of our very simple requirements for lexical processing, the coding difference seems to be enough to produce the observed dissociation.

7. Conclusions

We have briefly shown how basic neuroanatomical observations may be explored in the context of a connectionist account of unilateral visual neglect and of the closely associated neglect dyslexia. The neuroanatomical facts we chose – hemispheric division and coarse- *versus* fine-coding – are low-level differences that may be simply expressed but which have complex and unforeseen consequences for attentional processing. As such, they constitute parsimonious explanations for some of the more puzzling data in the neglect literature.

Acknowledgments

This work was supported in part by Wellcome Trust Grant 059080.

References

1. Andersen, R.A., Asanuma, C., Essick, G.K. & Siegel, R.M. (1985). Encoding of spatial location by posterior parietal neurons. *Science, 230*, 456-458.
2. Brown, H.D. & Kosslyn, S.M. (1993). Cerebral lateralization [Review]. *Current Opinions in Neurobiology, 3*, 183-186.
3. Chatterjee, A. (1995). Cross-over, completion and confabulation in unilateral spatial neglect. *Brain, 118*, 455–465.
4. Cohen, J.D., Romero, R.D., Farah, M.J. & Servan-Schreiber, D. (1994). Mechanisms of spatial attention: The relation of macrostructure to microstructure in parietal neglect. *Journal of Cognitive Neuroscience, 6*, 377–387.
5. Halligan, P.W. & Marshall, J.C. (1994). Toward a principled explanation of unilateral neglect. *Cognitive Neuropsychology, 11*, 167-206.
6. Jacobs, R.A. & Kosslyn. S.M. (1994). Encoding shape and spatial relations: the role of receptive field size in coordinating complementary representations. *Cognitive Science, 18*, 361–386.
7. Kinsbourne, M. (1977). Hemi-neglect and hemisphere rivalry. *Advances in Neurology, 8*, 41-49.
8. Ladavas, E., Shallice, T. & Zanella, M.T. (1997). Preserved semantic access in neglect dyslexia. Neuropsychologia, *35*, 257–270.
9. Mennemeier, M., Vezey, E., Chatterjee, A., Rapcsak, S.Z. & Heilman, K.M. (1997). Contributions of the left and right cerebral hemispheres to line bisection. *Neuropsychologia, 35*, 703-715.
10. Monaghan, P. & Shillcock, R. (1998). The cross-over effect in unilateral neglect: modelling detailed data in the line-bisection task. *Brain, 121*, 907–921.
11. Monaghan, P. & Shillcock, R. (*forthcoming*). Hemispheric asymmetries in cognitive modelling: A connectionist model of unilateral visual neglect and neglect dyslexia.
12. Patterson, K. & Wilson, B. (1990). A ROSE is a ROSE or a NOSE: A deficit in initial letter identification. *Cognitive Neuropsychology, 7*, 447–477.
13. Plaut, D.C., McClelland, J.L., Seidenberg, M.S. & Patterson, K. (1996). Understanding normal and impaired word reading: Computational principles in quasi-regular domains. *Psychological Review, 103*, 56–115.

14. Pouget, A. & Sejnowski, T.J. (1997). Lesion in a basis function model of parietal cortex: Comparison with hemineglect. In P. Thier & H.O. Karnath (Eds.). *Parietal Lobe Contributions in Orientation in 3D Space*. pp.521-538. Berlin: Springer.
15. Reggia, J., Goodall, S. & Shkuro, Y. (1998). Computational studies of lateralization of phoneme sequence generation. *Neural Computation, 10*, 1277–1297.
16. Riddoch,, J., Humphreys, G., Cleton, P. & Fery, P. (1990). Interaction of attentional and lexical processes in neglect dyslexia. Cognitive Neuropsychology, *7*, 479–517.
17. Schenkenberg, T., Bradford, D.C. & Ajax, E.T. (1980). Line bisection and unilateral visual neglect in patients with neurological impairment. *Neurology, 30*, 509–517.
18. Shillcock, R., Ellison, T.M. & Monaghan, P. (*in press*). Eye-fixation behaviour, lexical storage and visual word recognition in a split processing model. *Psychological Review*.
19. Shillcock, R.C. & Monaghan, P. (*in press*). The computational exploration of visual word recognition in a split model. *Neural Computation*.
20. Stone, S.P., Wilson, B., Wroot, A., Halligan, P.W., Lange, L.S., Marshall, J.C. & Greenwood, R.J. (1991). The assessment of visuo-spatial neglect after acute stroke. *Journal of Neurology, Neurosurgery and Psychiatry, 54*, 345-350.

Pseudopatterns and dual-network memory models: Advantages and shortcomings

Robert M. French, Bernard Ans, & Stéphane Rousset

Abstract

The dual-network memory model is designed to be a neurobiologically plausible manner of avoiding catastrophic interference. We discuss a number of advantages of this model and potential clues that the model has provided in the areas of memory consolidation, category-specific deficits, anterograde and retrograde amnesia. We discuss a surprising result about how this class of models handles episodic ("snap-shot") memory — namely, that they seem to be able to handle both episodic and abstract memory — and discuss two other promising areas of research involving these models.

1. Introduction

Neural networks typically store patterns in a single set of weights. This lack of modularity can mean that when new patterns are learned by the network, the new information may radically interfere with previously stored patterns. This problem, called catastrophic interference, was first brought to light by McCloskey and Cohen [12] and Ratcliff [15] and has been studied by numerous authors since (see [8] for a review). In this paper we will discuss a particular connectionist architecture designed to overcome this problem: the "dual-network" architecture. This model is loosely patterned after the "hippocampal-neocortical" architecture of the brain. In cognitive tasks, such as recall, pattern recognition, etc., humans do not experience catastrophic interference. McClelland, McNaughton and O'Reilly [11] suggested that the reason for this is the brain's bi-partite hippocampal-neocortical division of labor. New patterns are initially learned only by the hippocampus. The hippocampus then slowly trains the neocortex and, in this way, new patterns to do not interfere with already stored patterns.

There are a number of problems with this account; in particular, it does not explain why the transferred hippocampal patterns — regardless of the speed with they are learned by the neocortex — do not interfere with the patterns already in neocortex. Whether the new information "trickles" into the neocortex or is learned quickly is not the key issue here; new information, whether from the environment or from the hippocampus, can still overwrite old information in neocortex. French [6] and Ans & Rousset [1, 2] independently developed dual-network models that overcame this problem based on the use of pseudopatterns [17] to transfer information from one network to the other.

The key to these the dual-network models is pseudopattern transfer. In this paper we will discuss this crucial information transfer mechanism. We suggest that a clearer understanding of pseudopattern information transfer may be able to provide insights into the function of REM sleep, memory consolidation, category-specific deficits, anterograde and retrograde amnesia. We will discuss a number of potential problems with this type of mechanism and will suggest how the pseudopattern generation mechanism might be optimized.

2. Sensitivity and stability to new information

One of the most important problems facing connectionist models of memory — in fact, facing *any* model of memory — is how to make them simultaneously sensitive to, but not disrupted by, new input. One solution to this problem — arguably, the solution discovered by evolution for the human brain — is to have two separate storage areas: one for new information (the hippocampus), the other for previously learned patterns (the neocortex). The idea would be that the hippocampus gradually transfers the new patterns to the neocortex and, in this way, catastrophic interference would be avoided [11]. The major problem with this suggestion is that the rate, however slow or fast, at which new information is learned by the neocortex does not prevent the overwriting of previously learned patterns by the new patterns. The key to overcoming catastrophic interference is to *interleave* previously learned patterns (or some approximation of these patterns) with the new patterns during learning.

3. Dual-network memory models

To overcome the problem of catastrophic interference French [6] and Ans & Rousset [1, 2] independently proposed dual-network memory models. Even though their respective models differ in a number of respects, the essence of the two architectures is largely the same. The overall system consists of two separate, coupled networks. New information arrives at only one of the networks and is interleaved with information from the other network. The newly learned information is then passed from the first network to the second. Information is transferred from one network to the other by *pseudopatterns* [17].

Consider the following problem: A neural network has learned a number of input-output patterns corresponding to some underlying function f. We have no access to the network's weights, its connection topology, etc, and, most importantly, *the original patterns learned by the network are no longer available*. How can we, nonetheless, get an approximation of the original function f?

One solution to this problem consists of sending input (in the simplest case, random input) into the network and observing the output for each random input. We thus create a series of *pseudopatterns*, ψ_i, where each pattern ψ_i is defined by a random input and the output of the network after that input has been sent through it. This set of pseudopatterns reflects the originally learned function f. The greater the number of pseudopatterns, the better the approximation that can be obtained of the originally learned function f. Pseudopatterns were first introduced by Robins [17] to overcome catastrophic interference. Robins suggested that when a network had to learn a new pattern, a number of pseudopatterns be generated. Then, instead of learning just the new pattern, *P*, the network would be trained on the new pattern plus the set of pseudopatterns that reflected what it had previously learned. In this way, the new pattern would be interleaved with patterns that, even though they were not the originally learned patterns, nonetheless reflected the original function learned. Robins showed that his technique did, indeed, reduce catastrophic interference [4, 16, 17, etc.].

This technique was used successfully by French [6] and Ans & Rousset [1] as the main mechanism of information transfer between two separate networks. The idea is to have two networks (for simplicity we will call these two networks the "hippocampal" and the "neocortical" networks, although these designations should not be taken too literally) that exchange information by means of pseudopatterns (the neural correlate of pseudopattern information transfer might be experienced as dream sleep [18]). When a new pattern, *P*, is to be learned by the hippocampal network (new information from the environment is learned exclusively by the hippocampal network), there are (at least) three ways to proceed:

i) A set of neocortical pseudopatterns $\{\psi_i\}^{NEOCORTEX}$ is created. Hippocampal network learning continues until the network has learned the new pattern as well as the neocortical pseudopatterns to criterion [6].

ii) A set of neocortical pseudopatterns $\{\psi_i\}^{NEOCORTEX}$ is created. Hippocampal network learning continues only until the new pattern *P* falls below criterion, even though the network may or may not have learned the neocortical pseudopatterns to criterion [17].

iii) There is no fixed set of neocortical pseudopatterns. Rather, new neocortical pseudopatterns are continually generated while the hippocampal network gradually learns the new pattern, *P*. Hippocampal network learning continues until *P* falls below criterion [1, 2].

In Ans & Rousset's model the inputs used to produce the pseudopatterns are not simply random patterns, as they are in the other two models, but rather they result from a "reverberation" between the first and second layer of the network in their network of origin. The first and second layers of both networks in the Ans & Rousset model constitute an autoassociator. When a pseudo-input is given to the network, it is cycled from the first layer to the second to the first to the second, etc., until a maximum number of cycles has been reached. This "input attractor" (that has not necessarily reached a stable state) is then fed through the network to produce the pseudopattern that will be learned by the other network. Pseudopatterns are also used to transfer the newly learned information from the hippocampal network to the neocortical network. To do this, pseudopatterns $\{\psi_i\}^{HIPPOCAMPUS}$ are generated by the hippocampal network and are then learned by the neocortical network.

4. Advantages of the dual-network approach

It has been shown [1, 2, 6] that this type of approach does, indeed, eliminate catastrophic forgetting. The forgetting curves for this type of model are far more realistic than in any standard backpropagation or Hopfield network. For example, in one experiment French [6] compared a standard backpropagation network to a dual-network memory on a task that consisted of having the networks sequentially learn a total of 20 items. After learning the 20th item, the error performance is measured for each item in the 20-item list. For standard backpropagation, all 19 previous items were well above the 0.2 error threshold, whereas for the dual-network memory, forgetting was far more gradual (e.g., 8 of the most recently learned items were still at or below the 0.2 threshold). Ans & Rousset [2] have shown similar improvements in forgetting with their dual-network reverberating architecture.

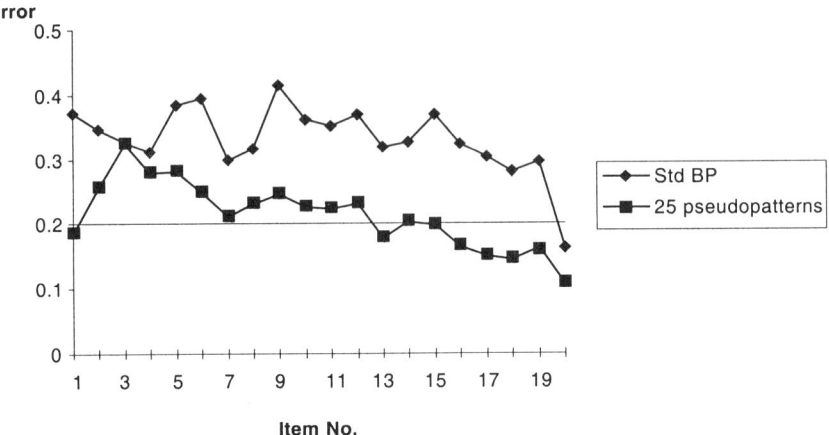

Figure 1: Amount of error for each of 20 items learned sequentially after the final item had been learned to a 0.2 criterion [6].

It also seems that this process of passing information back and forth between two networks using pseudopatterns may lead to representational compression over time, which may help explain rather puzzling category-specific losses observed in certain cases of anomia [5, 7].

Further, Robins & McCallum [16] suggest that this pseudopattern information transfer in this type of dual-network system is the primary means of long-term memory consolidation. These authors claim that memory consolidation by pseudopattern information transfer to the neocortex is "a computational model of sleep consolidation" that is supported by "psychological, evolutionary and neurophysiological data (in particular accounting for the role of the hippocampus in consolidation)." (For a review, see [20]). Related conclusions concerning "off-line" mutual reactivation of memory traces in both hippocampus and neocortex are discussed in [21]. Even though the contention that memory consolidation is contested by certain authors (e.g., [24]), the point remains that a mechanism resembling pseudopattern information transfer may well be responsible, at least in part, for LTM memory consolidation.

Ans & Rousset [2] have also argued that lesioning the flow of pseudopattern transfer from the "hippocampal" to the "neocortical" network should induce an anterograde amnesia behavior and, because pseudopatterns from the neocortical network would continue to refresh the hippocampal network, no severe retrograde amnesia be observed. In the other direction, damage to neocortical to hippocampal transfer would mean that pre-lesion information would continue to be consolidated, but there would be post-lesion catastrophic interference by new patterns of information already present in hippocampus.

5. Potential difficulties

Potential difficulties with this architecture fall broadly into three categories.

Episodic ("snap-shot") memory

It is well known that people can recall "snap-shot" episodes of experience, i.e., precise events that occurred at a specific time and a specific place. These so-called snap-shot memories are widely believed to be stored in the hippocampus, although undoubtedly prefrontal cortex also plays some lesser role in episodic memory [3, 19]. A common concern that has been voiced about the pseudopattern transfer mechanism involves its ability to preserve these snap-shot memories. This concern arises from the intuition that the random nature of pseudopattern input would tend to "blur to abstraction" the originally learned patterns. In other words, precise snapshot memories corresponding to particular patterns would be lost.

This important concern merits closer examination. In Ans & Rousset's dual-network model [1], the pseudopattern inputs are first reverberated between the input and hidden layers towards an attractor. Since the input-hidden layers of each network act much like the auto-associative "clean-up" units used in many current connectionist models, intuitively, this would seem to lead to one of the following situations:

- If the input attractor basins were very large, there might be a problem of coverage of the input space, i.e., even though in this case pseudopatterns would indeed tend to be the originally learned items (thereby satisfying the snap-shot memory criterion), those with small attractor basins could be systematically neglected for items with much larger attractor basins. This would mean that the neglected items would never be transmitted from one network to the other and would therefore be forgotten by the network;
- If the input attractor basins were relatively small, then many pseudo-inputs would not fall into attractor basins corresponding to previously learned input and the originally learned snap-shot memories would be lost.

In one of the most surprising results to come out of dual-network memory model research, it would seem that it is possible that neither of these situations occurs. The dual-network model would seem to provide the best of both worlds: abstraction occurs but, at the same time, specific memories are preserved.

In a preliminary study, Ans & Rousset [2] trained the hippocampal network on a set of 20 arbitrary input-output patterns, $\{P_i\}_{i=1}^{20}$. Then, in order to transfer this learning to the neocortical network, they produced 32-bit random pseudo-inputs and reverberated these inputs towards attractors in the hippocampal network in order to produce the hippocampal pseudopatterns that would subsequently be learned by the neocortical network. Each 32-bit pseudo-input was compared to the inputs of the 20 previously learned patterns P_i. A normalized Euclidean distance metric was defined on the input space as follows: $d(X,Y) = \left[\frac{1}{N} \sum_{i=1}^{N} (x_i - y_i)^2 \right]^{1/2}$, where N is the length of the input vector (in this case, 32) for each pattern. This metric keeps distance between any two input patterns between 0 and 1. Any pseudo-input that was within 0.5 of the input of any previously learned pattern, P_i, *was rejected*. This draconian filtering meant that only 13% of the total number of pseudo-inputs generated were actually used to train the neocortical network. In short, the neocortical network was trained only on patterns that specifically *did not resemble* the patterns previously learned by the hippocampal network. One might expect, because of the artificially forced dissimilarity between the hippocampal pseudopatterns and the patterns originally learned by the hippocampal network, that the original patterns, $\{P_i\}_{i=1}^{20}$, would not be accurately transferred to the neocortical network. Surprisingly, this proved not to be the case. In fact, all twenty of the originally learned patterns were successfully transferred to neocortical memory! This result has been tested numerous times and seems to be robust.

Although considerable research still needs to be done on this aspect of dual-network memory models, this result would suggest that this type of memory model not only accommodates the generally accepted idea that hippocampal memory is appropriately modeled by a sparse auto-associative network [13, 14, 23, etc.], thereby allowing it to precisely recover previously learned information, but simultaneously allows abstraction to take place in the neocortical network.

Contextualizing pseudopattern generation

When we encounter a new pattern — say, the first time we see a yo-yo — we must be able to learn this pattern without it interfering with other patterns we have already learned. In the dual-network model this is achieved by interleaving the new pattern with pseudopatterns that reflect previously learned patterns. But there is a problem with this — namely, that pseudopatterns reflecting *the undifferentiated contents* of neocortical memory are interleaved with each new pattern that must be learned in the hippocampal memory. This is surely not necessary to prevent catastrophic interference and, moreover, it is ludicrous to suppose that we need an approximate copy in hippocampus of everything that has ever been stored in long-term memory every time we learn a single new association.

In particular, in the case of our first encounter with a yo-yo, we can reasonably assume that our internal representations of patterns relating to sabre-tooth tigers and unicorns, computer memory chips, and the Crab Nebula will almost certainly be orthogonal to the representation we will develop of "yo-yo." This would mean that the internal representation for the new pattern "yo-yo" would not interfere with those representations for unicorns, computer chips and clusters of stars. But it certainly *could* interfere with our representation for concepts like "wheel," "pendulum", etc.

What needs to happen is that the input from the "yo-yo" pattern must activate similar concepts in long-term memory and that *these related concepts* need to be refreshed so as not to be overwritten. But how can this been done in a reasonable manner? Assume that the pattern to be learned by the hippocampal network is $P: i \to o$. One suggestion might be to give i as input to the neocortical network in order to produce a similar pseudopattern. This would, however, produce a neocortical pseudopattern $\psi: i \to \hat{o}$. The problem is that the hippocampal network (or any network) *cannot* simultaneously learn both P and ψ.

In fact, Ans & Rousset have shown that when neocortical pseudopatterns whose input resembles the input of the pattern to be learned are interleaved with the new pattern, the network frequently fails to converge. Performance is considerably better with pseudopatterns whose initial input (i.e., before reverberation) is random.

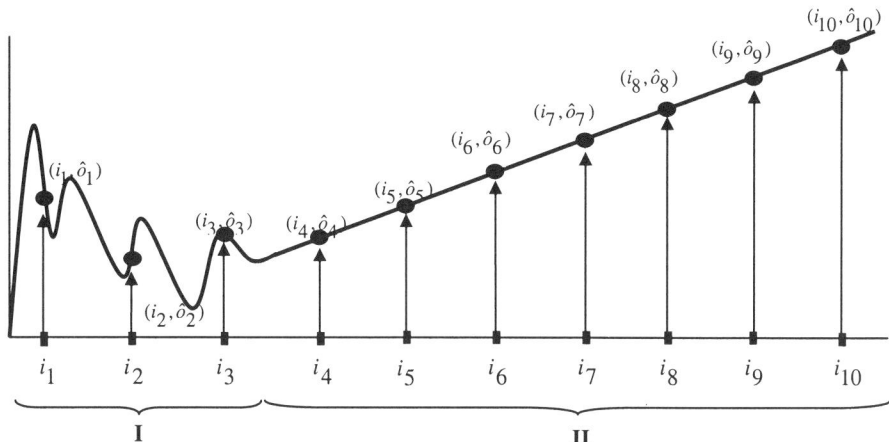

Figure 2: A uniform distribution of pseudo-inputs to generate pseudo-patterns leads to an over-exploration of Region II and an under-exploration of Region I.

Optimizing pseudopattern generation

The final issue that we will consider in this paper is closely related to the previous problem of pseudopattern contextualization. The question is: Can pseudopattern generation be optimized to improve recovery of the originally learned function and is the brain doing anything similar to this kind of optimization?

We begin this discussion with a problem first posed by John Holland [10]. Suppose we have a *two*-armed bandit which has two payoff arms. One arm pays off with a ratio of p, the other with a ratio of q where $p > q$, but we do not know which arm gives which payoff. We have N tokens and we wish to maximize our earnings. If we knew which arm was which, we would, of course, put all of our tokens in the arm with payoff ratio p. But we don't have this information, so we must "waste" some of our supply of tokens to try to determine which arm pays off more. One strategy might be to decide to allocate N/4 tokens to the first arm, N/4 tokens to the second and, then, whichever arm had produced the greatest payoff, put the remaining N/2 tokens in the slot corresponding to that arm. The problem with this strategy, of course, is that if the p ratio is 1000 and q is 1, then we would realize almost immediately which is the better arm and most of the N/4 tokens put in the arm corresponding to payoff q would be wasted.

Holland proved a theorem concerning the strategy to adopt. Basically, "the loss rate will be optimally reduced if the number of trials allocated [to the apparently most promising arm] grows slightly faster than an exponential function of the trials allocated [to the apparently least promising] arm." ([10], p. 83).

A related problem arises in pseudopattern generation. To see why, consider the following function that has been learned by a neural network. We wish to recover this function as best as possible using pseudopatterns.

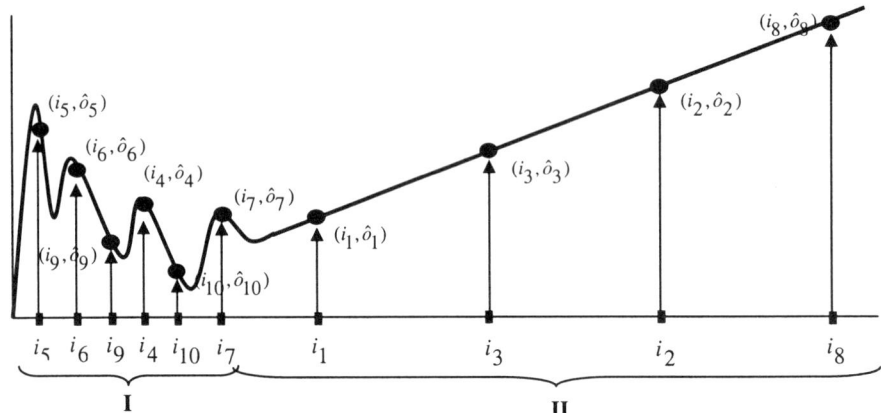

Figure 3: A better exploration of the space is obtained by using feedback from the patterns as they are generated. In this way, more of the patterns are concentrated in the more complex region of the function.

The set of pseudopatterns in Figure 4, $\{\psi_k : i_k \rightarrow \hat{o}_k\}_{k=1}^{10}$ does, indeed, approximate the originally learned function, but could we have done better by a more judicious choice of inputs to probe the network? Clearly, the answer is yes. In Figure 2, seven of the ten pseudopatterns fall in Region II, which does not need to be explored as carefully as Region I. Notice that nowhere have we used the information provided by the first pseudopatterns to determine the input to create the later ones. In a manner similar to the strategy for optimally allocating our tokens in the two-armed bandit problem, once we begin to be certain that we have a good representation of part of the function (Region II), we concentrate our remaining pseudo-inputs on other less well-understood regions (Region I).

If, for example, the network makes a linear interpolation prediction as to where the output from a given pseudo-input should fall, and this turns out to be correct (or close to correct), then this tells the system that it is probably in a well-understood area. In other words, in the first case of pseudo-input selection (Figure 2), we are not using the output of the network to modify subsequent pseudo-input selection. In the second case (Figure 3), as soon as the system begins to be able to accurately predict the outputs for a particular region of the function, then we can devote more of our pseudo-input "energy" to other regions of the function. This does not, of course, mean that we cease further exploration of the well-understood region. The function could be highly varying, in that region and we were simply lucky in our prediction.

But it does mean that we devote exponentially less of our pseudo-inputs to that area as our information of what the function in that region is likely to be improves.

If REM sleep and pseudopattern generation are related, we can probably conclude that pseudopattern generation is not random. It would seem likely that some kind of pseudopattern optimization is being carried on by the brain. What and how this works is an important question for further study. We suggest that perhaps there is some stochastic "2-armed bandit" optimization method being used by the brain in the type of noise that it "selects" to send through the system in order to improve the consolidation of new information in neocortex.

6. Conclusions

We have attempted to show that, while the dual-network memory model has great potential for explaining a number of phenomena related, in particular, to memory consolidation, there are still a number of problems that need to be considered. We have considered three of these problems relating to episodic memory, contextualization of pseudopattern generation and pseudopattern optimization. We have attempted to show why these areas pose problems for the dual-network model and have suggested ways in which these problems might be able to be overcome.

Acknowledgments

This research was supported in part by a research grant from the European Commission (HPRN-CT-1999-00065) and the Belgian government (IUPA P4/12).

References

1. Ans, B., & Rousset, S. (1997). Avoiding catastrophic forgetting by coupling two reverberating neural networks. *Academie des Sciences, Sciences de la vie, 320*, 989-997.
2. Ans, B., & Rousset, S. (2000). Neural Networks with a Self-Refreshing Memory : Knowledge Transfer in Sequential Learning Tasks without Catastrophic Forgetting. *Connection Sciences, 12, 1, 1-19*
3. Cohen, J. J., & Eichenbaum, H. (1993). *Memory, amnesia, and the Hippocampal System*. Cambridge: MIT Press.
4. Frean, M., & Robins, A. (1998). Catastrophic forgetting and "pseudorehearsal" in linear networks. In Downs T, Frean M., & Gallagher M (Eds.) *Proc. of the 9th Australian Conference on Neural Networks*, 173-178, Brisbane: U. of Queensland
5. French, R. M., & Mareschal, D. (1998). Could Category-Specific Semantic Deficits Reflect Differences in the Distributions of Features Within a Unified Semantic Memory? In *Proceedings of the Twentieth Annual Cognitive Science Society Conference*. NJ:LEA. 374-379.
6. French, R. M. (1997a). Pseudo-recurrent connectionist networks: An approach to the "sensitivity–stability" dilemma. *Connection Science, 9*(4), 353-379.
7. French, R. M. (1997b). Selective memory loss in aphasics: An insight from pseudo-recurrent connectionist networks. In J. Bullinaria, G. Houghton, D. Glasspool (eds.).

Connectionist Representations: Proceedings of the Fourth Neural Computation and Psychology Workshop. Springer-Verlag. 183-195.
8. French, R. M. (1999). Catastrophic Forgetting in Connectionist Networks. *Trends in Cognitive Sciences, 3*(4), 128-135.
9. Hebb, D. O. (1949). *Organization of Behavior.* New York, N. Y.: Wiley & Sons.
10. Holland, J. (1975). *Adaptation in natural and artificial systems.* Ann Arbor, MI: The University of Michigan Press.
11. McClelland, J., McNaughton, B., & O'Reilly, R. (1995). Why there are complementary learning systems in the hippocampus and neocortex: Insights from the successes and failures of connectionist models of learning and memory. *Psychological Review. 102,* 419–457.
12. McCloskey, M., & Cohen, N. (1989). Catastrophic interference in connectionist networks: The sequential learning problem. *The Psychology of Learning and Motivation, 24,* 109-165.
13. McNaughton, B., & Morris, R. (1987). Hippocampal synaptic enhancement and information storage within a distributed memory system. *Trends in Neurosciences, 10,* 408-415.
14. McNaughton, B., & Nadel, L. (1990). Hebb-Marr networks and the neurobiological representation of action in space. In M.A. Gluck, & D. Rumelhart (Eds.) *Neuroscience and Connectionist Theory.* Hillsdale, NJ: LEA, 1-63.
15. Ratcliff, R. (1990). Connectionist models of recognition memory: Constraints imposed by learning and forgetting functions. *Psychological Review, 97,* 285-308
16. Robins, A., & McCallum, S. (1998). Pseudorehearsal and the catastrophic forgetting solution in Hopfield type networks. *Connection Science, 10,* 121 - 135
17. Robins, A. (1995). Catastrophic forgetting, rehearsal, and pseudorehearsal. *Connection Science, 7,* 123 - 146.
18. Robins, A. (1996). Consolidation in neural networks and in the sleeping brain. *Connection Science, 8,* 259 - 275
19. Squire, L. (1992). Memory and the hippocampus: A synthesis from findings with rats, monkeys, and humans. *Psychological Review, 99,* 195-231.
20. Stickgold, R. (1999). Sleep: off-line memory reprocessing. *Trends in Cognitive Sciences, 2*(12), 484-492.
21. Sutherland, G., & McNaughton, B. (2000). Memory trace reactivation in hippocampal and neocortical neuronal ensembles. *Current Opinions in Neurobiology, 10,* 180-186.
22. Traub R., & Miles R (1991). *Neuronal networks of the hippocampus.* Cambridge, UK: Cambridge Univ. Press.
23. Treves A., Rolls E. (1994) Computational analysis of the role of the hippocampus in memory. *Hippocampus, 4,* 374-391.
24. Vertes, Robert P. and Eastman, K. E. (2000). The Case Against Memory Consolidation in REM sleep. *Behavioral and Brain Sciences, 23* (6).

A Learning Algorithm for Synfire Chains

Jacques Sougné

Abstract

Neurobiological studies indicate very precise temporal behavior of neuron firings. Abeles [1] has recorded spike timing of different cortical cells and, in particular, has observed the following level of precision: when a neuron A fires, neuron B would fire 151ms later and neuron C would fire precisely 289ms after that—with a precision across trials of 1 ms! Such long delays require dozens of combined transmission delays from the presynaptic neuron (A) to the postsynaptic neuron (C). The mechanism proposed by Abeles for generating such precise delayed synchronization has been called *synfire chains*. How could synfire chains develop? What learning procedure could generate such precise temporal chains? How could a connectionist network of spiking neurons learn synfire chains? An algorithm for a network of spiking neurons that learns synfire chains will be presented.

1. Introduction

The human brain is extremely sensitive to time. According to Abeles et al. [2], people hearing a sound from a source situated 1 degree to the right of the centerline of their face, at a distance of 1 meter, will receive the acoustic signal at their right ear 12 µs before the left ear. This means that detecting the source of a sound requires very accurate temporal detection. According to Calvin [4], throwing a projectile at a 20cm wide target, from a distance of 7 meters, requires a release time accuracy of less than 1 ms. Findings on synfire chains [1, 2, 17] indicate very precise temporal behavior of neuron firings. Researchers have recorded spike timing of different cortical cells in monkeys. In particular, they observed the following stimulus-dependent kind of pattern: when a neuron *a* fired, neuron *b* would fire 151ms later, and neuron *c* would fire 289ms later with a precision across trials of 1 ms! Delays of this duration would require dozens of transmission delays from *a* to *c* neuron. There are two hypotheses about how this phenomenon could occur. The first is based on an increase in a population rate which build excitation in another population which, in turn, increases its firing rate, etc. The second is the mechanism proposed by Abeles [1] which has been called *synfire chains*. Since cortical synapses are relatively weak, many inputs to cells must arrive at the same time for them to fire. Consequently, each step in the synfire chain requires a pool of neurons whose firings simultaneously raise the potential of the next pool of neurons to allow them to fire. In this mechanism each cell of the chain fire only once. Recent experiments [17] indicate that these precise firing sequences correlate more to behavior than rate modulation and does not seems to be a byproduct of rate modulation. These results seems to confirm the synfire chain hypothesis.

How could synfire chains develop? What learning procedure could generate such precise temporal chains? How can a connectionist network of spiking nodes learn and display synfire chains? This is the fundamental part of the present study. A network of spiking nodes (INFERNET) is provided with a learning algorithm that allows it to learn synfire chains.

2. Learning Synfire Chains

Previous work on learning synfire chains has focused on how they can develop from a chaotic net with an unsupervised Hebbian learning rule [3, 10, 11]. These studies involved an external stimulus which makes a pool of neurons fire at time 0. Subsequently a sequence of successive pools of neuron firings occurs engendered by the random connection weights of the network. Active connections are modified by a Hebbian learning rule. After learning, when the same stimulus is presented, the same chain of neuron firing is observed, thereby constituting a synfire chain. These studies show that these chains are stable, noise tolerant and that one network can store many different chains. Formal analysis showed that there is a relation between the network size and the length of learnable synfire chains [3, 10], and that the recall speed should be faster than the training speed [21].

This paper explores how a synfire chain can develop to link two pools of neuron firings provoked by two sequential external stimuli. After learning, the first external stimulus should activate the stored synfire chain.

3. INFERNET

INFERNET [18, 19, 20] is a network of spiking neurons [13]. In INFERNET, nodes can be in two different states: they can fire (on), or they can be at rest (off). A node fires at a precise moment and transmits activation to other connected nodes with some time course. When a node activation or potential $V_i^{(t)}$ reaches a threshold, it emits a spike. After firing, the potential is reset to some resting value V_r. Inputs increase the node potential, but some part of the node potential is lost at each time step. Spiking neuron models use a post synaptic potential (PSP) function.

INFERNET is not a fully connected network; its structure is organized by clusters of nodes which constitute subnets. Each subnet is fully connected. From each node of a subnet there is a connection to every other node within that subnet. Some subnet nodes have connections to external subnet nodes. It not only reduces the computational demands of the program, but also better corresponds to the actual organization of the brain. The brain is not fully connected; in particular, there are more intra-cortical connections than inter-cortical connections. Each connection is either excitatory or inhibitory. Excitatory connections increase the potential of postsynaptic nodes, while inhibitory connections decrease their potential. Two variables affect each connection: weight and delay. Each weight corresponds to the synaptic strength between a presynaptic and postsynaptic cell. The weight between a presynaptic node j and a postsynaptic node i is designated by w_{ij}. Noise is added to this value and the resulting noisy connection is denoted by $\hat{w}_{ij}$. The delay d of a connection determines when the effect of the presynaptic node firing will be maximum on the postsynaptic node. There is also a noise factor on the delay. The noisy delay is denoted by $\hat{d}$. This delay corresponds to the axonal, synaptic and dendritic delays of real neurons.

A signal, whether excitatory or inhibitory, will be affected by a leakage factor. When the signal has reached its maximum, at each following step of 1 ms, the signal will be divided by 2. Delays and leakage factors define the Excitatory Post Synaptic Potential (EPSP) or Inhibitory Post Synaptic Potential (IPSP) functions shown in Figure 1a. The y-axis refers to the postsynaptic node potential V_i. The x-axis is the time difference in ms between the time t and the time of the presynaptic node firing $t_j^{(f)}$. The resulting postsynaptic (PSP) equation $\varepsilon_{ij}(x)$ is given by:

$$\varepsilon_{ij}(x) = \frac{1}{2^x} \mathcal{H}(x) \qquad (1)$$

where: $\mathcal{H}(x) = \begin{cases} 1 \text{ if } x \geq 0 \\ 0 \text{ if } x < 0 \end{cases}$ (2)

and x is the difference between the time t, the time of the presynaptic node firing, and the noisy delay on the connection: $x = t - t_j^{(f)} - d$.

When a node potential V_i reaches a threshold θ, it emits a spike. Thereafter, the potential is reset to its resting value. Figure 1b illustrates a spike of an INFERNET node. After emitting a spike, a node enters a refractory period. This corresponds to the membrane resistance of real neurons which increases after a spike. In INFERNET, the refractory state of node i depends only on the last spike of the node i: $t_j^{(f)}$. A value dependent on the refractory state is subtracted from the node state value V_i. This value is denoted by $\eta_i(u)$, where u is the difference between the current time t and the time of the last spike of node i: $u = t - t_j^{(f)}$. The shape of this function is shown in Figure 1c.

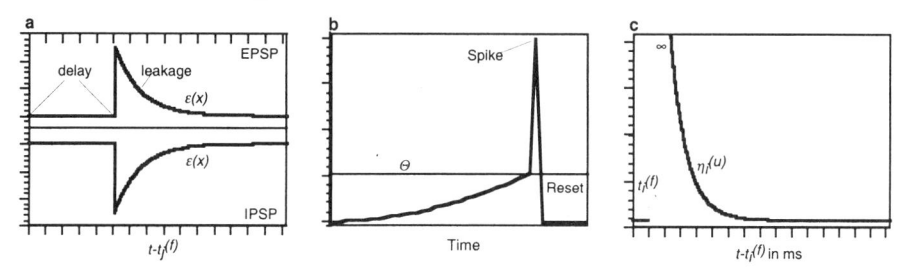

Figure 1, a: EPSP and IPSP function in INFERNET $\varepsilon(x)$. b: A spike in INFERNET. c: The function $\eta_i^{(u)}$ taking into account the refractory period of a node..

All variables affecting the potential of a node have now been defined. Equation (3) express how $V_i^{(t)}$, the potential or state of node I, is calculated at each time step.

$$V_i^{(t)} = \sum_{j \in \Gamma_i} \sum_{t_j^{(f)} \in F_j} \hat{w}_{ij} \, \varepsilon_{ij}(x) - \eta_i(u) \qquad (3)$$

Node i fires when its potential $V_i(t)$ reaches the threshold Θ. This potential is affected by connection weights $\hat{w}_{ij}$ coming from each presynaptic node j. The set of presynaptic connections to node i is given by $\Gamma_i = \{j | j \text{ is presynaptic to } i\}$. F_j is the set of all firing times of presynaptic nodes j: $t_j^{(f)}$. Noisy connection weights linking j node to i node are $\hat{w}_{ij}$. The equations $\varepsilon_{ij}(x)$ expresses the postsynaptic potential function as shown in Figure 1a. A comment is necessary about the threshold Θ.

Random noise is added to its value. Assuming a node is no longer in its refractory period, its firing requires simultaneous full excitatory input from more than one node. In agreement with real neural networks, a single synapse cannot provoke a postsynaptic action potential. A single cortical synapse raises the potential of a postsynaptic neuron by 0.1 to 1 mV [2].

3.1 Hebbian Learning

Long term potentiation (LTP) and depression (LTD) are the basic mechanisms of long-lasting modifications of synaptic efficiency. Hebb [9] postulated that when presynaptic activity coincides with postsynaptic activity, the connection between both neurons is strengthened. According to recent experiments, the modification of synaptic efficiency depends on precise timing of afferent signals (neurotransmitters binding to receptors) and the postsynaptic neuron spike. LTP seems to require that postsynaptic action potential be simultaneous or subsequent to EPSP [14, 22]. In short, when the signal from the presynaptic neuron firing arrives before, or during the spike of postsynaptic neuron, the synapse is strengthened (LTP). In contrast, when the signal issued from the presynaptic neuron firing arrives after the spike of postsynaptic neuron, the synapse is depressed (LTD).

3.2 Learning in INFERNET

The plasticity of a synapse w_{ij} is a function of three parameters: the firing time of the presynaptic neuron: $t_j^{(f)}$, the transmission delay between this firing and its effect on the postsynaptic neuron (d_{ij}), and the firing time of postsynaptic neuron $t_i^{(f)}$. Learning in INFERNET consists of modifying the weights of connections between nodes w_{ij} by a value Δw_{ij}. The INFERNET Hebbian learning function is shown in Figure 2. This function follows empirical studies [14, 22]. Similar functions were used in various simulation experiments [12, 16].

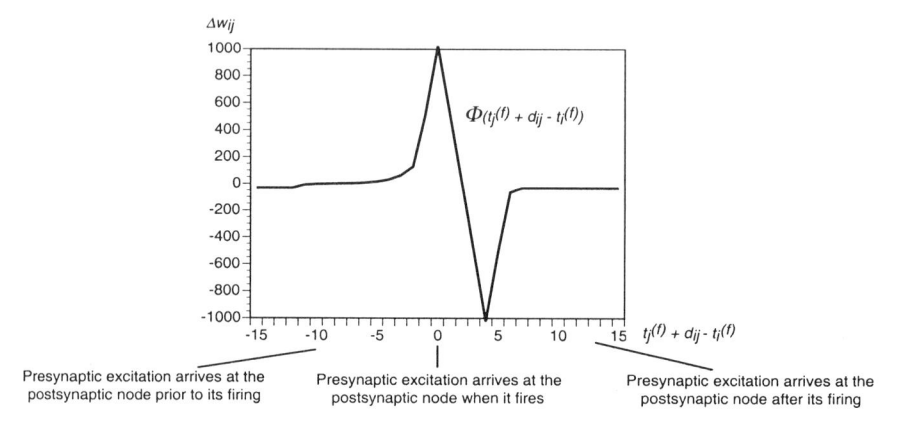

Figure 2: INFERNET Hebbian learning function: when the signal from the presynaptic neuron firing arrives before or during the spike of the postsynaptic neuron, the synapse is strengthened (LTP); when the signal arrives after the spike of the postsynaptic neuron, the synapse is depressed (LTD)

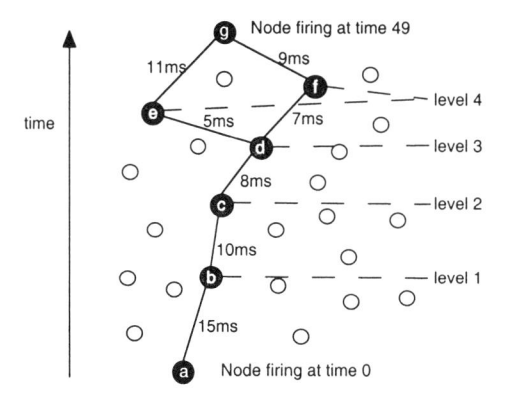

Figure 3: The chaining rule problem: Learning consists of finding a path that makes node g fire 49 ms later than nodes a and a'.

The learning algorithm's job is to reproduce the temporal relation between two successive inputs. This objective is quite difficult because two successive inputs can be separated by several tenths of a second and a single connection cannot alone be responsible for such long delays. A long chain of successive pools of node firings is therefore required. This problem is illustrated in Figure 3. The problem is linking nodes a and a' that fire at time 0 with node g firing at time 49. In the learning phase, only nodes a and a' then 49 ms later, g are externally stimulated. The system has to find a chain of node firing that makes the target node g fire at time 49 when the probe nodes a and a' are firing at time 0. This is the chaining rule problem. The level is defined as the number of steps (or pools of firing nodes) separating the input probe nodes' firing from the target nodes' firing.

3.3 The Chaining Rule

The chaining rule is based on the following assumption: We know the delay of signal propagation from a presynaptic node to a postsynaptic node. From the refractory state we know when a particular node fired, we can therefore detect which synapse can make a node fire at the right moment. In figure 3, one can detect which nodes make node g fire at the right moment (e and f). One can also determine which nodes cause the firing of node e at the right time (d and d'), and so on. In order to reduce combinatory explosion, only the n best contributing nodes are selected for the next level in this chaining rule. Connections between nodes will be modified according to equation (4):

$$\Delta w_{ij} = \Phi\left(t_j^{(f)} + d_{ij} - t_i^{(f)}\right) - \lambda \tag{4}$$

$$\text{where } \lambda = \begin{cases} -level & \text{if } \Phi\left(t_j^{(f)} + d_{ij} - t_i^{(f)}\right) \text{ is negative} \\ level & \text{otherwise} \end{cases} \tag{5}$$

This rule is based on the history of node firing and has some neurobiological justification. For example, the state of a synapse is indicative of its past activity

[15]: a synapse has a memory! Moreover, empirical studies [5] show that LTP also propagates from the originating synapse to neighbouring synapses, giving some plausibility to the chaining rule.

The learning algorithm is triggered only when external input is presented. We can imagine that external input provides a strong signal that triggers the chaining rule. Note that Hebbian learning does not seem to be dependent on this kind of signal and affects synapses after an action potential. Here, the target input is the signal to launch the chaining rule. The objective is to link the probe nodes' firing to the target nodes' firing and to avoid reinforcing other irrelevant firings.

3.4 Algorithm

For each input node firing $t_i^{(f)}$
    For each presynaptic node j
    Calculate Δw_{ij} and add it to w_{ij}
    Select the n best nodes $\{j'\}$
    For each node $j' \in \{j'\}$
        Set level to 1
        For each node j'' presynaptic to node j'
        Calculate $\Delta w_{j'j''}$ and add it to $w_{j'j''}$
        Select the n best nodes $\{j''\}$
        For each node $j'' \in \{j''\}$
            Set level to 2
            Etc. up to level 50.

4. Performance of the Algorithm

In the following experiments two inputs are presented, one (the probe) at time 0 ms and one (the target) some time later. The task for the network is to correctly reproduce the temporal association between these two inputs and therefore build a synfire chain between them. When trained, the network is able to trigger this synfire chain whenever the first input is presented. This ability was taken as a measure of the learning performance. The connections are randomly set and are modified by the learning algorithm, while the delays are randomly set (between 4 and 10 ms for intra subnet connections, and between 10 and 20 ms for inter subnet connections) but do not change. Noise affected thresholds and connection weights but not delays.

4.1 Learning Depends on the Lag between the Probe and the Target

In the first simulation, different lags between the probe and the target were tested. Results are shown in figure 4. Different measures were collected: the mean number of epochs (or association presentation) needed for the network to reproduce the association with a 1 ms precision. Figure 4a shows, not surprisingly, that as the lag between the probe and the target increases the number of presentations required increases. Figure 4b shows the proportion of successes on the 10 trials for different lag times. A success is counted if the network stabilize in a way that the target node fire at the right moment with a 1ms precision. Performance drops significantly for the 120ms group. Figure 4c shows the proportion of the successful trials in which some early and spurious firing occurred and was not removed by the learning algorithm. This emphasizes the difficulty of the task. The algorithm can correctly reinforce a connection (between nodes a and b) that leads to the probe node's firing at the right time, but could also cause the probe nodes to fire earlier if node a fires several times before.

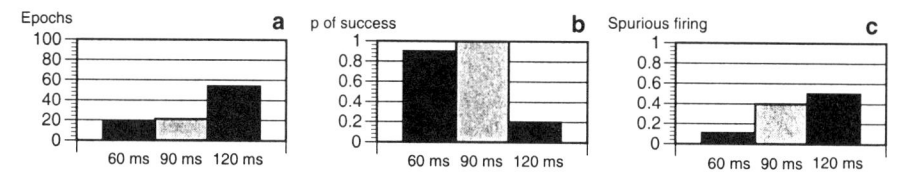

Figure 4: Learning synfire chains of different length. **a**: Number of epochs required for a synfire chain to be learned with a 1 ms precision. **b**: Proportion of learning sucesses. **c**: Proportion of spurious firing within successful cases.

4.2 Learning Depends on the Network Size

In this simulation, different network sizes that had to learn the same synfire chain (a 90 ms chain) were compared. The first network is composed of 800 nodes (50 subnets of 16 nodes), the second network contained 256 nodes (8 subnets of 32 nodes). Ten trials were run for each network. Results shown in figure 5 indicate a strong relation between the net size and the ability to learn synfire chains. The learning algorithm performs better the more nodes there are, providing many alternative paths between the probe nodes and the target nodes. This is consistent with formal analysis [3, 10].

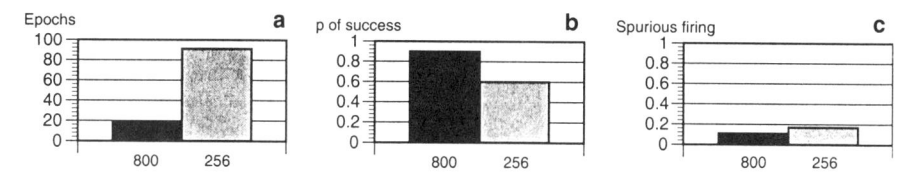

Figure 5: The performance of the algorithm with different network sizes. **a**: Number of epochs required for a synfire chain to be learned with a 1 ms precision. **b**: Proportion of learning sucesses. **c**: Proportion of spurious firing within successful cases.

4.3 The Need for Long Term Depression

This simulation compares the performance of the learning algorithm under three different conditions with ten trials per condition. In the first condition, LTP and LTD were used as described in figure 2, in the second condition LTD was removed but every connection was decayed by a fixed value at every epoch, the last condition only involved LTP. Results shown in figure 6 indicate a need for decaying connection weight. When no decay parameter is used all nodes began to fire too frequently. This is shown by the proportion of spurious firing shown in Figure 6c. Despite the approximately equal performance of the LTD and decay conditions, LTD is to be preferred since decay lowers connection weights blindly and may put the network in a state where it can no longer learn another synfire chain after learning the first one. This experiment seems to confirm data from Munro & Hernandez [16].

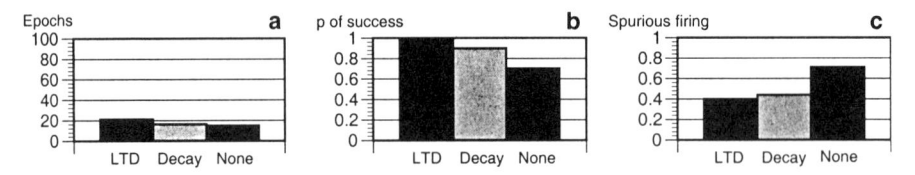

Figure 6: The effect of depression **a**: Number of epochs required for a synfire chain to be learned with a 1 ms precision. **b**: Proportion of learning sucesses. **c**: Proportion of spurious firing within successful cases.

4.4 Learning Depends on Sparseness of Connections

This simulation explores the effect of density of connection on the algorithm performance. INFERNET is organised in fully connected subnets that have a limited and tunable number of connections to other subnets. Two conditions were compared: the low connection density condition involved 50 subnets with 5 connections between each subnets and the high density condition involved 50 subnets with 6 connections between each subnet. Results shown in figure 7 indicate that the learning algorithm works better when the network is sparsely connected. When connections are too dense, target nodes fire too early and only rarely settle on the desired timing. This is probably due to the fact that there are too many direct paths between the probe nodes and the target nodes.

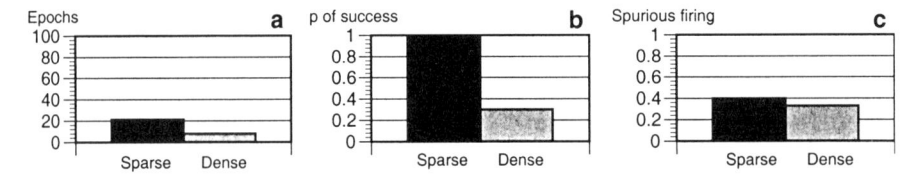

Figure 7: The effect of sparseness of connections. **a**: Number of epochs required for a synfire chain to be learned with a 1 ms precision. **b**: Proportion of learning successes. **c**: Proportion of spurious firings.

4.5 Catastrophic Interference

Catastrophic interference (for a review, see [7]) appears in neural networks when a particular learning experience is suddenly and completely destroyed by subsequent learning. A number of solutions to this problem have been proposed [6, 8]. This section explores catastrophic interference in the present network. In this simulation, involving 10 trials, the network had to learn sequentially 3 synfire chains. Thereafter, it was tested in order to see how much of each sequence it remembered. Results in Figure 8 show that the network remembered equally well all three synfire chains. These results seems to indicate that there is no catastrophic interference (the first learned chains are as well remembered as the last one). This lack of catastrophic interference could be due to the chaining rule in which only the most reinforced connections indicate which nodes can participate in the next level of the chaining algorithm.

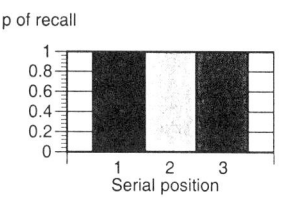

Figure 8: Three synfire chains were learned sequentially. The chart indicates an absence of catastrophic interference.

5. Conclusions

Human learning often involves relating two signals separated in time, or linking a signal, an action and a subsequent effect. These events are often separated in time, but nonetheless, humans can link them, thereby allowing them to accurately predict the right moment for a particular action.

There has been a recent surge in interest in the neurobiology of neuron spike timing. One of the major hypotheses concerning the timing behavior of neurons is synfire chains. Abeles and colleagues [1, 2, 17] have recorded neuron firing in the cortex of monkeys performing a task and have found stimulus dependent sequences of firing that were reproducible with a precision across trials of about 1 ms.

Synfire chains have been taken as a possible mechanism for representing relations between delayed events. This representation could enable the anticipation of an action or an effect following a signal. For example, a tennis player observing his opponent's serve, can anticipate the place where the ball will land and can prepare to return the ball. This is a learned ability. Consequently a set of synfire chain must be learnable and permit the linking of two events with a precise delay.

A learning algorithm based on a Hebbian learning rule has been presented in this paper. A number of simulations indicate that synfire chains can be learned, but that this learning is dependent on several factors, among them, the lag between the probe and the target stimuli, the size of the network (synfire chains require many nodes), the presence of long-term depression, and, finally the sparseness of connections between subnets. Furthermore, some presented results indicate that the learning algorithm presented here is not prone to catastrophic interference when it has to learn sequentially different synfire chains. Further studies are needed to explore how this learning algorithm could apply to more complex tasks and whether its behavior matches human data.

Acknowledgements

This research was supported by the Belgian PAI Grant p4/19 and the European Commission grant HPRN-CT-1999-00065. Special thanks to Robert French for his assistance in the work presented here.

References

1. Abeles, M. (1991). *Corticonics: Neural circuits of the cerebral cortex.* New-York: Cambridge University Press.
2. Abeles, M., Prut, Y., Bergman, H., Vaadia, E., & Aertsen, A. (1993). Integration, Synchronicity and Periodicity. In A. Aertsen (Ed.), *Brain Theory: Spatio-Temporal Aspects of Brain Function* (pp. 149-181). Amsterdam: Elsevier.
3. Bienenstock, E. (1995). A model of neocortex. *Network: Computation in Neural Systems, 6,* 179-224.
4. Calvin, W. H. (1983). A stone's throw and its launch window: timing precision and its implications. *Journal of Theoretical Biology, 104,* 121-135.
5. Engert, F., & Bonhoeffer, T. (1997). Synapse specificity of long-term potentiation breaks down at short distances. *Nature, 388,* 279-284.
6. French, R. M. (1997). Pseudo-recurrent connectionist networks: An approach to the 'sensitivity-stability' dilemma. *Connection Science, 9,* 353-379.
7. French, R. M. (1999). Catastrophic forgetting in connectionist networks. *Trends in Cognitive Sciences, 3,* 128-135.
8. French, R.M., Ans, B., & Rousset, S. (2001). Pseudopatterns and dual-network memory models: Advantages and shortcomings. In R.M. French, & J.P. Sougné (Eds.) *Proceedings of the Sixth Neural Computation and Psychology Workshop: Evolution, Learning, and Development.* London: Springer-Verlag.
9. Hebb, D. O. (1949). *The Organization of Behavior.* New York: Wiley.
10. Herrmann, M., Hertz, J.A., & Prügel-Bennet, A. (1995). Analysis of Synfire Chains. *Network: Computation in Neural Systems, 6,* 403-414.
11. Hertz, J.A., & Prügel-Bennet, A. (1996). Learning synfire chains by self organazation. *Network: Computation in Neural Systems, 7,* 357-363.
12. Levy, N., & Horn, D. (1999). Distributed synchrony in a Hebbian cell assembly of spiking neurons. In *Advances in Neural Information Processing Systems 11,* Cambridge Ma, MIT Press.
13. Maass, W., & Bishop, C.M. (1999). *Pulsed Neural Networks.* Cambridge, Ma: MIT Press.
14. Markram, H., Lübke, J., Frotscher, M., & Sakmann, B. (1997). Regulation of Synaptic Efficacy by coincidence of Postsynaptic Aps and EPSPs. *Science, 275,* 213-215.
15. Markram, H., Gupta, A., Uziel, A., Wang, Y., & Tsodyks, M. (1998). Information processing with frequency-dependent synaptic connections. *Neurobiology of Learning and Memory, 70,* 101-112.
16. Munro, P., & Hernandez, G. (1999) LTD facilitates learning in a noisy environment. In *Advances in Neural Information Processing Systems 11,* Cambridge Ma, MIT Press.
17. Prut, Y., Vaadia, E., Bergman, H., Haalman, I., Slovin, H., & Abeles, M. (1998). Spatiotemporal structure of cortical activity: Properties and behavioral relevance. *Journal of Neurophysiology, 79,* 2857-2874.
18. Sougné, J. P. (1996). A Connectionist Model of Reflective Reasoning Using Temporal Properties of Node Firing. In *Proceedings of the Eighteenth Annual Conference of the Cognitive Science* (pp. 666-671). Mahwah, NJ: Lawrence Erlbaum Associates.
19. Sougné, J. P. (1998). Connectionism and the problem of multiple instantiation. *Trends in Cognitive Sciences, 2,* 183-189.
20. Sougné, J. P. (1999). INFERNET: A neurocomputational model of binding and inference. Doctoral dissertation, Université de Liège.
21. Sterratt, D. C. (1999). Is biological temporal learning rule compatible with learning synfire chains? *Proceedings of the Ninth International Conference on Artificial Neural Networks.*
22. Zhang, L.I., Tao, H.W., Holt, C.E., Harris, W., & Poo, M. (1998). A critical window for cooperation and competition among developing retinotectal synapses. *Nature, 395,* 37-44.

Towards a Spatio-Temporal Analysis Tool for fMRI Data: An Application to Depth-from-Motion Processing in Humans

P.A. De Mazière and M.M. Van Hulle

Abstract

Statistical tools for functional neuro-imaging are aimed at investigating the relationship between the experimental paradigm and changes in the cerebral blood flow. They are usually based on univariate statistical techniques [3], however, since cognitive functions result from interactions, a number of new concepts and multivariate tools have recently been developed [5, 7, 12]. A different approach is to perform Input Variable Selection (IVS), a topic that has recently seen a resurgence of interest in the neural network modelling community. It boils down to the selection of a subset of "prototypical" signals that satisfy best a pre-specified criterion. We perform IVS by considering the data set as originating from a multivariate, attractor-based system, and select the prototypes that best approximate the attractor's dynamics [2, 6, 11]. In this way, we are able to consider both spatial and temporal correlations in one pass. As an example, we apply our technique to a fMRI-study concerning "Depth-from-Motion Processing in Humans" and compare our results with those obtained with the popular SPM technique.

1. Introduction

One of the major challenges in neuro-imaging is the development of powerful tools for analysing brain signals obtained with non-invasive scanners. The higher temporal resolution of fMRI and PET kinetic scanners [13] provides the neuro-scientist with huge amounts of data that need to be processed with fast, reliable analysis tools that are able to distinguish between signals of interest and noise, cardio-respiratory artifacts,...

Several research groups are involved in developing software packages, based on different views and theories on how to approach the issue, such as *Statistical Parametric Mapping (SPM)* (mainly based on Gaussian t-fields and the general linear model (GLM), [3, 4]), *Lyngby* (based on PCA, FCA and similar techniques, [7]), and *Analysis of Functional NeuroImages AFNI* (based on correlations, [1]).

Approaches based on artificial neural network (ANN) techniques have, up to now, largely been overlooked or, at least, not fully explored. The purpose of this contribution is to develop a tool for analysing fMRI data. The tool belongs on a class of techniques that have recently been introduced in the ANN field, namely Input Variable Selection (IVS). Since a periodic block-design has been adopted, the selection will be based on the expected periodic nature of the registered brain signals. In this way, active brain regions can be discerned from non-active or from noise and artifacts. The technique is derived from chaos theory. We compare our technique to other well-known ones such as PCA and correlation-based techniques. We will consider real fMRI data from a depth-from-motion study and compare our results to those reported in [10], which where obtained with SPM.

2. Preliminaries

Our technique exploits the apparent similarity between the expected periodic nature of the fMRI signals recorded in a block-design, and that of attractors in chaos theory. As an example of the latter, consider the Lorenz-attractor (Fig. 1, left panel): two "stable" states between which the system continuously oscillates. What is characteristic for this attractor is that similar patterns re-occur after a certain time, possibly with small deviations in amplitude.

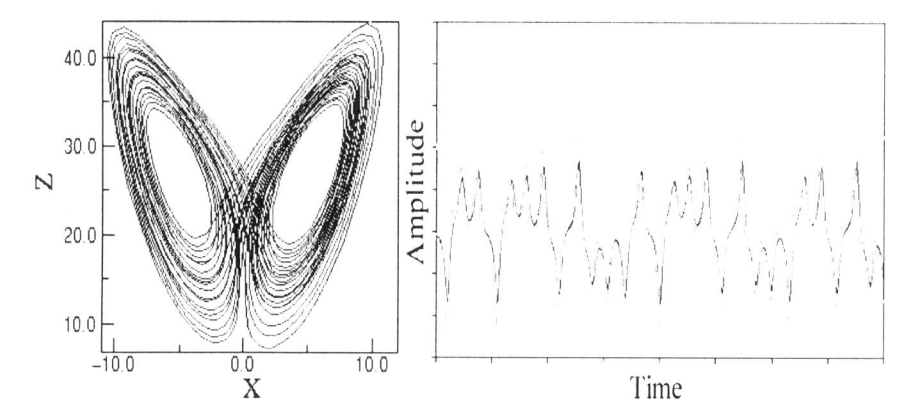

Figure 1: *Left panel:* A two dimensional graph of the three-dimensional Lorenz attractor. The two ring-shaped attractor states are clearly visible. *Right panel:* The temporal evolution of the amplitude along the three coordinate axes. Note the periodic nature of the three signals

During each run in a block-design, the subject is exposed to a periodic sequence of visual stimuli while successive scans of the brain are taken. The sequence consists of an alternation of two or more visual stimuli. Hence, if certain brain regions are responsive to stimuli in this sequence, we expect these regions to respond in a periodic manner. In other words, we expect them to correspond to different states of brain activity (Fig. 2). Noise, such as due to the cardio-respiratoric cycle, fMRI scanner drift, exogenous magnetic sources, motion artifacts,... do affect the registered fMRI signal, resulting in deviations in the attractor dynamics. In our case, each brain image is regarded as a point in D-dimensional space that defines the attractor; the images are sized 79 × 95 × 68 voxels (≈ 500k voxels) and each run consists of 120 brain images, i.e. 120 scans in time

The effective dimensionality of the attractor is defined by the theorems of Procaccia & Grassberger [6, 11]. It is aimed at characterising the attractor's spatio-temporal structure. It forms the cornerstone of our tool.

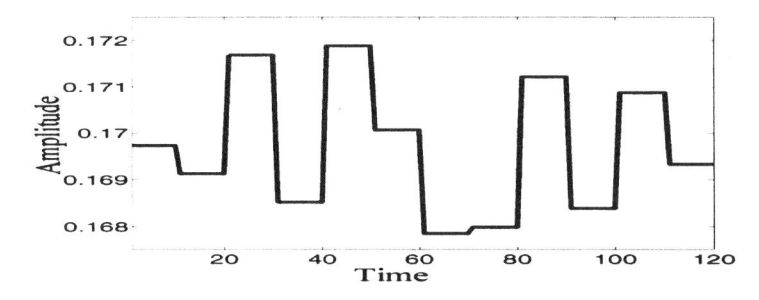

Figure 2: An averaged fMRI signal of a voxel, averaged within each block of the 120 sample run. Note the periodic nature of the signal. The entire run consists of two times 6 consecutive stimuli, each lasting for 10 images [10]

3. Methods

One run counts 120 3-dimensional brain images and each brain image consists of about 500k voxels. This results in a data set of more than 60M values, covering a 4 dimensional space (1 temporal and 3 spatial dimensions). The idea is to identify voxels, the spatio-temporal characteristics of which define brain-regions that code for the stimulus paradigm.

First, we represent the dataset in a high-dimensional space so that the temporal behaviour of each voxel can be explored with respect to that of the other voxels in the brain image. More in particular, each brain-image is represented as a vector in 500k-dimensional space, which allows us to study the temporal transitions in different brain signals.

In order to facilitate the data analysis, a dimensionality-reduction is performed. This is done by means of the Procaccia & Grassberger theorem. The latter states that the effective dimensionality is based on the spatio-temporal structure of a set of time tracks. As a result of this, time tracks that are periodic in nature will contribute to the effective dimensionality. Applied to our case, we determine the effective dimensionality of the dataset. The dimensionality-reduction is then achieved by selecting the voxels of which the fMRI signals contribute most to the effective dimensionality. This selection procedure ensures that only those voxels will be selected that are responsive to the alternating stimulus sequence. Noise sources, lacking this periodic feature, will therefore be ignored. Note that, since we in fact are selecting a subset of voxels, we perform IVS on the voxels

Since the Procaccia & Grassberger formula provides information about a collection of time tracks and not about a particular time track (contrary to, *e.g.*, an autocorrelation coefficient or the variance of that track), the voxel selection procedure can be regarded as an optimisation problem for which the selected voxels approximate the effective dimensionality of the entire brain within a given accuracy.

The selection procedure works as follows. First, a small subset of voxels is selected, mostly the ones that possess the largest signal variability: there is a good chance that they will contribute most to the attractor dynamics. We determine the effective dimensionality of this initial subset. Then, voxels are added to this initial subset. The effective dimensionality is determined and compared to the previous one and with that of the entire dataset. Based on the outcome, the added voxels are removed or left in the subset. We iterate this addition and removal of voxels until the desired accuracy is reached. Each voxel that is part of the final subset is called *prototype*. The ensemble of the time tracks of all prototypes characterises the entire dataset. The selection procedure is quite robust since starting from other subsets of voxels yields the same final set of prototypes.

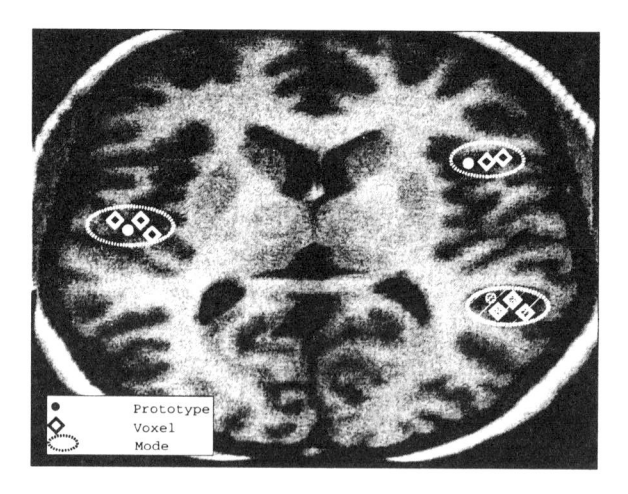

Figure 3: Three modes are drawn on top of a brain slice, each one of them consisting of one prototype (filled circle) and a few other voxels (diamonds). Voxels of the same mode are more similar in their temporal behaviour than voxels of differing modes

Each prototype is also the representative of a brain region, namely, a collection of voxels, in the vicinity of the prototype with the same temporal behaviour, up to some minor deviations ("noise"). Note that the effective dimensionality of the ensemble of the prototypes is not changed when a voxel is added the temporal behaviour of which is similar to one of the prototypes. We call the aforementioned collection of voxels a *mode*: its definition is based not only on the temporal similarity between the fMRI signals of these voxels and that of their common prototype, but also on the spatial proximity of these voxels so that they form a spatially connected set of voxels. An example of such modes is shown, in a schematic manner, in Fig. 3. Modes are likely to represent brain regions or parts of them.

Note that, contrary to the existing approaches, we have not used any information about the stimulus paradigm to identify active brain regions, except for the fact that we expect these regions to display a periodic behaviour.

The benefits are clear: no assumptions about the paradigm or about the haemodynamic response to the paradigm need to be made, no spatial or temporal smoothing is performed which could compromise the resolution, less parameters need to be specified by the user, etc...

Once the modes are determined, we relate the fMRI signals of the prototypes to the stimulus paradigm, by means of a regression function based on the haemodynamic response function (HRF)-convolved model, and which represents the stimulus paradigm. The resulting regression coefficients are combined in accordance to the paradigm, and result in t-values and corresponding p-values [8, 9]. The p-values then represent the statistical significance of the brain regions identified.

4. Comparison with PCA

PCA searches for an orthogonal set of axes for which the projections onto them code for the largest variance in the data set. The data set is then projected onto the subspace spanned by these axes. Hence, PCA is a projection-based technique that performs a dimensionality reduction. The orthogonal axes are combinations of the original coordinate axes: PCA does not select subsets of the original coordinate axes. Hence, our motivation is to develop a technique that selects dimensions rather than linear combinations of them.

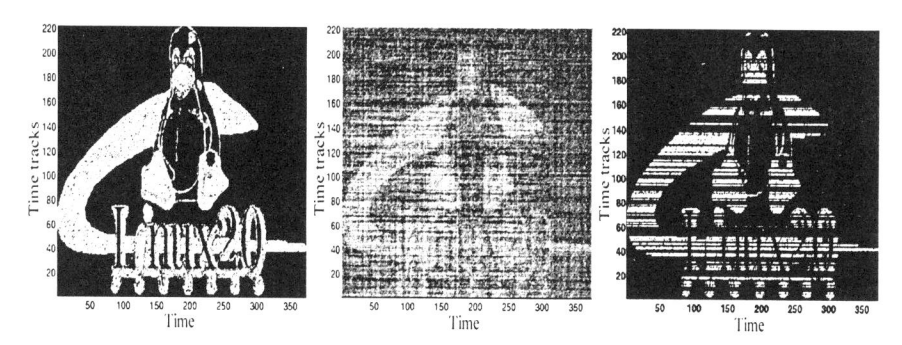

Figure 4: *Left panel:* The original data set. The horizontal axis represent the time axis; the vertical axis the 220 different signals. *Middle panel:* The result after a PCA transformation: the original signals are projected onto the subspace spanned by the first 79 principal components. *Right panel:* The result obtained with our selection technique. The original signal no. 44, its PCA-transformation, and the signal selected with our technique are illustrated in Fig. 5

In order to show the difference between the two techniques, consider the next experiment. We take a synthetic dataset, namely the well-known Linux-penguin (Fig. 4, left panel): the horizontal axis must be regarded as the time axis, while the vertical axis represents the distinct signals.

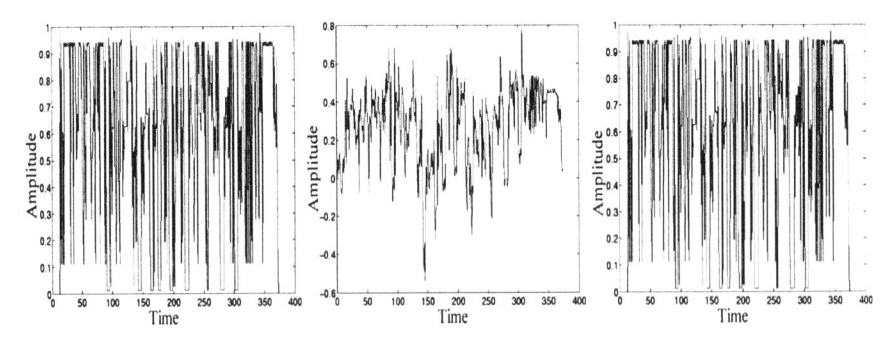

Figure 5: *Left panel:* Signal no. 44 taken from Fig. 4 (left panel). *Middle and right panels:* the PCA-transformed and the result obtained with our selection-based technique, respectively. The distortion introduced by PCA is apparent

Fig. 5 (left panel) shows an example signal taken from Fig. 4 (no. 44), Fig. 5 (middle panel) shows the result after projection onto the subspace spanned by the first 79 principal components (PCs). The number of PCs were chosen in such a manner that the projected set has the same effective dimensionality as the original set. Finally, Fig. 5 (right panel) shows the result obtained with our selection technique: of the original 220 signals, the 64 signals are retained that best approximate the effective dimensionality of the original set. Note that our technique, since it is selection-based, and not projection-based, does not distort the selected signals. Furthermore, both techniques yield signals sets with comparable effective dimensionalities but our technique seems to require less signals than PCs.

5. Statistical Parameter Mapping

Statistical Parameter Mapping (SPM) [3, 4], as the name suggests, is based on statistics and a general linear model (GLM) to represent the stimulus paradigm. The signals are temporally and spatially smoothened prior to the actual analysis. Note that these operations modify the original signals. In the actual analysis, the smoothened signals are fitted to the GLM. The resulting (3D) map of regression coefficients shows how well the signal of each voxel relates to the stimulus paradigm. By setting a user-specified significance level, the voxels, the regression coefficients of which exceed this level, will be further considered and identified with known brain regions.

The considerable number of parameters the user needs to specify in SPM is another disadvantage, besides the need for prior smoothing: its selection determines, *e.g.*, the extent of the spatial smoothing operation which, in turn, determines the size of the smallest brain region that can be detected. Prior knowledge of the brain region is often assumed in order to set the smoothing parameters correctly.

6. Results

6.1. Study

In this section, we will compare the active brain regions that have been identified by our tool to those identified by SPM. We use a data set obtained from a study performed by Orban and co-workers [10]. The goal of this study was to identify the brain areas involved in 3D structure from motion. The analysis was done using SPM and a factorial stimulus design was applied. Co-registration and normalisation towards Talairarch-Tournoux space were performed, and linear and second-order trends were removed by means of the GLM. The brain areas anterior to the centerplane in axial-coronal space were discarded. For more information concerning the experimental conditions, SPM-analysis and the results, we refer to [10]. Due to the factorial design, the effects of the stimuli can be compared. This comparison is performed in SPM by means of the GLM (regression analysis).

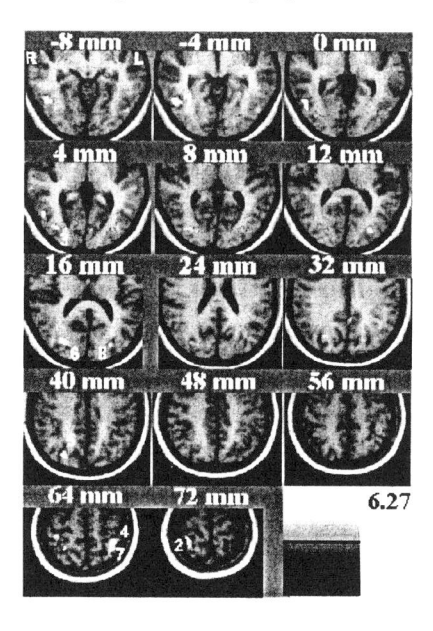

Figure 6: Axial view of the active brain regions for the 3D minus 2D stimuli case. Legend: 1 = right hMT/V5+, 2 = right DIPSA, 3 = right LOS, 4 = left DIPSA, 5 = right POIPS, 6 = right TRIPS, 7 = left DIPSL, 8 = left TRIPS (for the definitions of these abbreviations see [10]). The used scale represents a difference in z-score (A full-colour version of this paper is available on the internet: http://simone.neuro.kuleuven.ac.be/patrick/NCPW6.ps)

The main finding of this study was that the group analysis only yielded significant effects when the fMRI data were regressed with the stimulus paradigm "3D minus 2D stimuli". Brain region hMT/V5+ was detected as the most significant region [10], as shown in Fig. 6.

6.2. Comparison between SPM and our Tool

Our tool is now applied to a single subject of the same data set. First, as was the case for the SPM analysis, co-registration, Talairarch-Tournoux normalisation and re-alignment for head movements are performed, as well as a linear trend removal. However, contrary to the SPM case, no smoothing (temporal or spatial) is applied. Similarly to what is done in the pre-processing stage of SPM, in order to discard artifactual data, voxels of which the average temporal activation does not exceed 90% of the mean activation, are not included in the analysis. (Note that, in SPM, the user has no direct access to the number of voxels selected in the pre-processing stage.)

The accuracy with which the effective dimensionality is approximated is set to 3%. This results in 776 prototypes out of 306,204 voxels, *i.e.*, a 400-fold reduction. Starting from these prototypes, the brain regions ("modes") are determined with respect to the following constraints:

- a mode must contain at least 10 connected voxels;

- the signals of two voxels are regarded similar if their normalised Euclidean distance is less than 30% of the maximal distance between two prototypes;

- large modes ($> 1,000$ voxels) are discarded, since they are likely to be artifactual (*e.g.*, attributable to head movement). In this case, two large modes had to be removed, containing 2,597 and 1,020 voxels;

As a result, 82 modes or brain areas are retained, corresponding to 2,386 voxels in total. In the final step of the analysis, the representation for the stimulus paradigm is related to the 82 modes by means of the regression function, and the resulting coefficients are further evaluated using statistical procedures [8, 9] in order to obtain statistical significance values for the different modes. We used t-values for expressing the statistical significance of the relevant modes (Fig. 7). Note that we show more slices since the spatial resolution of our technique is higher than SPM's.

7. Conclusion

We have presented a new tool for analysing fMRI data based on a selection technique (IVS). The tool distinguishes itself from others by its higher spatial resolution and its ability to detect active brain areas, without prior knowledge about the stimulus paradigm, except for its periodic nature. In addition, the tool needs only a small number of parameters, such as those related to the physiological haemodynamic response function (HRF) and the physical time of registration (TR). Parameters needed for the analysis itself are the selection accuracy and the lower and upper limits on the amount of voxels in order for a brain region to be regarded as a valid one.

In addition to the absence of the need to smooth the data prior to the analysis, another advantage is that the ubiquitous linear and Gaussianity assumptions are only used at the end of the analysis, namely, for obtaining the significance levels (p-values) for the identified active regions.

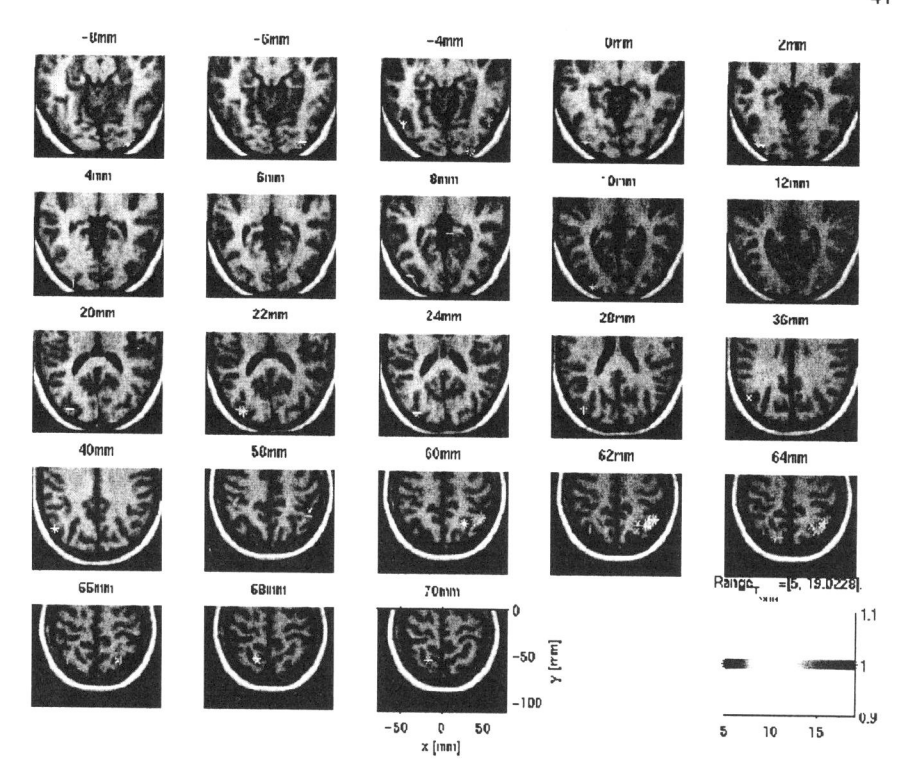

Figure 7: Axial view of the significant brain regions detected by our tool. Note that many of the regions found by SPM, are also found by our tool. Our tool also detects additional (smaller) regions, which is a result of the higher spatial resolution of our tool (Note that a full-colour version of this paper is available on the internet: http://simone.neuro.kuleuven.ac.be/patrick/NCPW6.ps)

Acknowledgements

We wish to thank J. De Brabanter, W. Van Assche and G. Verbeke, all from the K.U.Leuven, for fruitful discussions on the statistical aspect of this work, and G.A. Orban for kindly granting us permission to use his fMRI data. P.A.D.M. is supported by the Flemish Ministry for Science and Technology (VIS/98/012), and by the European Commission (Quality of Life and Management of Living Resources, QLG3-CT-2000-30161). M.M.V.H. is a research associate of the Fund for Scientific Research – Flanders (Belgium) and is supported by research grants received from the Fund for Scientific Research (G.0185.96), the National Lottery (Belgium) (9.0185.96), the Flemish Ministry of Education (GOA 95/99-06), and the Flemish Ministry for Science and Technology (VIS/98/012).

References

1. Cox, R.W. (1996). AFNI: Software for analysis and visualisation of functional magnetic resonance neuroimages. *Computers and Biomedical Research*, *29*, pp. 162-173.

2. Eckmann, J.-P. & Ruelle, D. (1985), Ergodic theory of chaos and strange attractors. *Reviews of Modern Physics, Vol. 57*, no. 3, part I, pp. 617-656.

3. Frackowiak, R.S.J., Friston, K.J., Frith, C.D., Dolan, R.J., & Maziotta, J.C. (1997), Human Brain Function. *Academic Press*

4. Friston, K.J., Holmes, A.P., Worsley, K.J., Poline, J.P., Frith, C.D., & Frackowiak., R.S.J. (1995). Statistical Parametric Maps in Functional Imaging: A General Linear Approach *Human Brain Mapping 2*, pp. 189-210.

5. Gold, S., Christian, B., Arndt, S., Zeien, G., Cizadlo, T., Johnson, D.L., Flaum, M. & Andreasen, N.C. (1998). *Human Brain Mapping, 6*, 73-84.

6. Grassberger, P. & Procaccia, I. (1983), Measuring the strangeness of strange attractors. *Physica, D9*, pp. 189-208.

7. Lars Kai Hansen, Finn °Arup Nielsen, Peter Toft, Matthew G. Liptrot, Cyril Goutte, Stephen C. Strother, Nick Lange, Anders Gade, David A. Rottenberg, Olaf B. Paulson (1999), "lyngby" - a modeler's Matlab toolbox for spatio-temporal analysis of functional neuroimages. *NeuroImage, 9*(6, part 2), pp. S241.

8. Montgomery, D.C. (1997). Design & Analysis of Experiments, 4th edition. *John Wiley & Sons*

9. Montgomery, D.C., & Runger, G.C. (1999). Applied Statistics and Probability for Engineers, 2nd edition.*John Wiley & Sons*

10. Orban, G.A., Sunaert, S., Todd, J.T., Van Hecke, P., & Marchal, G. (1999). Human Cortical Regions Involved in Extracting Depth from Motion. *Neuron, Vol. 24*, pp. 929-940.

11. Procaccia, I. (1984), The static and dynamic invariants that characterise chaos and the relations between them in theory and experiments. *Proceedings of the 59th Nobel Symposium (Phys. Scr. T9, 40)*.

12. Tononi, G., McIntosh, A.R., Russell, D.P. & Edelman, G.M. (1998). Functional clustering: identifying strongly interactive brain regions in neuroimaging data. *Neuroimage, 7*, 133-149.

13. Worsley, K.J. (1999). Statistics of Brain Mapping. 52nd *Session of the International Statistical Institute*, Helsinki, Finland.

A Simple Model Exhibiting Scalar Timing

J. L. Shapiro, John Wearden, and Rossano Barone

Abstract

Animal data on delayed-reward conditioning experiments shows a striking property — the data for different time intervals collapses into a single curve when the data is scaled by the time interval. This is called the scalar property of interval timing. Here a simple model of a neural clock is presented and shown to give rise to the scalar property. The model is an accumulator consisting of noisy, linear spiking neurons. It is analytically tractable and contains only three parameters. When coupled with simple reinforcement learning it simulates peak procedure experiments and temporal bisection.

1. Introduction

Numerous experiments have investigated the ability of animals to estimate time intervals and wait appropriately for delayed rewards. A typical type of experiment is called the peak procedure. In this, the animal is trained to respond after a given time interval has elapsed. There is no penalty for responding before the time has passed, but the animal only receives a reward at the first response after the time interval has passed. On most trials, the first response ends the trial.

On some trials, however, no reward is given even when the animal responds appropriately. This is to see when the animal stops responding. What typically happens in non-reward trials is that the animal starts responding at some time, responds for a period, and then stops responding. Responses averaged over many trials give a smooth curve. The highest response is at the time interval, and there is variation around this. Typical curves are illustrated in figure 1. This has been seen in many species, including rats, pigeons, turtles; humans will show similar results if the time intervals are short or if they are prevented from counting through distracting tasks. For reviews of interval timing phenomena, see [8] and [6].

A striking feature of the data on peak procedure and other experiments on time intervals is the so called "scalar property". In its weak form, this states that the variation in the response of the animal, as measured by the standard deviation of the response time, scales with the length of the time interval. Since the mean response time also scales with the length of the time interval, the ratio of the standard deviation to the mean response time (the *coefficient of variation*) is a constant independent of the time interval. This is shown in the left side of figure 2. The right side of figure 2 shows a stronger form of the scalar property. When the response probability is multiplied by the time interval and is plotted against the relative time (time divided by the time interval), the data from different time intervals collapse on to one curve.

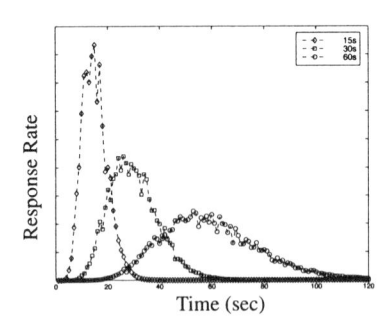

Figure 1: The typical appearance of the average response probability during non-reward trials in a peak procedure experiment as a function of time for three different time intervals (15, 30, and 60 seconds). This figure and following one are schematic views of the data in [5].

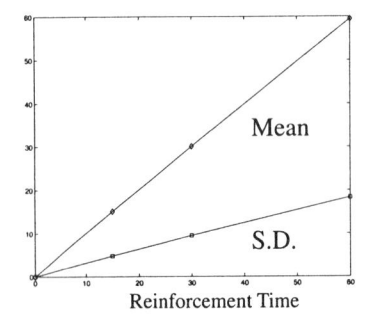

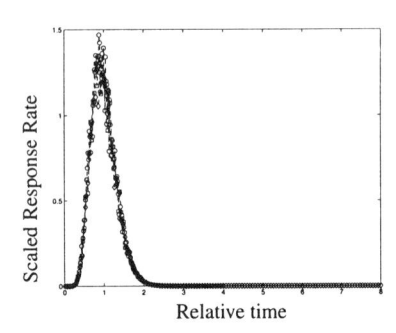

Figure 2: Two illustrations of the scalar property. The figure on the left shows that during non-reward trials the mean response time and its standard deviation are proportional to the time interval. The figure on the right illustrates that the data collapses when the response probability scaled with the time interval and plotted against relative time.

This strong form of the scalar property can be expressed mathematically as follows. Let T be the actual time since the start of the trial and $\tilde{T}$ be subjective time. Subjective time is the time duration which the animal perceives to have occurred, (or at least appears to perceive judging from its behaviour). The experiments show that $\tilde{T}$ varies for a given T. This variation can be expressed as a conditional probability, the probability of acting as though the time is $\tilde{T}$ given that the actual time is T, which is written $P(\tilde{T}|T)$. The fact that the data collapses implies this probability depends on T and $\tilde{T}$ in a special way,

$$P(\tilde{T}|T) \approx \frac{1}{T} P_{\text{inv}}\left(\frac{\tilde{T}}{T}\right). \qquad (1)$$

Here P_{inv} is the function which describes the shape of the scaled curves. Thus, time acts as a scale factor. This is a strong and striking result.

A key question which remains unanswered is: what is the origin of the scalar property. Since the scalar property is ubiquitous across experiments and species, it may be revealing something fundamental about the nature of animals' internal clocks. But it has been difficult to think of processes which produce it. It is well known that any model based on the accumulation of independent errors, such as a clock with a variable pulse-rate, does not produce the scalar property. In such a model it would be the ratio of the *variance* to the mean response time which would be independent of the time interval (a consequence of the law of large numbers). In section 3, a simple stochastic process will be presented which gives rise to scalar timing.

In section 5, we show that reinforcement learning coupled to our model of scalar timing simulates experiments on peak procedure and another experiment called temporal bisection.

2. Previous Work

One way to generate the strong scalar property is to assume that there is a source of multiplicative noise. If one assumes that subjective time $\tilde{T}$ is related to actual time T through a noise source ξ

$$\tilde{T} = \xi T, \qquad (2)$$

then equation (1) immediately follows, with P_{inv} equal to the noise distribution of ξ. A standard information processing model called Scalar Expectancy Theory [4] posits that multiplicative random variables intervene between the value of the count accumulated from a clock reinforcement time and retrieval time (see, for example [5] [6]). The clock is assumed to be an accurate time-keeper. The multiplicative noise is interpreted as motivational and attentional factors, and coding and decoding. It has been argued that it is possible to manipulate this noise source using drugs [9]. However, there does not appear to be any proposed mechanism for the noise, or why the accurate clock is not accessible. The scalar property is put in at the outset.

A connectionist model of scalar timing was proposed by Church and Broadbent [1]. This consists of a number of oscillators of differing frequencies, an associative memory to store time represented as the phases of the oscillators, and a comparator for decision making. The scalar property is built into the model through the assumption that the frequency of each oscillator is multiplied by a random variable generated from the same noise source. The model does simulate a number of interval procedures qualitatively [14], but does not provide quantitative predictions. A similar model used oscillators coupled to additive Gaussian noise in which the noise variance was proportional to the time interval [3]. This model described acquisition and particularly extinction between trials in rats, but again fails to explain the origin of the scalar property, as that is put into the model at the outset.

Another connectionist model was proposed by Grossberg and Merrill [7]. This model also consists of numerous clocks in the form of a number of neurons each of which is tuned to respond after a different time delay. This model also addresses the reinforcement learning problem. At reinforcement the response is associated with the clock nodes. Although the scalar property is not directly tested in this model, each clock node appears to have a tuning curve which has increasing width with increasing delay and the response appears to reproduce the scalar property qualitatively. This is a consequence of the dynamical equations used to described them. Grossberg argues that these equations describe plausible dynamics for real neurons. Thus, this model does seem to propose an explanation of the origin of the scalar property — it originates from the dynamics of neurons. However, the model does require many self-similar clock nodes, and contains a number of parameters which must be set to get an appropriate range of tuning curves of appropriate shapes.

A set of models which is very effective at modelling Pavlovian eye-blink experiments has been proposed by Desmond and Moore [10] based on the model of Sutton, Barto, and Desmond [11]. This model represents time using a tapped delay line; at each time-step, a different node in the delay line is activated. Time acts as one of the conditioned stimuli which using temporal difference (TD) reinforcement learning is associated with the response through the unconditioned stimulus. This models elucidates the reinforcement learning problem very clearly and simulates many experiments, but does not attempt to model the scalar property; time is represented accurately by the system.

3. A Simple Stochastic Network Which Produces the Scalar Property

We now consider a very simple connectionist model which reproduces the scalar property. It consists of a network of spiking neurons which are noisy and linear. The network encodes time as the quantity of activity, i.e. the number of spikes in the network. This grows over time.

The network consist of N identical neurons. Any neuron can potentially be connected to any other; the actual connectivity pattern is defined by a connection matrix **C**, where C_{ij} is 1 if neuron j receives input from neuron i and 0 otherwise. The connection strength is the same between all connected neurons and is taken to be 1. We assume that there is a characteristic time τ (10mS, say) required for a spike produced by a neuron to invoke a spike in a connected neuron. In principle, the spikes should be modelled as stochastic in time, e.g. via a Poisson process. However, we use strict, discrete units of size τ. There is no limit to the spiking rate of these neurons, so saturation effects are ignored.

The basic assumptions concerning the interactions between neurons are as follows,

Assumption 1 *There is a probability γ of a spike crossing a synapse,*

Assumption 2 *Spikes do not interact at synapses; they behave independently.*

Thus, these neurons act like noisy linear neurons. The expected number of spikes produced by a neuron is just γ times the number of pre-synaptic spikes. In other words, if $x_i(t)$ denotes the number of spikes produced by neuron i at time t, and $h_i(t)$ is the number of spikes feeding into that neuron,

$$h_i(t) = \sum_j C_{ji} x_j(t), \qquad (3)$$

then $x_i(t+\tau)$ is drawn from a binomial distribution with mean $\gamma h_i(t)$ and variance $\gamma(1-\gamma)h_i(t)$.

Since the expected number of spikes produced by the neuron is less than the number exciting it, there is a loss of network activity when spikes cross synapses. However, the number of spikes also increases due to the fan-out of each neuron. Let C_i denote the fan-out of the ith node, $C_i = \sum_j C_{ij}$. To simplify the discussion, assume that all neurons have the same connectivity, $C_i = C$ for all i, (This is only necessary to hold on average and that is how the simulations in the next sections are done.) Under this assumption, the total activity of the network $n(t)$ can be expressed in terms of that in the previous time-step, $n(t-\tau)$, because $n(t)$ is drawn from a binomial distribution with mean $\gamma C n(t-\tau)$ and variance $\gamma(1-\gamma)C n(t-\tau)$.

At each time-step, the number of spikes will grow due to the fan-out of the neurons; each neuron can excite several other neurons. At the same time, the number of spikes will shrink due to the fact that a spike invokes another spike with a probability less than 1. An essential assumption of this work is that these two processes balance each other. This is required to get the mean activity to be proportional to the time interval.

Assumption 3 *The amplification of neuron activity due to each neuron exciting multiple other neurons exactly balances the diminution of activity due to the finite probability a neuron exciting other neurons.*

In other words,

$$C\gamma = 1. \qquad (4)$$

Finally, in order for this network to act as an accumulator, it must receive input during the time interval which is being perceived. The input is assumed statistically the same at each time-step.

Assumption 4 *During a perceived time interval, the network receives constant or statistically stationary input.*

These assumptions modify the dynamics for those neurons connected to the inputs. Let $I(t)$ denote the number of spikes externally excited in the network at time t. The equation for the total neuron activity is,

$$n(t+\tau) = \sum_{\alpha=1}^{Cn(t)} \eta_\alpha + I(t). \tag{5}$$

where the η's are random variables — they are 1 with probability γ and 0 with probability $1-\gamma$ It is this equation which gives rise to the main features of scalar timing.

4. Discussion of the Model

This simple network is sufficient to give rise to scalar timing. Elsewhere [12] we show that for large time t, the network activation obeys the scalar property. Let $P(n|t)$ be the probability of network having total activity n at time t. The result is,

$$\lim_{t\to\infty} P(n|t) = \frac{1}{t}P_{\text{inv}}(n/t), \tag{6}$$

with corrections of order $1/t$. Thus, if an interval of time is measured through the activity of this network, the perception of time will obey the strong scalar property (equation 1).

For example, the coefficient of variation is

$$Q(t) = \frac{\sqrt{1-\gamma}}{\sqrt{2}m_I}\left[1+\frac{\tau}{t}\left(1+\frac{2\sigma_I^2}{1-\gamma}\right)\right] + O(\frac{\tau}{t})^2. \tag{7}$$

where m_I is the mean external excitation, and σ_I is the variance in the external excitation. This goes to a constant for large t. More generally, the scalar distribution, P_{inv} is a gamma-distribution with mean m_I/τ and variance $(1-\gamma)m_I/(2\tau^2)$. These two combinations of the three parameters of the model fully describe the system for large t.

5. Learning Behaviour

The above model represents a way of representing and measuring time. How would this be used to produce behaviour? Here a simple model is considered in which time is converted into a spatial code which is then associated with time intervals. Other methods of encoding time using this clock and more details will be presented elsewhere [12].

5.1 Learning Peak Procedure

In this model, the accumulator network feeds into a set of nodes which encode the activity of the accumulator. Each of these node is tuned to respond to a single value of accumulator activity. I.e. for any accumulator network output there is a single node which responds. These nodes for a spatial encoding of time.

For each possible response there is a response node. Each of the spatial code nodes feed into response nodes. The connection between the the ith spatial node and the jth response node is w_{ij}; it is these connections which learn to associate a response with an appropriate time delay. There is also a connection between the stimuli and the response nodes. It is the combined influence of the stimulus and the spatial nodes which cause the response node to fire. The association between stimulus and response is not modelled, but could follow a standard TD learning rule.

This model uses the TD learning equations of Desmond and Moore [2] as described in [11]. The spatial nodes play the role of the tapped delay-line nodes in that model. However, here they are stimulated by the accumulator rather than each other, and they will follow a stochastic trajectory due to the fluctuating nature of the accumulator. Learning uses an eligibility trace for the spatial nodes, and learning is via a simple association rather than a TD rule.

The model as been used to simulate peak procedure. In the simulations, the model is forced to respond for the first set of trials (50 trials in the simulations); otherwise the model would never respond. This could represent shaping in real experiments. After that the model learns using reward trials for an additional number of trials (100 trials in these simulations). The model is then tested on a number of non-reward trials.

Figure 3 shows the average responses for non-reward trials, averaged over 100 runs. The figure also shows that the model produces the scalar property quite strongly.

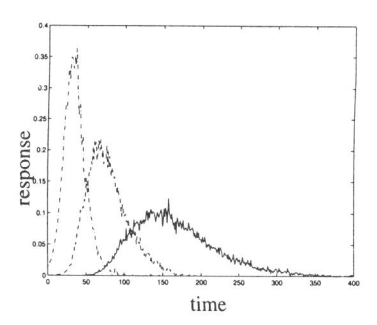

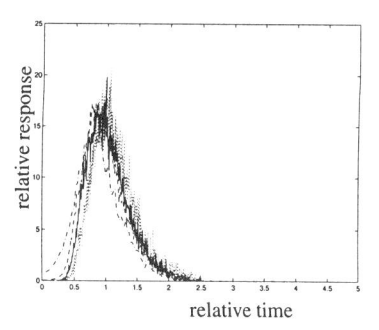

Figure 3: Left figure: Average response of the spatially encoded network for non-reward trials. The time intervals are 40τ (dot-dashed line), 80τ (dashed line), and 160τ (solid line). 100 non-reward trials have been averaged. Right Figure: The response probabilities scaled with the time interval and plotted against relative time. Same details as previous figure with addition of time interval of 320.

5.2 Learning Temporal Generalization

Another type of procedure which can be modelled is temporal bisection. In these experiments, a stimulus for a short duration S results in rewards for one action; the same stimulus for a longer duration L results in rewards for a different action. After an animal has been trained on the two intervals, intermediate intervals are used in order to see for what duration the animal produces either action with equal likelihood, the so-called bisection point. It has been observed that the bisection point is below the arithmetic mean and is closer to the geometric mean.

The model has two different response nodes, one trained on an interval of 40 time-steps and the other trained on longer intervals (60, 80, 100, 120, 160, and 320). The bisection point is interpreted as the point where the probabilities of producing each action are equal. In these simulations, there is nothing to stop the network from making both actions simultaneously or make no action. Thus, the two actions are independent, and the bisection point is essentially where curves in figure 3 cross. The results of this experiment are shown in figure 4. The bisection points are systematically below the geometric mean, although the difference is small. One way to estimate the bisection point is to take were the perceived time is the same. That is, the value of the network activity which is equally likely to be generated by time interval L as from time interval S. This is not the geometric mean, but is given by

$$\text{bisection point} = \frac{LS}{L-S} \log \frac{L}{S} \qquad (8)$$

This is systematically less than the geometric mean, but the relative difference is less than 10% unless the long interval is more than 5 times greater the short one. The simulation results are consistent with this value.

6. Discussion

Previous models of interval timing fail to explain its most striking feature — the collapse of the data when scaled by the time interval. We have presented a simple model of an accumulator clock based on spiking, noisy, linear neurons which produces this effect. It is a simple, analytically tractable model. The parameters are: τ — the time for a spike on one neuron to excite spikes on connected neurons, m_I — the average number of spikes excited externally at each time step, and the variance of the branching process, which in this model is $1 - \gamma$. A major weakness of this model is that it requires fine-tuning of a pair of parameters, equation (4), so that the expected number of spikes grows linearly during a measured interval.

Once a scalar clock is produced, simple reinforcement learning can be used to associate the clock signal with appropriate responses. An set of intermediate nodes which encodes time spatially was used here. Simple reinforcement learning between the intermediate nodes at reinforcement and an eligibility trace simulates peak procedure and bisection experiments, although the bisection point is below the geometric mean.

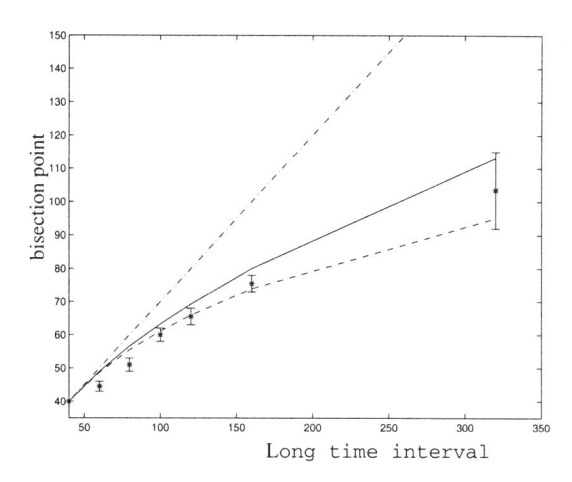

Figure 4: The results of bisection experiments. The short time interval is 40τ; the bisection point is shown for long intervals of 60τ, 80τ, 100τ, 120τ, 160τ, and 320τ. The dot-dashed line shows the arithmetic mean, the solid line shows the geometric mean, and the dashed line is the point where the clock output are as likely to have come from either interval, which is given by equation (8)

The next step is to draw connections to psychology and neurology. Because there are only a few parameters in this models, there will be relationships between measurable quantities. For example, it is not clear whether there is sufficient freedom to describe the range of pharmacological manipulations which have been carried out. Until we find what this relationships are in this model and test them experimentally, it will remain unclear how relevant this model is interval timing phenomena.

References

1. Church, R., & Broadbent, H. (1990). Alternative representations of time, number, and rate. *Cognition*, 37:55–81.

2. Desmond, J. E. (1990). Temporally adaptive responses in neural models: The stimulus trace. In Michael Gabriel and John Moore, editors, *Learning and Computational Neuroscience: Foundations of Adaptive Networks*, A Bradford Book, pages 421–456. The MIT Press.

3. French, R. M., & Ferrara, A . (1999). Modeling time perception in rats: Evidence for catastrophic interference in animal learning. In *Proceedings of the 21st Annual Conference of the Cognitive Science Conference*, pages 173–178. LEA.

4. Gibbon, J . (1977). Scalar expectancy theory and WeberÕs law in animal learning. *Psychology Review*, 84:279–325.

5. Gibbon, J. (1992). Ubiquity of scalar timing with a Poisson clock. *Journal of Mathematical Psychology, 35,* 283–293.

6. Gibbon, J., & Church, R. M. (1990). Representation of time. *Cognition, 37,* 23–54.

7. Grossberg, S., & Merrill, J. W. L. (1992). A neural network model of adaptively timed reinforcement learning and hippocampal dynamics. *Cognitive Brain Research,1,* 3–38.

8. Hinton, S. C., & Meck, W. H. (1997). How time flies: Functional and neural mechansims of interval timing. In C. M. Bradshaw, & E. Szadabi, (Eds.), *Time and Behaviour: Psychological and Neurobehavioural Analyses.* Amsterdam: Elsevier Science.

9. Meck, W. H. (1996). Neuropharmacology of timing and time perception. *Cognitive Brain Research, 3,* 227–242.

10. Moore, J. W., Desmond, J. E., & Berthier, N. E. (1989). Adaptively timed conditioned responses and the cerebellum: A neural network approach. *Biological Cybernetics, 62 ,* 17–28.

11. Moore, J. W., Berthier, N., & Blazis, D. (1990) . Classical eye-blink conditioning: Brain systems and implementation of a computational model. In M. Gabriel, & J. Moore, (Eds.), *Learning and Compu-tational Neuroscience: Foundations of Adaptive Networks.* (pp. 359–387). Cambridge, Ma: The MIT Press.

12. Shapiro, J., Wearden, J., & Barrone, R. (In preparation). *Simple models of scalar timing.*

13. Sutton, R. S., & Barto, A. G. (1990).Time-derivative models of pavlovian reinforcement. In M. Gabriel, & J. Moore, (Eds.), *Learning and Computational Neuroscience: Foundations of Adaptive Networks.* (pp. 497–537). Cambridge, Ma: The MIT Press.

14. Wearden, J. H., & Doherty, M. F. (1995). Exploring and developing a connectionist model of animal timing: Peak procedure and fixed-interval simulations. *Journal of Experimental Psychology, 21,* 99–115.

Modularity and Specialized Learning in the Organization of Behaviour

Joanna Bryson and Lynn Andrea Stein

Abstract

Research in artificial neural networks (ANN) has provided new insights for psychologists, particularly in the areas of memory, perception, representation and learning. However, the types and levels of psychological modelling possible in artificial neural systems is limited by the current state of the technology. This chapter discusses modularity as illuminated from research in *complete agents*, such as autonomous robots or virtual reality characters. We describe the sorts of modularity that have been found useful in agent research. We then consider the issues involved in modelling such systems neurally, particularly with respect to the implications of this work for learning and development. We conclude that such a system would be highly desirable, but currently poses serious technical challenges to the field of ANN. We propose that in the mean time, psychologists may want to consider modelling learning in specialised hybrid systems which can support both complex behaviour and neural learning.

1. Introduction

Research in artificial neural networks (ANN) has provided new insights for psychologists, particularly in the areas of memory, perception, representation and learning. However, the types and levels of psychological modelling possible in artificial neural systems is limited by the current state of the technology. We are only beginning to be able to model neurally many of the complex structures and interactions we know to exist in animal brains. In particular, modularity is a much-discussed feature of the brain, but we have only rudimentary models of it in neural networks.

This chapter discusses modularity as illuminated by research in *complete agents*, such as autonomous robots or virtual reality characters. We use this research to identify the sorts of modules that have been found useful in our field. We then describe a possible mapping between these established AI modules and the modularity present in mammalian brains. We conclude with a discussion of the implications of this work for learning, development, and the future of ANN and agent models of animal intelligence.

2. Modularity in AI and Psychology

The extensive use of modularity in complete agents was popularised in the mid 1980's with the establishment of *behaviour based artificial intelligence* (BBAI). Behaviour-based AI refers to an approach inspired by Minsky [22], where many small, relatively simple elements of intelligence act in parallel, each handling its own area of expertise [3, 19]. In theory, these component elements are both easier to design and more plausible to have evolved than a single complex monolithic system to govern all of behaviour. In the modular theory of intelligence, the apparent complexity of intelligent behaviour arises from two sources: the interactions between multiple units running in parallel, and the inherent complexity of the environment the individual units are reacting to.

The behaviour-based approach generated significant advances in mobile robotics [4] and has come to dominate both the fields of robotics and virtual reality [17, 27]. It has lead to a revolution in the way computation is thought about [29]. Nevertheless, it has not been entirely successful. Advances in the development of humanoid agents still come disappointingly slowly. Further, there is no single dominant behaviour-based architecture that is used by even a large percentage, let alone a majority, of complete agent developers.

The engineering advantage of modularity is simple: it decomposes the problem of intelligence into manageable chunks. In essence, it is a design advantage. Modularity is a form of hierarchy. Each module handles a portion of the agent's overall problem space, leaving the complete agent with an exponentially reduced space of behaviour options to consider. However, the modular approach also introduces a number of problems. First, there is the question of how to decompose an apparently coherent intelligence into modules. What state and/or behaviour belongs together, and what apart? Even more problematic, once behaviours have been separated into at least semi-autonomous modules, how can overall behavioral coherence be reestablished? A modular approach is of no advantage if the problem of integrating behaviour leads to a greater problem of design than the decomposition originally avoided.

From a psychological perspective, interest in modularity dates back to Freud [12], and even to Hume [16]. A modular architecture is motivated not only by the inconsistencies of human behaviour, but also by neuroscience, which has shown a diversity of organs in the human brain. Here the same questions emerge: how is the brain modularised? To what extent are the modules encapsulated — that is, how strictly are they separated? Which modules communicate, and at what level of abstraction? What functions do the various organs perform? And how are their parallel operations coordinated into fluid, largely coherent behaviour? In the following sections, we hope to provide answers or at least hypotheses for these questions based on our work in complete agent architectures.

3. Modularity in Complete Agent Architectures

In this section we describe the sorts of modules that are requisite for making a complete agent function. We also show how these required modules relate to animal intelligence. To begin though, we delineate two different sorts of modularity: architectural modules vs. skill modules.

3.1 Types of Modularity

In thinking about the organisation of agent behaviour, we must consider two different sorts of modular decomposition. One is decomposition of generic function, where modules might include planning, vision, motor control, and such-like. The second is decomposition by task, where modules might include walking, sleeping, hunting, grooming and so forth. Initially, the behaviour-based revolution was about moving from the former, generic sort of modularity to the latter, task-oriented sort of modularity. Although the major benefits of BBAI come from the special-purpose, task-oriented sort of modularity, the experience of the last 15 years has shown that some generic modules are also necessary in an architecture in order for that architecture to be useful and usable.

One of our hypotheses (described further below) is that this more generic modularity is analogous to the modularity by organ in the brain, e.g. the roles of the cerebellum or the hippocampus. We call this sort of modules *architectural modules*, both because in the brain it often characterised by different underlying neurological architectures, and because in agent organisation it contributes architectural features. The task-specific modules we hypothesise to be more analogous to the within-organ modularity exhibited in the brain, by neural assemblies differentiated only by physical space and connectivity. We will call this sort of modules *skill modules*. The emphasis in this chapter will be primarily on different sorts of architectural modules, since this is the emphasis of our own research. The reader should be aware, however, that most work in BBAI concentrates on the development of skill modules.

3.2 Architectural Modularity in Complete Agents

Bryson [5] describes the emergence over the last 15 years of three sorts of architectural requirements for complete autonomous agents capable of complex, scalable behaviour. That article reviews the literature in four separate agent traditions: BBAI; the hybrid or multi-layer community, which combines behaviours with more conventional planning; the Procedural Reasoning System (PRS) / Beliefs, Desires and Intentions (BDI) communities, which have adapted conventional planning and representation in response to the success of BBAI and the demands of real-time, embedded systems; and the Soar / ACT-R communities, which have been working with their distributed representation cognitive agent architectures for many years. In this chapter, we will not reiterate that review, but will only describe the results.

Briefly, there are three sorts of architectural modules that seem necessary. First, nearly all autonomous agents consist primarily of a system of skill modules. These are often referred to as *behaviours* due to BBAI. Despite their name, "behaviours" actually *generate* behaviour; they are not a description of expressed behaviour. There is no one-to-one correspondence between a behaviour module and an expressed behaviour. Much of expressed behaviour is supposed to *emerge* from the interaction of two or more skill-module behaviours operating at the same time. The second sort of architectural module performs *action selection*, which provides for coherent behaviour in a distributed system. Finally, because action selection generally works by focusing attention on one subset of possible behaviour, there needs also to be a dedicated *environment monitoring* module. This module switches the action-selection attention in response to salient environmental events.

Nearly all autonomous agent architectures today use a modular skill structure. They incorporate a set of primitive behaviours that no other part of the system needs to understand the workings of. This reflects the combinatorial advantage of modularity referred to in the introduction. It comes at a cost of less fine-grained control and the inefficiencies of not being able to combine the outputs of related actions or motions. But for a resource-constrained agent working in a dynamic, real-time environment where responses must be quick and appropriate, the advantage of being able to quickly activate pre-compiled skills outweighs these sorts of costs.

Some researchers try to reduce some of the complexity of coordinating skill modules by using homogeneous behaviour representations or coordinated output formats for all the modules (see for examples Arkin [1] and Tyrrell [30] respectively). We feel these strategies overconstrain the sorts of computation and representation in a skill module. Skill modules should contain not only motor actions, but also whatever perceptual skills are necessary in order to support them. Perception is not just sensing. Sensory stimuli are often ambiguous and often require both recent context and longer-term experience and expectations to discriminate. The state needed to learn or tune skills, or to disambiguate or categorise the perception, should also be a part of the skill module. Actions may also vary significantly depending on context, but a designer may want to encapsulate all these expressions as a single behaviour for simplicity. For this reason, we prefer using objects specified in object-oriented programming languages to represent skill modules.

This representation of skill modules has advantages for two reasons. From an engineering perspective, it allows for the behaviour decomposition problem to be addressed with techniques similar to those developed for object decomposition (e.g [9]) and allows us to quickly develop more powerful, complex behaviour [6]. From a cognitive modelling perspective, this level of abstraction is more closely analogous to the level of description usually used for ascribing functions to cortical areas. Although early BBAI eschewed variable state, in general the notion of empowered, perceiving semi-autonomous behaviours is more in keeping with this work than the trend in hybrid and BDI architectures to making the skill modules into simple motor primitives. It is also very like the recent contributions of the multi-agent systems (MAS) community [31].

The next architectural module is hierarchically structured plans for action selection. "Action selection" is the term applied to the ongoing problem of determining exactly what an agent is going to do next. The BBAI literature often refers to this problem as "behaviour arbitration", in MAS it is called "agent coordination". When all behaviours are running in parallel, each may have an action it is currently attempting to express. If these conflict, some method of coordination must maintain the coherence of the complete agent's behaviour.

Plans in the context of agent action selection are structures which indicate, given a particular environmental context and decision history, what to do next. BBAI originally strongly resisted the use of plan-like structures for coordination, because they were felt to lose most of the advantages of modularity by imposing a form of centralised control. However, these systems proved useful not only for coordination, but also as simple memory. A system using plans to coordinate behaviour does not need complex or perfectly ordered episodic memory in order to disambiguate where it is in a multi-step plan, even if the consequences of its previous actions are not present in its environment.

For example, consider, a robot tidying an office. If it has a fixed order for tidying drawers it neither needs to repeatedly inspect closed drawers it had already tidied, nor to remember the complete list of drawers already visited each time it selects the next one. It only needs to remember its agenda, and its current place in its agenda. Notice the agenda does not need to specify any details about how a drawer is cleaned. The drawer cleaning behaviour is free to chose a methodology appropriate to its perceptions, or to determine whether a drawer is already "clean enough."

Although such structures for supporting action selection are referred to as *plans*, this does not necessarily imply that they are created by *planning*, at least not by the agent. Most complete agents exploit plans provided by their designers, rather than engaging in the slow and unreliable constructive planning process. Choosing what to do next this way, without deliberation, is often called *reactive planning*. When plan-like structures are provided to facilitate this process, they are referred to as *reactive plans*.

Hierarchical structures for action selection allow for the focus of attention on a particular set of behaviours that are likely to be applicable in a particular circumstance. The hierarchies are parsed by the recognition of circumstance, and can be reparsed arbitrarily frequently in order to ensure the applicability of current behaviour. This observation leads to the third architectural module found in all successful agent architectures: a mechanism for monitoring the environment and realizing that the agent should attend to new goals. This sort of system is sometimes referred to as an *alarm system* (e.g. [28]). Having a parallel system is necessitated by hierarchical control: any agent in a dynamic environment that may have its attention focussed on a particular task needs a system to guarantee that it notices salient events in the environment. Otherwise, the agent may overlook both dangers and opportunities. The alarm system must necessarily be relatively simple, requiring no cognitive overhead that would distract from the primary task. It is usually a case of pattern matching, of recognising salient indicators in the environment and then switching cognitive attention to analysing and coping with such situations.

3.3 Equivalent Modularity in Animal Brains

In summary of the last section, complete agent architectures have converged on three sorts of architectural modules for supporting complex, dynamic behaviour. These are a system of skill modules, structured action selection, and an environment monitoring system. If such an organisation is necessary or at least very useful for controlling intelligent agents, then it is also likely to be present in animals, since they have evolved to face similar problems of information management. In this section we speculate about what the analogous modules might be.

First, the skill modules we believe are reflected in within-organ modularity in the brain. This is not to imply that they are represented only within one organ or region: many skills necessarily combine vertically a large number of brain areas and organs, for example the retinas, visual cortex, associative cortices, motor pre-planning and motor coordination. To some extent, each of these areas is architecturally specialised, but the combination of a particular ensemble of simultaneous activations across these structures might model the various skill sets modularised in agent architectures.

Next, Prescott et al. [24] postulate that the basal ganglia is the architectural module responsible for action selection (see also [21]). These researchers focus primarily on the problem of action selection as behaviour arbitration, but the basal system is also well integrated with some of the mid-brain systems that have been implicated in species-typical action patterns as well as the cortical systems which might hold the perceptual skill modules required to discriminate context.

Finally, the behaviour of the alert system is analogous to behaviour attributed to the limbic system. In particular, the amygdala has been implicated in learning to recognise and attend to salient situations [8]. This attention takes the form of emotional responses, which are characterised by the selective activation of appropriate behaviours and expectations.

Thus it is plausible, if far from proven, that the sorts of modularity used in complete autonomous agents are also present in animals and humans. This reenforces our proposal that the agent platform could be useful for psychological modelling.

4. Learning in Behaviour Oriented Architectures

A particular modular organisation provides structure not only for how an intelligence operates in particular situations, but also for how it can learn. We assume that learning happens *within* modules, not across or outside them. So, for the organisation of modules described in the previous section, there are only certain places learning can take place. There can be specialised or perceptual learning within the skill modules. There can be learning of new skill modules within the architectural module of skill modules. There can be the learning of new reactive plans within the action selection module, and there can be learning of selection rules for priorities and alarms. Notice that the sorts of perceptual learning that takes place in skill-modules can also affect the execution of reactive plans and of attention switching. In our model, learning new categories, both new discriminations and new generalisations, is the sort of perceptual learning that should take place in a skill module.

Not all complete agent architectures perform learning in this way. Some which are more closely aligned traditional planning maintain a single database of "beliefs" [15]. Production-based agents [18] often have both a database of beliefs and a database of *productions*: if-then rules governing intelligent control. These productions, analogous to behaviours in their modularity, are not as complex as the sorts of skill modules we have been describing. For the remainder of this chapter, we will be emphasising complete agent architectures that are *behaviour oriented*. Behaviour-oriented systems differ from being strictly behaviour-based because they have structured action selection, but are otherwise near to traditional BBAI systems in their emphasis on relatively autonomous behaviours capable of perception and action (see Figure 1).

Agents tend to be created for particular environments and tasks; in other words they are niche specific. Consequently, the kinds of things they are likely to need to learn can generally be determined in advance. This is a currently popular view of animal learning [13, 25], and we have made it central to our methodology for developing behaviour

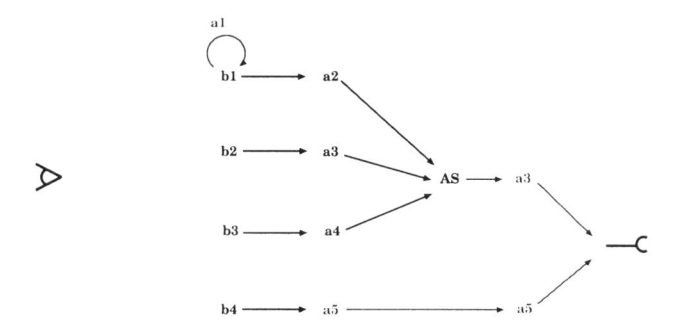

Figure 1: Behaviour-oriented systems have multiple, semi-autonomous skill modules or *behaviours* ($b_1 \ldots$) which generate actions ($a_1 \ldots$) based on their own perception (indicated by the eye icon). Actions which affect state outside their generating behaviour, whether internal to the agent or external, are generally subject to arbitration by an action selection (AS) system.

oriented agents, which we call behaviour oriented design (BOD) [6]. We determine the modular decomposition for a set of skills around the kinds of perception and memory those skills need to operate appropriately. Thus each skill module contains specialised representations and perception and action routines for maintaining those representations, as well as control for the skilled actions that the agent applies to its environment.

The second most common form of learning in complete agents is the learning of new reactive plans. This is usually done in one of two ways, either by reasoning or planning a new plan (e.g. [2, 14]), or more recently by social learning such as imitation or receiving instruction (e.g. [10, 26]). However, in practice, a surprisingly large number of agents use plans programmed by their designers. So far, this is still the fastest and most reliable way to get appropriate behaviour from an agent.

5. Learning New Behaviours

In our earlier discussion of skill modules, we claimed that behaviour oriented design requires the use of complex algorithms and specialised representations, and that modules are therefore better represented in object oriented languages than in current ANN. However, there is at least one reason to favour an ANN representation of skill modules. That is the problem of designing an agent that can learn or develop *new* skill modules. This is clearly desirable, and has been the focus of significant research (see [10] for a recent example and review.) However, to date, most efforts on these lines would qualify as specialised learning *within* a single skill module / representation system from the perspective of behaviour oriented design.

The reason that we would like to be able to represent behaviours in terms of ANN is as follows. Consider Figure 2. In this figure, representation of the skill modules has been split into two functional modules: the Behaviour Long Term Memory (BLTM) and the Working Memory (WM). The working memory allows for rapid, short term changes not only for perceptual memory, but also in the representation of the behaviours. The BLTM provides a relatively stable reference source for how these modules should appear when

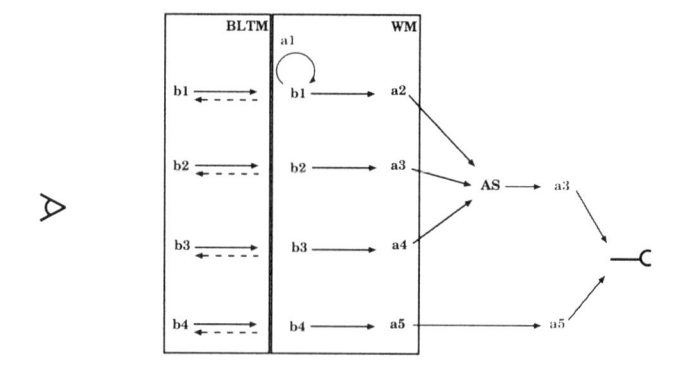

Figure 2: A system capable of learning behaviours must 1) represent them on a common substrate and 2) allow them to be modified. Here behaviours are represented in a special long term memory (BLTM) *and* in a plastic working memory (WM) where they can be modified. During consolidation (dashed lines) modifications may either alter the original behaviours or create new ones.

activated. Skill representations might be modified due to particular circumstances, such as compensating for tiredness or high wind, or responding to a novel situation such as using chopsticks on slippery rice noodles for the first time.

In this model, the adjustments made in plastic, short term memory also affect the long term memory. This sort of dual- or multi-rate learning is receiving a good deal of attention in ANN currently (see [7, 11, 20]). Depending on long term experience, we would like this consolidation to have two possible effects. Let's imagine that b_2 has been modified in working memory in order to provide an appropriate expression of a_2. If the same modifications of b_2 prove useful in the near future, then they will be present for consolidation for a protracted period, and likely to effect the permanent representation of b_2. However, if the modifications are only sometimes applicable, we would like a new behaviour b_2' to become established. This process should also trigger perceptual learning, so that the two behaviours can discriminate their appropriate context for the purpose of action selection. Also b_2 and b_2' would now be free to further specialise away from each other.

6. Conclusions and Future Directions

In this chapter we have shown that complete agents can provide higher-level AI models for psychologically important concepts such as modularity and specialised learning. We have also described the sorts of systems that are the current state-of-the-art in complete agent architectures. We have shown that using ANN to represent at least some parts of a complete agent might be highly desirable, but unfortunately we have also argued that the complexity of the algorithms and the specialised representations are not yet met in current models of ANN modularity (e.g. [23]).

On the other hand, we believe that combining research in these two fields might be a highly useful direction. Complete agent researchers are already experimenting with ANN for learning and simple control, but most such systems are not yet advanced enough to be interesting to psychologists. However, the psychologically interesting ANN systems currently under development such as those described in this volume might be furthered by embedding them in a complete agent. In this way, conventional programming can be used to provide for appropriate experimental platforms to test such systems. A complete agent can provide realistic inputs, and test outputs in realistic settings. Thus, we hope to see a further uniting of these two fields in the future.

Acknowledgements

Thanks to David Glasspool for his comments and suggestions.

References

1. Arkin, R. C. (1998). *Behavior-Based Robotics*. MIT Press, Cambridge, MA.
2. Bonasso, R. P., Firby, R. J., Gat, E., Kortenkamp, D., Miller, D. P., and Slack, M. G. (1997). Experiences with an architecture for intelligent, reactive agents. *Journal of Experimental & Theoretical Artificial Intelligence*, 9(2/3):237–256.
3. Brooks, R. A. (1991). Intelligence without representation. *Artificial Intelligence*, 47:139–159.
4. Brooks, R. A. and Stein, L. A. (1993). Building brains for bodies. Memo 1439, Massachusetts Institute of TechnologyArtificial Intelligence Lab, Cambridge, MA.
5. Bryson, J. (2000a). Cross-paradigm analysis of autonomous agent architecture. *Journal of Experimental and Theoretical Artificial Intelligence*, 12(2):165–190.
6. Bryson, J. (2000b). Making modularity work: Combining memory systems and intelligent processes in a dialog agent. In Sloman, A., editor, *AISB'00 Symposium on Designing a Functioning Mind*.
7. Bullinaria, J. (2000). Exploring the Baldwin effect in evolving adaptable control systems. In this volume. Springer.
8. Carlson, N. R. (2000). *Physiology of Behavior*. Allyn and Bacon, Boston.
9. Coad, P., North, D., and Mayfield, M. (1997). *Object Models: Strategies, Patterns and Applications*. Prentice Hall, 2nd edition.
10. Demiris, J. and Hayes, G. (1999). Active and passive routes to imitation. In *Proceedings of the AISB'99 Symposium on Imitation in Animals and Artifacts*, Edinburgh. AISB.
11. French, R., Ans, B., and Rousset, S. (2000). Pseudopatterns and dual-network memory models: Advantages and shortcomings. In this volume. Springer.
12. Freud, S. (1900). *The Interpretation of Dreams*. Avon, New York.
13. Gallistel, C. R. (1990). *The Organization of Learning*. MIT Press / Bradford Books, Cambridge, MA.
14. Gat, E. (1991). *Reliable Goal-Directed Reactive Control of Autonomous Mobile Robots*. PhD thesis, Virginia Polytechnic Institute and State University.
15. Georgeff, M. P. and Lansky, A. L. (1987). Reactive reasoning and planning. In *Proceedings of the Sixth National Conference on Artificial Intelligence (AAAI-87)*, pages 677–682, Seattle, WA.

16. Hume, D. (1748). *Philisophical Essays Concerning Human Understanding*.
17. Kortenkamp, D., Bonasso, R. P., and Murphy, R., editors (1998). *Artificial Intelligence and Mobile Robots: Case Studies of Successful Robot Systems*. MIT Press, Cambridge, MA.
18. Laird, J. E. and Rosenbloom, P. S. (1994). The evolution of the Soar cognitive architecture. Technical Report CSE-TR-219-94, Department of EE & CS, University of Michigan, Ann Arbor. also in *Mind Matters*, Steier and Mitchell, eds.
19. Matarić, M. J. (1997). Behavior-based control: Examples from navigation, learning, and group behavior. *Journal of Experimental & Theoretical Artificial Intelligence*, 9(2/3):323–336.
20. McClelland, J. L., McNaughton, B. L., and O'Reilly, R. C. (1995). Why there are complementary learning systems in the hippocampus and neocortex: Insights from the successes and failures of connectionist models of learning and memory. *Psychological Review*, 102(3):419–457.
21. Mink, J. W. (1996). The basal ganglia: focused selection and inhibition of competing motor programs. *Progress In Neurobiology*, 50(4):381–425.
22. Minsky, M. (1986). *The Society of Mind*. Simon and Schuster, New York, NY.
23. Plaut, D. C. (1999). Systematicity and specialization in semantics. In Heinke, D., Humphreys, G. W., and Olson, A., editors, *Connectionist models in cognitive neuroscience: Proceedings of the Fifth Annual Neural Computation and Psychology Workshop*, New York. Springer-Verlag.
24. Prescott, T. J., Gurney, K., Gonzalez, F. M., and Redgrave, P. (to appear). The evolution of action selection. In McFarland, D. and Holland, O., editors, *Towards the Whole Iguana*. MIT Press, Cambridge, MA.
25. Roper, T. J. (1983). Learning as a biological phenomena. In Halliday, T. R. and Slater, P. J. B., editors, *Genes, Development and Learning*, volume 3 of *Animal Behaviour*, chapter 6, pages 178–212. Blackwell Scientific Publications, Oxford.
26. Schaal, S. (1999). Is imitation learning the route to humanoid robots? *Trends in Cognitive Sciences*, 3(6):233–242.
27. Sengers, P. (1999). *Anti-Boxology: Agent Design in Cultural Context*. PhD thesis, School of Computer Science, Carnegie Mellon University.
28. Sloman, A. (2000). Models of models of mind. In Sloman, A., editor, *AISB'00 Symposium on Designing a Functioning Mind*.
29. Stein, L. A. (1999). Challenging the computational metaphor: Implications for how we think. *Cybernetics and Systems*, 30(6):473–507.
30. Tyrrell, T. (1993). *Computational Mechanisms for Action Selection*. PhD thesis, University of Edinburgh. Centre for Cognitive Science.
31. Weiss, G., editor (1999). *Multiagent Systems: A Modern Approach to Distributed Artificial Intelligence*. MIT Press, Cambridge, MA.

Modeling Modulatory Aspects in Association Processes

Jairo Diniz-Filho & Teresa Bernarda Ludermir

Abstract

It is thought that the amygdala and the orbitofrontal cortex are involved in learning and memory systems and that groups of cholinergic and adrenergic neurons may function as modulators in the activity of those systems [9]. It is also believed that association learning is very important in the control of motivational and emotional behaviours [8]. Furthermore, it has been suggested that neurons involved in homeostatic regulation mechanisms (evolutionarily old structures in the brain) are related to neocortical neurons (evolutionarily modern sectors) via emotion [3]. On the other hand, modularity has been often considered in systems simulating brain activity [5, 1]. In this work, we propose a modular system with a particular module able to evaluate some variables reflecting the own system functions, in order to simulate internal states. According to that, this module has a modulatory role on the other modules' computations. Our aim is to show the properties of the system proposed under the influence of this particular module.

1. Introduction

There is support with experimental evidence for a conception of brain organization as processing streams [4]. Research interests are directed now to the way these streams are organized and how they interact. Some evidence is coming in favor of the idea that these processing streams compute complementary properties [4]. On the other hand, considering that there are multiple stages, it is reasonable to conceive that groups of cells embedded in these streams are organized as processing modules (not completely independent ones), and that interaction intra and intermodules may take place [7]. There are several models of neural networks based on modularity [5, 1, 2]. Commonly, these modular neural networks (MNNs) are designed as a collection of independent modules. To a lesser extent, they are conceived as MNNs with intra and intermodular interactions, beyond those of simply transfering results from one module's output to the other module's input. Besides the interaction between modules directly involved in main processing streams for diverse forms of perception, it has been also suggested in the literature that brain modules of more general nature may be involved as a reference to the reaction of the organism as a whole. According to these suggestions, whatever the information that is reaching the central nervous system, before it can take subsequent effects, there is a primary evaluation, from the point of view of internal survival parameters, in order to preserve homeostatic equilibrium [3]. This evaluation is taken as mainly innate, although some improvement may be acquired through experience, and is linked to the evolutionarily older components of brain structure, which are known to be involved in the development of feelings and emotion [3]. One of the important functions of emotion is to provide flexibility to behaviour, since the organism, while aware of the emotional state in response to a stimulus, can flexibly choose the appropriate response among the many possible. That is in contrast with fixed behavioural response to a stimulus. Among the influences of emotion, are also facilitations of memory storage, which may be implemented by relatively non-specific projecting systems, like cholinergic and noradrenergic pathways [9]. Backprojections widely spread from amygdala to cortex association areas in the ventral visual system have been demonstrated, which could be involved in the neural substrate for important functions such as recall, attention, constraint satisfaction and priming, through pattern association [9]. Both cholinergic and noradrenergic pathways and backprojections pathways are compatible with modulatory activities of general nature.

We are more precisely interested in the aspects of modulation and flexibility described above, since we detected a weak concern in neural networks literature with respect to those issues. Although we are not presently able to incorporate a more precise picture of the complex playing factors, we decided to contribute with an effort in the direction of modeling. We present a modular system to which a special component is incorporated, with the aim to represent this kind of parallel influence.

2. Architecture and Operation

Descriptions of associative memories vary according to modeler's knowledge background and interests [6, 10]. We worked with an ensemble of pattern association memories. We assume that these memories play the role of channels in an intermediate stage of potentially longer processing streams. As they represent the first step in the system's reaction (within a certain stage), we called them Reaction Networks (ReactNs). Each ReactN is given a learning rule that operates independently, although it is of the same type for all ReactNs. In Figure 1, it can be seen the basic organization of each ReactN. Figure 2 gives an idea of each individual ReactN memories' place in the general picture of the system. During the learning phase, only the ReactN components are trained. Details about Training and Recall are given below. Other components of the system are taken as modifyers placed along the processing streams. We assume that during Recall a general operation (called Ponderation) is performed over the ReactNs' outputs, and that it is performed by the PondN component. To perform that, PondN uses weight values produced by a Modulator component (called Modtr). To produce the weights, Modtr seeks predefined rules. They are simple rules that represent our assumptions about the system's reaction to more general features of information, e.g., to the order by which input examples are presented. An example of the system will be given in section 3. PondN and Modtr are not allowed to learn, since they are taken as 'innate' capabilities of the system, according to the above considerations.

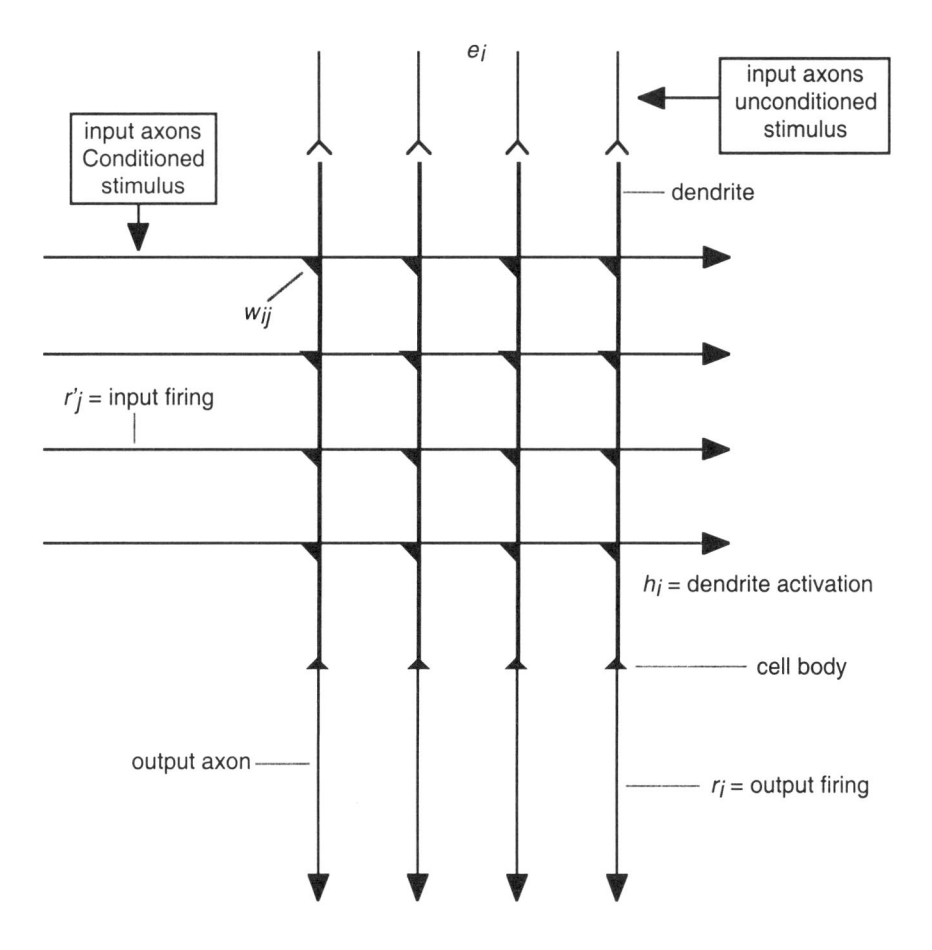

Figure 1: Basic organization of pattern association memory. Unconditioned stimulus produces activity firing rates e_i for the *ith* neuron, which forces output activity firing r_i. Conditioned stimulus activity firing r_j can be used to modify weights w_{ij}.

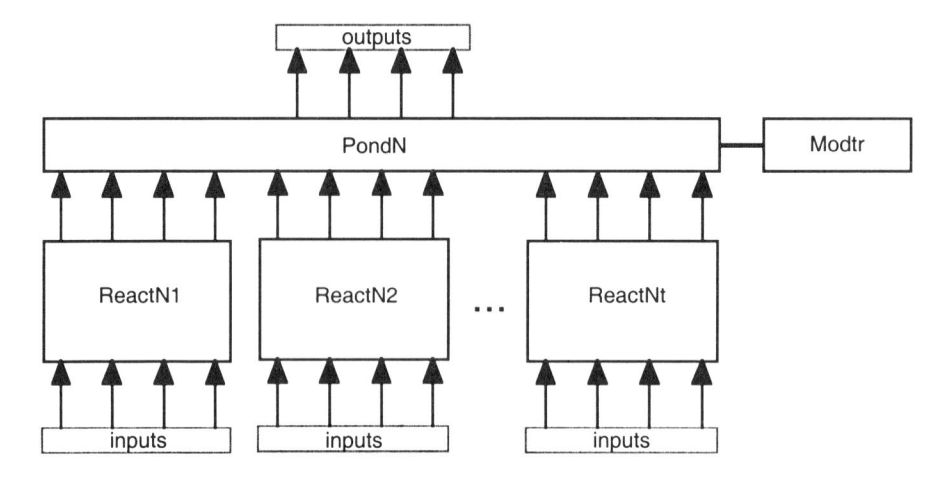

```
                    ┌─outputs─┐
                    ↑ ↑ ↑ ↑ ↑
        ┌───────────────────────────────┐  ┌──────┐
        │           PondN               │──│ Modtr│
        └───────────────────────────────┘  └──────┘
         ↑↑↑↑   ↑↑↑↑       ↑↑↑↑
        ┌────┐ ┌──────┐   ┌──────┐
        │ReactN1│ │ReactN2│ ...│ReactNt│
        └────┘ └──────┘   └──────┘
         ↑↑↑↑   ↑↑↑↑       ↑↑↑↑
        ┌────┐ ┌──────┐   ┌──────┐
        │inputs│ │inputs│  │inputs│
        └────┘ └──────┘   └──────┘
```

Figure 2: Components and relationships shown schematically. During the learning phase, input examples are passed through each ReactN, in order to be learned. PondN and Modtr (not allowed to learn) act according to predefined rules. These rules represent assumptions about the system's reaction to general features of the undergoing processes. During Recall, PondN performs a general (weighted) operation over the ReactNs' outputs. Thick line between Modtr and PondN represents the use of weights produced by Modtr. Modtr's output is a vector with values that change according to the rules.

2.1 Learning

During Learning, a set of examples containing associated patterns, prepared previously, is passed through a central channel (one of the ReactNs), following each example individually, by the order in which it appears in the example set. This same set is also passed through some neighbour (to the central) channels, giving rise to redundance in the representation.

At the same time, individual examples from other, previously modified, copies of this first set are then prepared to proceed with learning in the other memories. Set modifications are restricted to the unconditioned part of the examples set, as follows. Some ReactN receive a learning set that has an exceeding number of 1 (ones, in binary representation) incorporated to the right extremity of the unconditioned stimuli, for the same conditioned stimuli present in the central ReactN's learning set. Note that, for each example passed through the central ReactN, there is a copy of it that is also passed through the other ReactNs (not in the neighbourhood of the central ReactN), to which only the unconditioned part of the example is changed. This is to represent our assumption that, in an intermediate stage of processing streams, there are different representations to unconditioned stimuli associated in previously processed stages. Learning rule is local, taking the form of a Hebbian learning version in each of the referred memories. Following the diagram of Figure 1, we denote a conditioned stimulus as a vector **r'** of firing rates, and use the index j in the range of the elements $r'_1, r'_2, ..., r'_j$, representing the firing rates of individual neurons. In a similar way, we use the index i in the range of the unconditioned stimulus vector, denoted **e**, with elements $e_1, e_2, ..., e_i$ representing the inputs for individual neurons. Unconditioned stimulus pattern is applied through unmodifiable synapses, forcing the firing rate of the output neurons. The firing output vector, which we denote **r**, has elements $r_1, r_2, ..., r_i$ and represents the firing rates of individual output neurons. Putting into mathematical form, the forced firing rate for each output neuron i is given by

$$r_i = f(e_i), \qquad (1)$$

where the function f is the activation function. The expression of the Hebbian learning we used is

$$\delta w_{ij} = k r_i r'_j, \qquad (2)$$

where δw_{ij} is the change to be applied to the synaptic weight w_{ij}, and k is a learning rate.

2.2 Recall

During Recall, all ReactN are involved, each one contributing with its own associated pattern for a more general answer given by the whole system. For each ReactN, Recall is made individually through the following expression:

$$h_i = \sum_j r'_j w_{ij} \qquad (3)$$

where h_i is the total activation of an individual output neuron i, and Σ_j represents the sum of all activations produced under the influence of both the active neuron r'_j and its respective strengthened synapse w_{ij}.

The result for each h_i is then passed through a threshold function, which gives the values for the outputs:

$$r_i = f1(h_i) \qquad (4)$$

From the point where each ReactN gives its outputs for the current inputs on, in the system architecture, the results elicited by the inputs during Recall enter a subsequent module (PondN), where they are re-operated. This also gives rise to the effect of a parallel module (Modtr), in order to mimic the proposed parallel influence, like is thought to occur in an organism internal evaluation. The work of Modtr consists in sensing the result of the first Recall step just described, submiting it to previously defined rules designed to represent a more general system's reaction in the present stage. As an example of a mechanism to stablish a rule, used in the section 3 demonstration, Modtr can observe the order used to present the pattern examples in the Recall phase, compare with predefined parameters, and put the result in the form of a vector **m**, to be used in the subsequent step. Each element of **m** is associated to its respective ReactN. So, the answer from Modtr assume the form of a vector **m**, with elements m_t, with t equal to the number of ReactN modules. The subsequent step is performed by PondN, able to combine the results of previous modules in a particular way. For this paper, we consider that activation of each individual input i in PondN is given by the following expression:

$$a_i = \sum_t r_{it} m_t \qquad (5)$$

where r_{it} means the firing rate of the individual neurons for each ReactNt and m_t is the corresponding value in the **m** array.

Constraints can be associated with the modulatory component, Modtr, and thus they are taken as a reference to the functioning state of the system. This reference is such that, once a specific input evokes undesirable effects (predefined in the system), although it can be learned, the system signals with a reaction answer. This answer can be used further in the processing streams as an indication of the inputs' fitting with respect to the "organism" affairs.

A subsequent step is a calculation of PondN outputs. The value of a_i for every output neuron is divided by the sum of the values in the elements of **m**.

$$a_i' = a_i / (m_1 + m_2 + ... + m_t) \qquad (6)$$

After that, these values are passed individually through a second threshold (step) function, $f2$, giving the final r_i outputs:

$$r_i = f2(a_i') \qquad (7)$$

Both step functions, $f1$ and $f2$, are such that $f(x) = 0$, if $x <=$ *threshold;* and $f(x) = 1$, otherwise.

3. A Three ReactNs Case

We implemented the above structure as a simple instance containing three ReactNs. A demonstration of its operation is summarized in Figure 3 for the Recall phase. In this case, the central ReactN is considered to be ReactN$_2$, and ReactN$_3$ has a similar behavior to it, that is, both use the same training set.

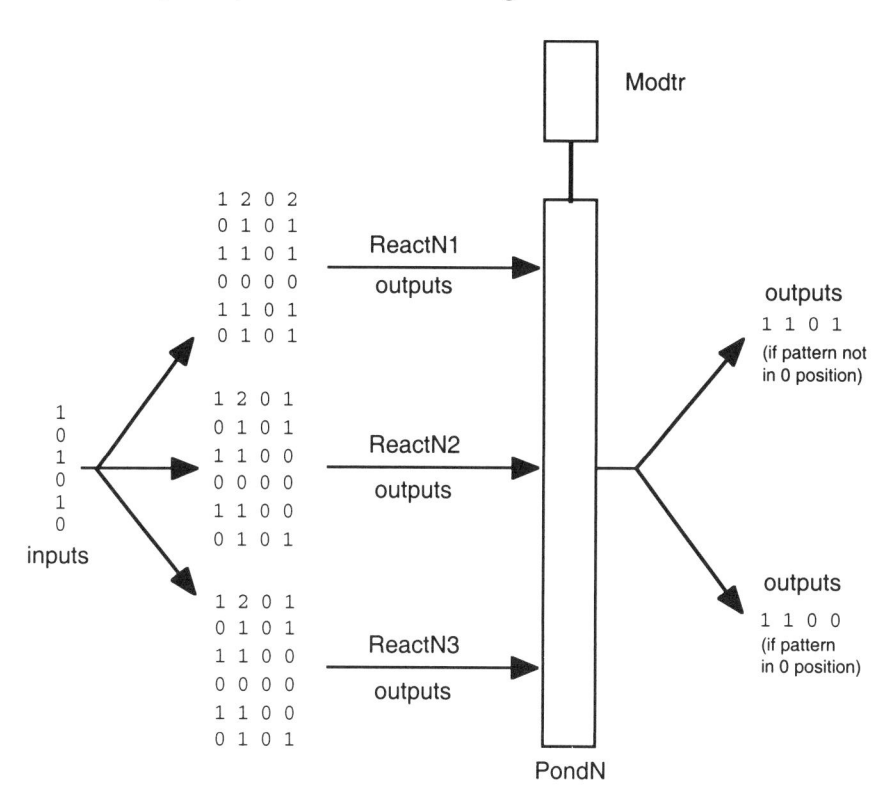

```
                              Modtr

          1 2 0 2
          0 1 0 1
          1 1 0 1     ReactN1
          0 0 0 0     outputs              outputs
          1 1 0 1                          1 1 0 1
          0 1 0 1                          (if pattern not
                                           in 0 position)
   1      1 2 0 1
   0      0 1 0 1
   1      1 1 0 0     ReactN2
   0      0 0 0 0     outputs
   1      1 1 0 0
   0      0 1 0 1
 inputs
                                           outputs
          1 2 0 1
          0 1 0 1                          1 1 0 0
          1 1 0 0     ReactN3              (if pattern
          0 0 0 0     outputs              in 0 position)
          1 1 0 0
          0 1 0 1
                              PondN
```

Figure 3: A three ReactNs case. During Recall, the example with inputs 101010 is passed through the matrices with learned weights in each of the three ReactNs (as ReactN$_1$ was trained with a different example set, its matrix presents modifications in some of its weights, with respect to both ReactN$_2$ and ReactN$_3$ matrices). In a subsequent phase, the outputs from each ReactN are passed through PondN, which presents results under the influence of Modtr, according to the following: if the example is in the first position in the Recall input set, PondN decides through a combination of ReacN$_2$ and ReacN$_3$ matrices (default, inferior branch, in the figure, giving the result 1100). If not, the final outputs are decided through a combination of ReacN$_1$ and ReacN$_2$ matrices (superior branch, resulting 1101).

3.1 Learning

During the Learning phase, conditioned stimuli were presented individually as six inputs to each ReactN. For the inputs represented by the vector 101010, the associated unconditioned stimulus was 1101 (last value changed to 1) for $ReactN_1$, but it was 1100 for both $ReactN_2$ and $ReactN_3$. Similar relationships were conserved for subsequent training with a second (different) stimulus. The three matrices are shown in the Figure 3, with their respective values changed after training. They are $ReactN_1$ (top), $ReactN_2$ (middle) and $ReactN_3$ (bottom). Note that values in the third column of the $ReactN_1$ matrix are modified in relation to the other matrices, as expected, since $ReactN_1$ was submitted to a particular training set, different from that used by $ReactN_2$ and $ReactN_3$. Remember that both PondN and Modtr are prewired and their operations only take part during the Recall phase.

3.2 Recall

During the Recall phase, the 101010 inputs were presented to all ReactNs, as indicated by the arrows. The outputs from each ReactN are passed through PondN, which produces results under the influence of Modtr, as in the general case, discussed in section 2 (Architecture and Operation).

In this particular case, the rule by which Modtr changes PondN decision depends strictly on the order used to present the examples during the Recall phase. For the first presented example (0 position in the Recall set) only, critical values are produced at Modtr's output. These values leave PondN free to decide through a combination of $ReactN_2$ and $ReactN_3$ (inferior branch, in Figure 3, considered to be the default answer of the system). For all other examples in the Recall set, the final PondN's outputs are decided through a combination of $ReactN_1$ and $ReactN_2$ (superior branch).

Threshold functions (*f1* and *f2*) are all step functions. The threshold for function *f1* was 2 and for *f2* was 0.5. The value for the learning rate *k* was 1.

As can be seen in Figure 3, for the same input vector (101010), there are two possibilities for the outputs, 1100 or 1101. The precise result depends on the decision made by PondN, as explained above, as well as on Modtr's role, that is, if it forces or not a change to the superior branch.

4. Conclusions

In our model, we make use of different channels for the inputs, which give rise to multiple processing streams. Some experimental evidence in natural systems suggests the existence of lateral connections between processing streams [4]. Although lateral connections remain to be appropriately characterized, they serve as support to our interconnected modular proposition. The nature of intermodular connections in our model is not simply that of communicating the results of computations between modules, by which mean outputs in one stage are transfered to inputs in the other. The intermodular connections were conceived to incorporate the notion that in the intermediate stages of main processing streams, primarily concerned with the inputs' features being analyzed, there may be the influence of parallel processing systems. Those parallel factors presumably reflect the reactions of other sectors of nervous system, which besides being aware of these inputs, are allegedly used as a reference to what is proper to the organism as a whole, since they have contact with homeostatic parameters [3]. In this sense, our proposition is divergent from a number of published MNNs, since we consider intermediate structures, where interaction take place between different modules, while stream is being processed, instead of simply transfering results. We made the parallel evaluation system seem of a more general instead of localized nature, in order to take into account the general nature of modulatory activities refered in section 1. Thus, we coupled its effect as an operation over groups of neurons, giving rise to a mechanism of flexibility, instead of defining it as isolated operations with particular effects over each neuron. Our example demonstration shows that the result of the computation using the isolated capacity of independent modules (interpreted as the default, or inferior branch in Figure 3) is different from that presented when the intermediate modulatory component forces the system through the superior branch in Figure 3. With the proposed structure, we incorporated some flexibility to the network function, according to the evaluation of the Modtr module. The states of Modtr module work as a selector function for possible outputs, thus representing the possibility of a choice in behavioural responses.

References

1. Ando, H., Suzuki, S., Fujita, T. (1999). Unsupervised Visual Learning of Three-dimensional Objects Using a Modular Network Architecture. *Neural Networks, 12,* 1037-1051

2. Caelli, T., Guan, L., & Wen, W. (1999). Modularity in Neural Computing. *Proceedings of the IEEE, 87,* 1497-1518

3. Damasio, A. R. (1994). *Descartes' Error. Emotion, Reason and the Human Brain.* Avon Books Inc, New York

4. Grossberg, S. (2000). *The Complementary Brain. A Unifying View of Brain Specialization and Modularity.* Technical Report CAS/CNS-TR-98-003.

5. Körner, E., Gewaltig, M. O., Körner, U., Richter, A., Rodemann, T. (1999). A Model of Computation in Neocortical Architecture. *Neural Networks, 12,* 989-1005

6. Kulkarni, S. R., Lugosi, G., & Venkatesh, S. S. (1998). Learning Pattern Classica-tion - A Survey. *IEEE Transactions on Information Theory, 44,* 2178-2206.

7. Ozawa, S., Tsutsumi, K., & Baba, N. (1998). An Artificial Modular Neural Network and its Basic Dynamical Characteristics. *Biological Cybernetics, 78,* 19-36.

8. Rolls, E. T. (1990). A Theory of Emotion, and its Application to Understanding the Neural Basis of Emotion. *Cognition and Emotion, 4,* 161{190

9. Rolls, E. T., & Treves, A. (1998). *Neural Networks and Brain Function.* Oxford University Press, New York

10. Schwenker, F., Sommer, F. T., & Palm, G. (1996). Iterative Retrieval of Sparsely Coded Associative Memory Patterns. *Neural Networks, 9,* 445-455.

Recognition of Novelty Made Easy: Constraints of Channel Capacity on Generative Networks

A. Lőrincz, B. Szatmáry, G. Szirtes, and B. Takács

Abstract

We subscribe to the idea that the brain employs generative networks. In turn, we conclude that channel capacity constraints form the main obstacle for effective information transfer in the brain. Robust and fast information flow processing methods warranting efficient information transfer, e.g. grouping of inputs and information maximization principles need to be applied. For this reason, indepent component analyses on groups of patterns were conducted using (a) model labyrinth, (b) movies on highway traffic and (c) mixed acoustical signals. We found that in all cases 'familiar' inputs give rise to cumulated firing histograms close to exponential distributions, whereas 'novel' information are better described by broad, sometimes truncated Gaussian distributions. It can be shown that upon minimization of mutual information between processing channels, noise can reveal itself locally. Therefore, we conjecture that novelty - as opposed to noise - can be recognized by means of the statistics of neuronal firing in brain areas.

1. Introduction

1.1 Generative Networks

We conjecture that the sensory information is processed by generative networks in the brain. Generative networks work by bottom-up processing of the input providing an internal representation and then top-down processing of the internal representation by means of the long-term memory [16,14,32,27,29]. Our hypothesis on generative networks is motivated by the desire to resolve the homunculus fallacy (see, e.g., [34]). This fallacy says that an internal representation is still meaningless unless someone can read it. Alas, the better the internal representation, the more elaborate is the necessary reader (interpreter). However, where is that reader? Is the reader using a representation? What sort of representation is used? In turn, an infinite regression follows. The fallacy can be shortcut [26] by considering (a hierarchy of) generative systems. In this case, the reader is 'making sense' of the representation by top-down generation of the input: The input (!) 'makes sense' if it can be derived from the internal representation. The representation of the internal representation is the reconstructed (internally generated) input. In turn, the fallacy disappears.

All this means that sensory processing is not simply coding, but also involves decoding (i.e., reconstruction). Such connectionist systems are called reconstruction networks [14,15,13]. The internal representation is used to generate the reconstructed (expected) input and error is produced between the expected input and the actual input. Finally, this error is used to correct the values of the internal representation. The main objection against such iterative schemes is that those may not be fast enough [23]. Experiments on humans show that frontal cortex can analyze complex scenes within 150 ms [36]. According to [23] this data is in favor of feedforward networks and may represent a challenge for iterative schemes.

On the other hand, iteration is not an obstacle for processing data if decoding is perfect. This is the case when coding and decoding invert each other [29]. Reduced to triviality, the steps of the iteration are as follows:
- Network starts with zeroed internal representation and with zeroed reconstructed input.
- Input is provided.
- Error becomes equal to the input.
- Coding gives rise to the internal representation.
- Reconstruction inverts (decodes) the internal representation.
- Error disappears.

In turn, feedforward networks and generative networks will have the same one-step delay in the forming of the internal representation.

If we consider generative schemes then the optimization information transfer[1] is of immediate consequence. Loss of information in bottom-up processing gives rise to deteriorated reconstruction. Efficient use of the channel capacity in the bottom-up (coding) and top-down (decoding) processes is a necessity in reconstruction networks.

1.2 Efficient Coding for Generative Networks

Efficiency is related to the possible speed of the whole coding-decoding process and is influenced by the capacity of the channels. Channel capacity measures the maximum rate (pulse/s, bit/s) for the given channel. In most cases channel capacity forms a hard constraint (the increase of the number of channels is costly), so we need other tools to make information transfer efficient. According to Shannon [35,8] optimal coding can be achieved by grouping the atomic units of the inputs first and by coding these new blocks instead of coding the atomic units themselves.

We argue that concatenation of disjoint inputs may be a smart trick applied by the brain. For example, inputs from different modalities, or temporal sequences, or both, i.e., spatio-temporal sequences can be grouped to improve the efficiency of coding.

The efficiency can also be improved by minimization of the mutual information beetween the processing channels. It is well known (see, e.g., [8]) that for statistically independent components the maximization of information transfer

---

[1] Considerations on the origin of optimization of information transfer have a long history. See, e.g., [1,3,10,30] and references therein.

induces the minimization of mutual information between processing channels. Thus, we ought to consider the minimization of mutual information between processing channels, and processing needs to be shaped by independent component (IC) analysis (ICA) [4,21,19]. IC analysis in the linear case corresponds to a matrix transformation. Given a set of inputs, the components of the transformed input set, i.e., the components of the bottom-up processed input, will be as independent as possible in statistical sense. It is worth noting that ICA can be formed by means of local (Hebbian) learning rules, which makes it appropriate for neuronal modeling [18,22].

2. Modeling Methods

We shall examine how the bottom-up processed input looks like upon minimization of mutual information. We shall use temporal sequences as inputs. A neural network accepts vector-valued inputs. We interpret 'sequences' by concatenating a few (say k) temporally subsequent inputs into a (k times) longer input vector. The method will be called embedding and the depth of the concatenation (i.e., k) will be the embedding dimension. After embedding, we present this new vector to the network that performs independent component analysis. Rows of the derived matrix correspond to vectors in 'sequence space'. These rows will be called temporal independent components (TICs) and the algorithm will be called TICA[2]. Computation of the outputs of the matrix for a given input data will be called 'testing'. We make the following distinction between 'familiar' and 'novel' inputs. Inputs (examples) are divided into categories. For each category, respective bottom-up processing matrices are developed by means of TICA. TIC analysis in each category is performed by means of the training samples, a randomly selected subset of the examples of the category. Testing is performed by means of the examples not used for training. To each bottom-up processing matrix, we have within-category testing samples that will be called 'familiar samples'. We can also test samples from other categories that will be called 'novel samples'.

In our computer studies, the FastICA [18] software package was used because FastICA has no adjustable parameters. It is of equal importance that FastICA makes use of negentropy, a quantity, which is invariant for invertible linear transformations. Upon minimization of mutual information, directions with roughly optimized negentropy values are provided. In larger simulations, PCA preprocessing was used for dimensionality reduction [12].

3. Results

TICA was performed on temporal sequences taken from different data sources. Three examples are shown in Figs. 1-3.

---

[2] TICA is not unknown in the literature. Hateren and Ruderman have shown [11] that filters produced by TICA are in better agreement with receptive fields of the primary visual cortex than those of ICA when natural image sequences (instead of individual images) are considered.

The first example is a simple artificial labyrinth (Figure 1). We intended to imitate the visual inputs of a rat moving on a fixed path in a labyrinth (see the caption of the picture for further details). The inputs were concatenated and ICA was performed. After learning, inputs underwent testing. In this particular example, all inputs were used for training: Testing of familiar inputs, in turn, correponds to testing on the training set. Figure 1.B depicts the histograms of the transformed activities taken in the learned (familiar) labyrinth (Fig. 1.B.a) and in a novel labyrinth (Fig. 1.B.b). In all cases, the histograms were computed by taking the absolute values of the activities. For the learned cases, an exponential curve fits well, whereas the novel cases are better described by a truncated Gaussian function.

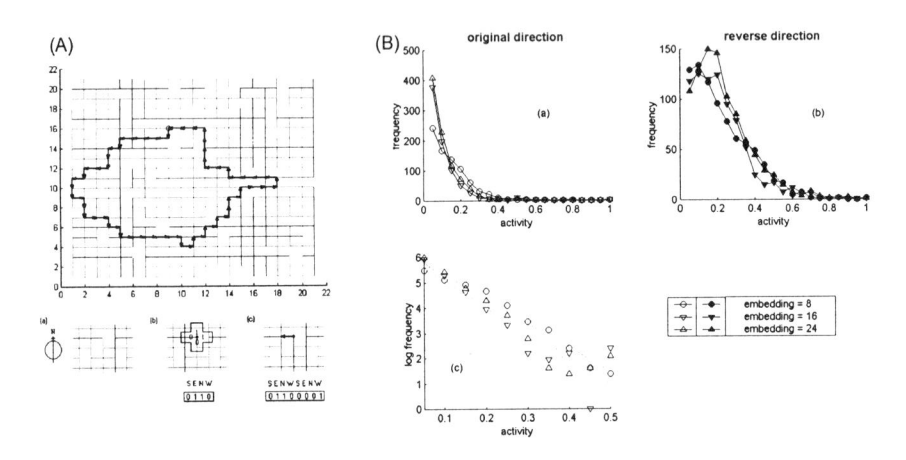

Figure 1: The labyrinth example. **(A)** Arrows show the randomly generated circular path in the labyrinth. If a direction is not available, then the 'animal' sees a wall in that direction ('1'), otherwise it does not ('0'). The processed sensory information is generated from the surrounding 'walls' in every position of the path (a). The first four digits show the presence or the absence of the walls at the south, east, north, and west side of the 'animal'. The presence (absence) of a wall makes the respective input '1' ('0') (b). The second four digits mark the direction of the path in the given position (c). The path is 58 step long. Input sequence is built from the adjacent inputs on the path. **(B)** TIC output distributions. Learning on the input set of the labyrinth (A) formed a TICA matrix. Inset (a) shows the histograms of the activities in the case, when familiar inputs were tested. Linear fits to the logarithm of the histograms are shown in inset (c). In the next case another input set was made on a different labyrinth (or in the same labyrinth walking in reverse direction) and was tested on the TICA matrix. Inset (b) shows the histograms for this case of 'novel' inputs. Histograms were calculated for embedding depths of 8, 16, and 24.

In the second example a movie database was used (Figure 2). The movies were taken from a car in traffic. We used two input sets cut from the same movie in different positions. Upon TIC analysis, histograms of familiar outputs are narrower (sparser) than histograms on novel outputs. This observation is numerically confirmed by the calculated negentropy values on the rectified distributions.

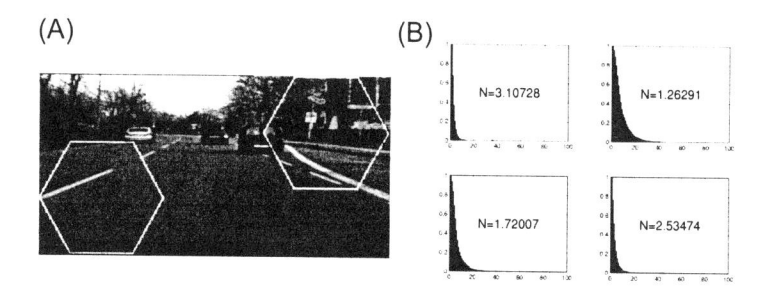

Figure 2: Traffic example. Hexagonal windows of 169 pixels in different positions (hexagonal areas in subfigure **A**) were used for creating the familiar and unfamiliar data sets. **(B)** Distributions of TICA outputs. 7 inputs were concatenated to form an embedded input. Diagonal (off-diagonal) histograms describe testing for familiar (novel) inputs. The negentropy (N) values [8] of the histograms (discretizations of the rectified distributions) are given within each subfigure.

In the third example acoustic signals were used (Figure 3).

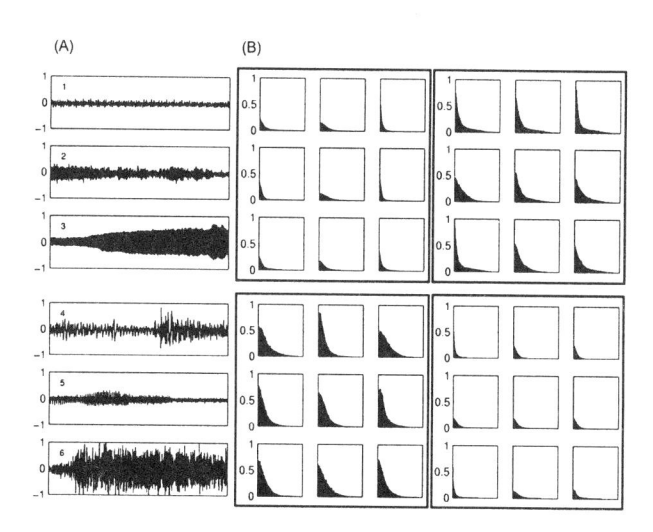

Figure 3: The case of acoustic signals. **(A)** Samples are about 250 ms in length and are from different sources (e.g. music, sounds in a forest or sounds of a whale). **(B)** TIC output distributions. The training set for the TICA matrix was created from a mix of three samples out of the six signals. Mixing of the other three samples made the 'novel' inputs. Embedding depth is 16. The number of TICs is thus 3x16=48. Here - in contrast to Figure 1 and 2 - the histograms of randomly selected individual outputs are shown. Diagonal (off-diagonal) blocks of histograms represent familiar (novel) tests.

4. Discussion

Individual histograms on familiar inputs are most reminiscent to findings in the prefrontal cortex, the inferior temporal visual cortices (see, e.g., [2,37] and references therein), and the hippocampus. Individual histograms sometimes show both broad and sharp parts. Those are best characterized by mixed Gaussian and exponential distribution. Qualitatively similar histograms have been reported in [37].

We voted for generative networks as opposed to feedforward schemes to save the 'homunculus' without falling into trap of the 'homunculus fallacy'. Ordinary generative networks, however, are poor candidates for 'making sense' of the reading. This can be seen by the following argument: Assume that the internal representation has the same dimension then the input. Assume further that bottom-up and top-down processing are linear and are of full rank. Then, a generative scheme can reconstruct every possible input with zero error. In other words, every input 'makes sense' for an ordinary generative network. How could then a generative network 'recognize' that the actual input 'does not make sense'? The first step of 'making sense' is to recognize 'familiarity' vs. 'novelty'. Novelty and noise are also to be distinguished. In a hierarchical system, noise at one level may be grouped into novel information at another level. In fact, the optimization of information transfer requires that frequent symbols need short codes (low level in the hierarchy), whereas less frequent symbols need longer codes (higher levels in the hierarchy). A less frequent symbol may seem nothing but noise at low levels. ICA and TICA are optimal to make this distinction. Denoising (noise filtering) of IC outputs can be approximated by diminishing small amplitude outputs [17]. In turn, noise filtering is local. Reconstruction is performed by high activity IC outputs. Noise appears in the reconstruction error. This reconstruction error can be analyzed by higher levels, as suggested in [32]. Recognition of novelty is the task of a higher level and works on large reconstruction errors (large residuals) in a given area. Our simulations imply that the recognition of novelty is 'easy' and can be instantaneous in optimized generative networks. We found on different examples, that a novel (not yet seen, not yet optimized) input gives rise to distinctly different statistical properties at the level of internal representation.

Computational considerations, such as entropy minimization, factor analysis, or sparse coding lead to representational schemes that optimize information transfer [1,3,9,10]. Factor analysis and ICA have been considered as the main strategy of encoding sensory information for a long time [1,3,10]. Computer simulations using independent component analysis provide good agreement with receptive field properties of neurons in the primary visual cortex [30,5]. TICA has been shown superior to ICA in this respect [11]. Temporal firing properties of LGN [38] are such that TICA processing is possible in the primary visual cortex.

In this paper, we put forth the idea that long-term memory is represented by connection strengths of top-down processing. In turn, channel capacity constraint forms the underlying reason for maximization of information transfer and that independent component analysis of (spatial, temporal, modularity-wise) sequences of symbols or symbol sets need to be considered. Let us consider the question whether ICA/TICA processing is possible in the neocortex.

Interestingly, the concept that ICA processing is performed in a generative network allows the 'derivation' of several important details of the complex hippocampal-entorhinal region [26,27,28,29,7]. Processing of temporal sequences in the hippocampus, the focal point for the learning of long-term explicit memory, has been suggested in the literature [25,24]. According to simulations of CA3 principal cells of the hippocampus, information from remote synapses on the dendritic tree is not damped compared to synapses close to the soma [13,20] and thus TICA is possible in the hippocampus. The hippocampal output via the entorhinal cortex may serve as a training signal for neocortical areas as follows. Neocortical areas may report novelty by their firing patterns. The hippocampus analyzes the novel data, concatenates the codes spanning different time domains and different areas, and promotes the learning process in the reporting areas. Then 'collaboration' between direct backward projections from the hippocampus and hierarchical backward projections between adjacent cortical areas (see, e.g., [30] and references therein) should play a role. This consequence is supported by the findings that there are as many backward projections between adjacent cortical areas as forward connections [37]. It may be worth mentioning that the concept of TICA can be extended to positive coding networks [6].

We stress two things. The emphasis is on forming blocks of symbols, i.e., concatenating in time, grouping in space, and grouping between different modalities. Secondly, none of the test samples was used during training, but some represented similar, whereas others represented dissimilar statistics. ICA is working on the higher order moments of the activity distribution and tends to make this distribution narrower (sparser). The learned statistics is thus expected to be sparser. Statistical properties can form the basis of novelty recognition in optimized generative networks. This feature is not specific to generative schemes [31]. A feedforward network that optimizes information transfer will behave the same way. The assumption that 'making sense' is top-down 'inferencing' via reconstruction in a recurrent network is the novelty in our work. Maximization of information transfer, grouping of inputs into blocks, fast novelty recognition, recognition of familiarity, and sparsity of the representation are consequences.

5. Conclusions

We assumed that the brain makes use of generative networks. Then, however, channel capacity constraint and coding efficiency form the main obstacle for top-down filtering of the inputs of these networks. We have argued that coding efficiency can be optimized by grouping blocks of inputs to maximize information transfer. Computer experiments were run by making blocks of inputs on the temporal domain. Independent component analysis was performed and ICA transformation was computed. Familiar and novel blocks of inputs underwent this ICA transformation. Upon transformation, novel inputs can be distinguished from familiar ones simply on the bases of neuronal firing distribution. This information is immediately available in perfect reconstruction (coding and decoding) networks and thus, in these structures, the recognition of novelty/familiarity can be as fast as recognition itself.

Acknowledgements

This work was partially supported by OTKA Grant No. 32487. Enlightening discussions with Professor Irving Biederman are gratefully acknowledged. Traffic database was kindly provided by Honda Future Technology Research, Offenbach, Germany with permission from Prof. Werner von Seelen, Ruhr-Universität Bochum, Institut für Theoretische Biologie.

References

1. Attneave, F. (1954). Some informational aspects of visual perception. *Psychological Review, 61,* 183-193.
2. Baddeley, R. J., Abbott, L. F., Booth, M., Sengpiel, F., Freeman, T., Wakeman, E. A. & Rolls, E. T. (1997). Responses of neurons in primary and inferior temporal visual cortices to natural scenes. *Proc. Roy. Soc.* London *B264,* 1775-1783.
3. Barlow, H. B. (1961). In: *Sensory communication.* Rosenblith, W. A. (ed.). Cambridge, MA: MIT Press.
4. Bell, A. J. & Sejnowski, T. J. (1995). An information-maximization approach to blind separation and blind deconvolution. *Neural Computation, 7,* 1129-1159.
5. Bell, A. J. & Sejnowski, T. J. (1997). The 'independent components' of natural scenes are edge filters. *Vision Research 37,* 3327-3389.
6. Charles, D. & Fyfe, C. (1998). Modelling multiple cause structure using rectification constraints. *Neural Systems, 9,*167-182.
7. Chrobak, J. J., Lőrincz, A. & Buzsáki G. (2000). Physiological patterns in the hipocampus-entorhinal cortex system. *Hippocampus, 10,* 457-465.
8. Cover, T. & Thomas, J. (1991). Elements of information theory. New York, USA: John Wiley and Sons.
9. Dong, D. W. & Atick, J. J. (1995). Temporal decorrelation: a theory of lagged and nonlagged responses in the lateral geniculate nucleus. *Network, 6,* 159-178
10. Field, D. J. (1987). Relations between the statistics of natural images and the response properties of cortical cells. *Journal of the Otical Society of America, A4,* 2379-2394.
11. Hatteren, J. H. & Ruderman, D. L. (1998). Independent component analysis of natural image sequences yields spatio-temporal filters similar to simple cells in primary visual cortex. *Proc. Roy. Soc. London, B265,* 2315-2320.
12. Haykin, S. (1999). Neural Networks: A comprehensive foundation. New Jersey, USA: Prentice Hall.
13. Henze, D. A., Cameron, W.E. & Barrionuevo, G. (1996). Dendritic morphology and its effects on the amplitude and rise-time of synaptic signals in hippocampal CA3 pyramidal cells. *J. Comp. Neurology, 369,* 331-344.

14. Hinton, G. E. & Ghahramani, Z. (1997). Generative models for discovering sparse distributed representations. *Philosophical Transactions of the Royal Society B, 352*, 1177-1190.
15. Hinton, G. E. & Zemel, R. S. (1994). Autoencoders, minimum description length, and Helmholtz free energy. In: J. D. Cowan, G. Tesauro, & J. Alspector (ed), *Advances in neural information processing systems, 6,* 3-10. San Mateo, CA: Morgan Kaufmann.
16. Horn, B.K.P. (1977). Understanding image intensities. *Artifical Intelligence, 8,* 201-231.
17. Hvärinen, A., Hoyer, P., & Oja, E. (1999) Sparse code shrinkage: Denoising by maximum likelihood extimation. In: Kearns, M., Solla, S. A., & Cohn, D. (ed), *Advances in Neural Information Processing Systems, 12*, MIT Press, Cambridge MA.
18. Hvärinen, A. & Oja, E. (1997). A fast fixed-point algorithm for independent component analysis. *Neural Computation, 9,* 1483-1492.
19. Hvärinen, A. & Oja, E. (2000). Independent component analysis: Algorithms and applications. *Neural Networks, 13(4-5)*:411430.
20. Jaffe, D. B & Carnevale, N. T. (1999). Passive normalization of synaptic integration influenced by dendritic architecture. *Journal of Neurophysiolog, 82,* 3268-3285.
21. Jutten, C. & Herault, J. (1991). Blind separation of sources, Part I: An adaptive algorithm based on neuromimetic architecture. *Signal Processing, 24,* 1-10.
22. Karhunen, J. Oja, E. Wang, L. Vigário, R. & Joutsensalo, J. (1997). A class of neural networks for independent component analysis. *IEEE Transactions on Neural Networks, 8,* 486-504.
23. Koch, C. & Poggio, T. (1999). Predicting the visual world: Silence is golden. *Nature Neuroscience, 2,* 9-10.
24. Lisman, J. E. (1999). Relating hippocampal circuitry to function: Recall of memory sequences by reciprocal dentate-CA3 interactions. *Neuron, 22,* 233-242.
25. Lisman, J. E. & Idiart, M. A. P. (1995). A mechanism for storing 7±2 short-term memories in oscillatory subcycles. *Science, 267,* 1512-1514.
26. Lőrincz, A. (1997). Towards a unified model of cortical computation II: From control architecture to a model of consciousness. *Neural Network World, 7,* 137-152.
27. Lőrincz, A. (1998). Forming independent components via temporal locking of reconstruction architectures: A functional model of the hippocampus. *Biological Cybernetics, 79,* 263-275.
28. Lőrincz, A. & Buzsáki, G. (1999). Computational model of the entorhinal-hippocampal region derived from a single principle. In: *Proceedings of IJCNN, Washington, July 9-16,* JCNN2136.PDF IEEE Catalog Number: 99CH36339C, ISBN: 0-7803-5532-6
29. Lőrincz, A. & Buzsáki, Gy. (2000). Two-phase computational model training long-term memories in the entorhinal-hippocampal region. In: Scharfman, H. E., Witter, M. P., Schwarcz, R. (ed), *The parahippocampal region: Implications for*

neurological and psychiatric diseases, 911, New York Academy of Sciences, 83-111.
30. Olshausen, B. A. & Field, D. J. (1996). Emergence of simple-cell receptive field properties by learning a sparse code for natural images. *Nature, 381,* 607-609.
31. Parra, L., Deco, G. & Miesbach, S. (1995). Statistical independence and novelty detection with information preserving nonlinear maps. *Neural Computation, 8,* 260-269.
32. Rao, R. P. N. & Ballard, D.H. (1999). Predictive coding in the visual cortex: A functional interpretation of some extra-classical receptive-field effects. *Nature Neuroscience, 2,* 79-87.
33. Rolls, E. T. & Treves, A. (1998). Neural networks and brain function. Oxford, UK: Oxford University Press.
34. Searle, J. R. (1992). The rediscovery of mind. Cambridge, MA: Bradford Books, MIT Press.
35. Shannon, C. E. (1948). A mathematical theory of communication. *AT&T Bell Labs. Tech. J., 27,* 397-423.
36. Thorpe, S. Fize, D. & Marlot, C. (1996). Speed of processing in the human visual system. *Nature, 381,* 520-522.
37. Treves, A., Panzeri, S., Rolls, E.T., Booth, M. & Wakeman, E. A. (1999). Firing rate distributions and efficiency of information transmission of inferior temporal cortex neurons to natural visual stimuli. *Neural Computation, 11,* 601-631.
38. Wimbauer, S., Wenish, O. G., Miller, K. D. & van Hemmen, J. L. (1997). Development of spatio-temporal receptive fields of simple cells: I. Model formulation. *Biological Cybernetics, 77,* 456-461.

A Biologically Plausible Maturation of an ART Network

Maartje E.J. Raijmakers & Peter C.M. Molenaar

Abstract

We present a numerical bifurcation analysis of a shunting neural network (SNN). Fold bifurcations appear to occur in the plane of parameters that are subject to post-natal maturation: the range and strength of lateral connections. The SNN is implemented as a content addressable memory (CAM) in Exact ART, a complete implementation of Adaptive Resonance Theory. The stability and functionality of Exact ART with the CAM in different dynamic regimes is maintained. Moreover, through a bifurcation, the learning behavior of Exact ART changes from forming local representations to forming distributed representations. Presently, we extend Exact ART by adding biologically plausible evolution equations for activation dependent maturation of lateral connections in a SNN [14]. The resulting model is an epigenetic model of both morphological development and cognitive behavioral development.

1. Introduction

Artificial neural network simulation usually is carried out with configurations that are, from a developmental-biological point of view, in a mature state. That is, the contours of neural field organization are given and simulation usually pertains to the adaptation of interconnections within and between the given fields. Recently, in cognitive science, constructive neural networks of development, model maturation by the addition of structure during and in addition to learning (e.g., [3, 9]). In these cases, more inspired by computation theory than by biology, conceptual resources (hidden units or recurrent calculation depth) are added to the network during or even as a function of learning. It is, however, also possible to study the morphogenesis of neural networks by introducing biologically plausible continuous epigenetic rules according to which neuronal morphology and network development take place [14]. One of our objectives is to insert Van Ooyen's et al., [14], maturation equations of a shunting neural network (SNN) into an Adaptive Resonance Theory (ART, [5]) model of learning recognition codes. In this case, the maturation process of a cognitively applied neural network is modeled by ordinary differential equations that describe the evolution of real valued parameters as a function of activation. As will be shown, in the plane of these parameters local bifurcations of activation dynamics occur. That is, parameter changes trigger qualitative changes in network dynamics and, as a consequence, qualitative changes in the cognitive network behavior as well. The latter induction of qualitative new cognitive behavior without adding conceptual resources is a consistent interpretation of epigenetic theories of cognitive development, like Piaget's stage theory [8, 10]. The induction of qualitative new cognitive behavior by means of parameter changes (i.e., without adding conceptual resources) would be a consistent interpretation of epigenetic theories of cognitive development, like Piaget's stage theory [8, 10]. The results presented in this article are an important step towards such a developmental model of transitions in development.

A second important aim of the described simulation of network ontogenesis is to study the effects of cloning a network germ into an ensemble of many genetically identical copies and studying the epigenetic trajectories of this ensemble to full network development. It is expected on the basis of theoretical quantitative arguments that the endstates within such an ensemble of genetically identical clones will show considerable phenotypical variation due to the nonlinearities inherent in the epigenetical processes [7]. This conjecture is currently being tested for an ensemble of simple ART network clones with respect to several phenotypical characteristics, including morphological (neuronal morphology) and behavioral (pattern classification) phenotypes.

The present article mainly concerns a bifurcation analysis of a SNN that appears as the category representation field in Exact ART [12], which is an Adaptive Resonance Theory model [5]. We show the occurrence of codimension-one bifurcations in the distance-dependent on-center off-surround SNN with fixed external input under variation of the four parameters that define the shunting connections in the network. Furthermore, we show that the equilibrium states that occur in different dynamic regimes of the SNN are stable. The mentioned morphological simulation of Van Ooyen et al. [14] concerns the evolution of just these parameters. To examine the resulting cognitive behavior of the SNN in different dynamic regimes we study the classification behavior of Exact ART with a category representation field in different dynamic regimes that occur with different values of the bifurcation parameters. To this end, we changed the dynamic equations of the adaptive weights in accordance with Carpenter [1], such that distributed representations can be learned without catastrophic forgetting.

In the next section we introduce the on-center off-surround SNN and we describe its basic properties. Section 3 presents the numerical bifurcation analysis of the SNN: the applied techniques and the results. Additionally, for the examined ranges of parameters we present a qualitative description of network equilibria, which is related to the nature of category representations formed by Exact ART. Section 4 describes the learning behavior on a simple classification task of Exact ART with the F2-field in several dynamic regimes. Raijmakers, van der Maas and Molenaar [13] describe all the details of this study.

2. On-center Off-Surround Shunting Neural Networks

Grossberg [4] introduces several strongly related models of on-center off-surround SNN's. Such a network consists of mutually connected neurons with activities that obey the nerve-cell-membrane equation. The activities of units j are described by (1).

$$\frac{dx_j}{dt} = -Ax_j + (B - x_j)(\sum_{k=1}^{M} C_{kj} f(x_k) + I_j) - (x_j + D) \sum_{k \neq j}^{M} E_{kj} f(x_k), \qquad (1)$$

$$C_{kj} = \hat{C} \exp(-\sigma_e^{-2}|k-j|^2), \hat{C} \text{ is set such that } \sum_{k=1}^{M} C_{kj} = d_e \quad (1a)$$

$$E_{kj} = \hat{E} \exp(-\sigma_i^{-2}|k-j|^2), \hat{E} \text{ is set such that } \sum_{k \neq j}^{M} E_{kj} = d_i, E_{jj} = 0 \quad (1b)$$

j=1, 2, ..., M. x_j is the activity of the j-th neuron (population) in the SNN. The vector of activities x_j will be referred to as **x**. A is the passive decay parameter of the activity. I_j is the external input. B and D are the parameters that limit activities x_j from respectively above and below. Parameters C_{kj} and E_{kj} define the lateral excitatory and inhibitory connections within the SNN. Parameters C_{kj} and E_{kj} are defined by (1a) and (1b). C_{kj} and E_{kj} are Gaussian distributed along the distance between units j and units k. σ_e and σ_i determine the range across which lateral connections exist. Parameters d_e and d_i determine the strength of lateral connections.

3. Numerical Bifurcation Analysis

The SNN activity, **x**, has many equilibrium states which depend, among other factors, on the activity vector at time zero, **x**(0). If we study the change of equilibria we can not examine all the coexisting equilibria, because they are too numerous. Therefore, the numerical bifurcation analysis of the SNN defined by (1) only concerns those settings of the network that might occur in Exact ART. That is, equilibrium states of which the zero vector ($x_j = 0$, j † M) is part of the basin of attraction. Consequently, we can not apply the current continuation algorithms for numerical bifurcation analysis.

We first briefly explain the methods we used for numerical bifurcation analysis and the qualitative description of equilibria. A complete description is given in [13]. Second, we present the bifurcation diagrams and the qualitative changes of equilibria. Additionally, we discuss how bifurcations relate to transitions in the qualitative description of equilibria.

3.1 Method

The qualitative description of network equilibria is based on the application of the SNN in Exact ART. An equilibrium state is an M-dimensional (M = 25) vector, which we describe by two numbers: the number of units with an activity above threshold (N_u) and the number of clusters of units with activity above threshold (N_c). The threshold is the activity value above which learning takes place in Exact ART, and is fixed arbitrarily ($\theta_y = .35$). A winner-takes-all network, for example, has one unit above threshold, N_u would equal 1, and this unit forms also the only cluster of units above threshold, N_c would equal 1. An activity vector which follows a bell-shaped curve as a function of unit number (1   25), for example, has one cluster of active units, N_c would equal 1, and possibly several active units, $N_u > 1$. Since the changes of equilibria are mostly abrupt, as we will show below, the precise value of θ_y (within certain bounds) is not critical for N_c (but sometimes it is for N_u).

The numerical bifurcation analysis concentrates on fold-bifurcation and Hopf-bifurcations. These bifurcation points are characterized by special conditions of the linearization matrix. The linearization matrix is the Jacobian matrix, DF, of an equilibrium state of the system.

$$DF = \left[\frac{\partial F_j}{\partial x_k}\right] \tag{2}$$

is the Jacobian matrix of the first partial derivatives of the function $F = (F_1, .., F_j, .., F_M)^T$. For (1), elements DF_{jk} of the Jacobian matrix are defined by (3a) and (3b):

$$DF_{jk} = (B - \bar{x}_j)(2C_{kj}\bar{x}_k) - \bar{x}_j(2E_{kj}\bar{x}_k), j \neq k \tag{3a}$$

$$DF_{jj} = -A - (\sum_{k=1}^{M} C_{kj}\bar{x}_k^2 + I_j) + (B - \bar{x}_j)(2C_{jj}\bar{x}_j) - (\sum_{k \neq j}^{M} E_{kj}\bar{x}_k^2) \tag{3b}$$

where $\bar{x}_j$ and $\bar{x}_k$ are vector elements of the equilibrium state $\bar{x}$ of system (1).

Fold-bifurcation points occur at those parameter values for which one eigenvalue of the linearization matrix of the equilibrium state becomes zero. Hopf bifurcation points are characterized by a simple pair of pure imaginary eigenvalues and no other eigenvalues with zero real part. This implies that we can first detect the bifurcation points with zero real part and then, depending on the imaginary part, we determine whether the point is a fold or a Hopf bifurcation.

To detect bifurcation points, we search for zero eigenvalues of the linearization matrix in a two-dimensional grid of parameters, but in the direction of one parameter simultaneously. At the beginning of each step we set x_j, all $j \dagger M$, to zero and we integrate the system defined by (1) until equilibrium is reached. LSODAR [6] is used as integration method. Then, we calculate the maximum eigenvalue of the linearization matrix. If a local maximum appears in this series of maximum eigenvalues, we search for a zero point by means of a one-dimensional search method: Golden Section Search. A zero eigenvalue indicates a bifurcation point. The stability of equilibria depends on the eigenvalues of the Jacobian matrix. If all eigenvalues are negative, the equilibrium is stable. Otherwise the equilibrium is unstable.

3.2 Results

We performed numerical bifurcation analysis of the SNN in regions of parameters that are subject to post-natal maturation, the range (σ_i and σ_e) and strength (d_i and d_e) of lateral connections. The parameter regions that we examined are $\sigma_i \times \sigma_e$ ([.6, 10.5] x [.1, 2.06]) and $d_i \times d_e$ ([.12, 23.88] x [.1, .99]). The results are based on a fixed input vector. However, it appeared from repeated simulations with different random input vectors that the overall picture of the qualitative description of the equilibrium behavior is a general result. Figure 1a is a bifurcation diagram, which also shows a qualitative description of the equilibrium states of (1) in the $\sigma_e \times \sigma_i$ plane ($d_i = 24.0$, $d_e = 1.0$). The gray tones denote both N_u and N_c for each combination of σ_e and σ_i (see Figure 1c). It appears that only isolated activity peaks, instead of clusters of activity peaks, are found in the equilibrium behavior (i.e., $N_u = N_c$). Most changes of equilibrium patterns due to variation of σ_i and σ_e consist of either the appearance or the disappearance of peaks in the activity of units. More specifically, with a decreasing range of inhibitory connections, i.e. decreasing σ_i, the number of active units increases. The variation of σ_e causes less changes in the equilibrium state of the SNN than σ_i does, but its effect is considerable: if the range of excitatory connections is too broad none of the activities comes above threshold.

The gray tones in Figure 1b present a qualitative description of equilibrium states in the $d_i \times d_e$ plane ($\sigma_i = 9$, $\sigma_e = .6$). With large values of d_i, that is with strong inhibitory feedback, only isolated activity peaks ($N_c=N_u$) and zero activity peaks ($N_c=N_u=0$) occur as we saw in the examined $\sigma_i \times \sigma_e$ plane. With smaller values of d_i also broader peaks, that is clusters of active units, occur. In Figure 1b this is reflected by striped areas ($N_u \neq N_c$). The area with clusters of active units instead of isolated activity peaks ($N_u \neq N_c$) is relatively small. Very small values of d_i result in uniform activity ($N_c = 0$, $N_u = 25$).

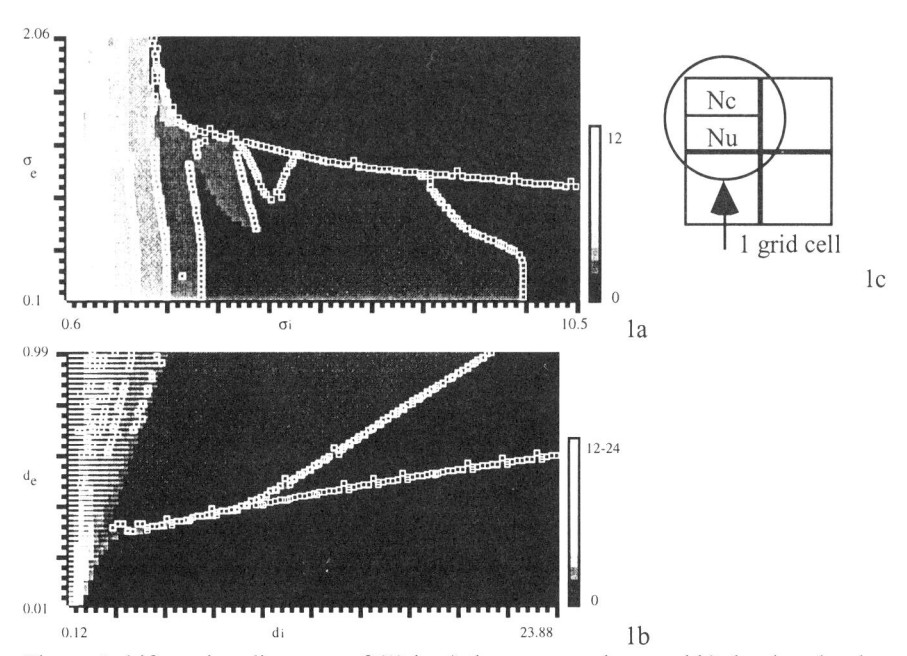

Figure 1: bifurcation diagrams of (1) in a) the $\sigma_i \times \sigma_e$ plane and b) the $d_i \times d_e$ plane. Gray tones denote the number of clusters of active units, N_c, and the number of active units, N_u, in each grid cell conform 1c. The white squares denote the Fold-bifurcation points that were found by the grid search. Colored versions of these figures can be found at: http://macnet007.psy.uva.nl/Users/Raijmakers/chapter4.html

In general we can conclude that the area with a winner-takes-all dynamics ($N_c=N_u=1$), i.e. one peak, is limited and that there exist a large area, in both the σ_i

x σ_e and the d_i x d_e plane, within which no units becomes active above threshold. However, if σ_e is below 1, activity patterns with various numbers of active units occur. Clustered peaks also occur, but only exist for a rather small range of d_i. Simulations with different random input vectors show equivalent results. In section 4 we will examine the learning behavior of Exact ART with a category representation field in different dynamic regimes.

In Figure 1a and Figure 1b, the white squares reflect bifurcation points, which all appear to be fold-bifurcation points. No Hopf-bifurcation points and no stable limit cycles were found. It appears that if a boundary between different regimes does not coincide with a bifurcation point the parameter change causes the starting state (i.e. zero activities, $\mathbf{x}(0) = \mathbf{0}$) to become a member of a different basin of attraction. Also the reverse situation occurs: Fold-bifurcation points are found that do not change the qualitative behavior of the SNN in terms of N_c and N_u. In that case, activity peaks appear and disappear simultaneously. The significance of these results with regard to developmental hypotheses depends on the application of a SNN in a neural network model of cognitive behavior. Therefore, we apply the SNN to an Adaptive Resonance Theory network.

4. Exact ART Under Various Dynamic Regimes

ART-networks contain a SNN as the category representation field, F2, where it obeys a winner-takes-all dynamics. The most common implementations of ART, viz. ART1, ART2, and ART3, are such that minimal computation time is needed. For that reason, several aspects of these models, among which the SNN, are implemented by equilibrium behavior only. Exact ART [12], in contrast, is a complete implementation of an ART network by a system of differential equations.

4.1 Exact ART

Exact ART bears a large resemblance to ART2 [2]. As ART2, Exact ART is built of several subsystems: F0 is the input transformation field, F1 combines bottom-up input and top-down expectations, F2 is the category representation field, bottom-up and top-down weights between F1 and F2 form an adaptive filter, and the orienting subsystem resets F1 and F2 if it detects a mismatch between bottom-up input (F1) and top-down expectations (F2). The functions of these different parts in learning stable recognition codes and the functional interactions of the different parts are equivalent to those of ART2. The main difference with ART2 is that Exact ART is completely defined by a system of ordinary differential equations. Furthermore, the F2-field is implemented by a gated dipole field, which is based on [5]. In addition, the adaptive weights of Exact ART are replaced by adaptive thresholds as described in [1], to allow for forming distributed representations without catastrophic forgetting. Calculation of activities and weights takes place by integrating the complete system of differential equations simultaneously during the presentation of external input. A complete description of the model and mathematical proofs of its fundamental characteristics are given in [11, 12].

4.2 The Shunting Neural Network in Exact ART

In ART, categories are represented by the F2-field. In Exact ART, the units of the F2-field are gated dipoles. The gated dipoles of F2 constitute a SNN as described by (1). In most implementations of ART the F2-field is supposed to follow a winner-takes-all dynamics: The activity of the unit with the highest input becomes maximal, and the activities of the other units decrease to zero (winner-takes-all dynamics imply that $N_c = N_u = 1$). This is established in (1) if the positive feedback signals are self-excitatory only, and the inhibitory feedback signals are uniform, without self-inhibitory feedback. Weights between an F2-unit j and all units of layer F1 only adapt if in equilibrium the activity of the F2 unit j is above a threshold ($\theta_y = .35$). For that reason we classified the behavior of the SNN by counting the number of units that are activated above θ_y, i.e. N_u. As we showed in the section 3.2, the number of F2-units that become active simultaneously in Exact ART can be manipulated by parameter σ_i. In the next section we will examine the classification behavior of Exact ART with an F2-field in several dynamic regimes: with 2, 4, 5, and 6 simultaneously activated units. In the following simulations, the number of F2 units, M, is 25. Default parameter values can be found in [13].

4.3 Classification Behavior of Exact ART

An ART network performs an unsupervised classification of input patterns (but can be part of a supervised network ARTMAP). During the classification process, prototypes of the formed classes of patterns are preserved in the adaptive weights. We describe three aspects of the performance of Exact ART. First, it is necessary for ART in learning a stimulus set, that the match between each input pattern and its representation is sufficiently high (i.e. the mismatch R should be below 1-ρ, ρ is the so called vigilance parameter). Second, after learning, the F2 category representation of an input pattern should be stable, which means that a particular input pattern is represented by the same F2 code after each presentation. These two criteria, which are both used to study ART2, indicate whether the network learns a classification task at all. Thirdly, the kind of category representation that is formed during the learning process, being local or distributed, is of importance. The classification task we use to test Exact ART with different F2 dynamics is the same task as in [2] used to test the stability of learned representations in ART2. The task consists of four different input patterns that are presented in a fixed order: A, B, C, A, and D. This sequence is presented repeatedly for many trials. As is shown in [12] Exact ART with winner-takes-all dynamics learns the simple classification task in a stable manner. With ρ equal to .7, Exact ART with a winner-takes-all dynamics of F2 forms four different categories for the four different input patterns.

The simulation study with different values of σ_i shows that the category representations of input patterns also stabilize with 2 ($\sigma_i = 7.5$), 3 and 4($\sigma_i = 3.0$), 4 and 5 ($\sigma_i = 2.2$), 6 and 7 ($\sigma_i = 1.3$) simultaneously active units in F2. Moreover, the mismatch between input and learned categories is in all these cases sufficiently low, that is R < 1 - ρ. From this we can conclude that the F2-field of ART is robust. For up to 5 simultaneously active F2 units, local representations are formed. That is, each input pattern is represented by a set of units and these sets are disjunctive. If we decrease σ_i such that the number of simultaneously active F2 units (i.e., 7) times the number of different input patterns (i.e., 4) exceeds the total number of F2 units (i.e., 25), distributed representations are formed. That is, each input pattern has a unique representation (consisting of a set of F2 units), but some F2 units are active during the presentation of more than one input pattern. Representations of input patterns C and D, which are most alike, share five out of seven F2-units. The representations are again stable and the match between input and learned representation is sufficient for each input pattern. Nevertheless, the match is no longer perfect: R equals .01, .02, .27, and .19 for inputs A, B, C, and D respectively. In contrast to an F2-field with a winner-takes-all dynamics, now unique representations can be formed for each input pattern while at the same time similarities between patterns are represented by the recognition code.

5. Conclusion

Numerical bifurcation analysis of a shunting neural network (SNN) with fixed external input shows that fold-bifurcation points occur in the plane of parameters that represent the range and the strength of lateral connections. These structural properties of the SNN are known to vary by maturation. The SNN with winner-takes-all dynamics is, for example, as a content addressable memory applied to Adaptive Resonance Theory [5]. Since bifurcations are found in the dynamics of the SNN in the mentioned parameter plane, two questions arise: First, is the functionality of the SNN maintained in various dynamic regimes? Second, increases the functionality of the system such that the bifurcation is related to learning or development? We studied the functionality of the SNN by means of the classification behavior of Exact ART. The answer to the first question is positive: The functionality of Exact ART is maintained in different dynamic regimes of the SNN. The most striking result is, however, that with short range inhibitory connections the representation of input patterns appears to be distributed. From this we can conclude that the functionality of the category representation field is improved due to bifurcations in the SNN. On a biological level the change is related to development. On a cognitive behavioral level, however, the relation to development is not yet shown to exist. The induction of qualitative new cognitive behavior by means of parameter changes (i.e., without adding conceptual resources) would be a consistent interpretation of epigenetic theories of cognitive development, like Piaget's stage theory [8, 10]. The results presented in this article are an important step towards such a developmental model of transitions in development.

In the presented study, the change of the structural properties of the SNN is not a function of learning or activity within the network. We manipulated the structural parameters by hand. The simulation study of Van Ooyen et al. [14] concerns an activity dependent development of the lateral connections within the same SNN. Their model is fully based on biological properties of network development. Currently, we are incorporating their equations into the system of differential equations defining the category representation layer of Exact ART. We hope that the resulting epigenetic model of both behavioral and morphological development, first, can still learn stable recognition codes, and, second, can learn distributed representations which improve in representing and detecting similarities. If available the results of the ongoing study will, in addition to the above results, be presented at the conference.

Acknowledgment

This research was supported in part by the Dutch Organization of Scientific Research (NWO) Grant PPS 98016.

References

1. Carpenter, G. (1997). Distributed learning, recognition, and prediction by ART and ARTMAP neural networks. *Neural Networks, 10* (8), 1473-1493.
2. Carpenter, G., Grossberg, S. (1987). ART 2: Self-organization of stable category recognition codes for analog input patterns. *Applied Optics, 26* (23), p.4919-4930.
3. Elman, J.L. (1993). Learning and development in neural networks: The importance of starting small. *Cognition*, 48 (1), 1993, 71-99.
4. Grossberg, S. (1973). Contour Enhancement, Short Term Memory, and Constancies in Reverberating Neural Networks. *Studies in Applied Mathematics, 52* (3), p.213-257.
5. Grossberg, S. (1980). How does a brain build a cognitive code. *Psychological Review, 87*.
6. Hindmarsh, A. C., Odepack, a systematized collection of ode solvers. In: Stepleman, R. S. et al. (eds.) Scientific Computing. North Holland Amsterdam, pp. 55 — 64.
7. Molenaar, P.C.M., Boomsma, D.I. & Dolan, C.V. (1993). A third source of developmental differences. *Behavior Genetics, 23*, 519-524.
8. Molenaar, P.C.M. (1986). On the impossibility of acquiring more powerful structures: A neglected alternative. *Human Development, 29*, 245-251.
9. Quartz, S.R. (1993). Neural networks, nativism, and the plausibility of constructivism. *Cognition, 48*, 223-242.
10. Raijmakers, M.E.J. (1997). Is the learning paradox refuted? *Behavioral and Brain Sciences*, 20(4).
11. Raijmakers, M.E.J. (1998). Distributed learning in Exact ART. *Technical Report*, Department of Psychology, University of Amsterdam.
12. Raijmakers, M.E.J. & Molenaar, P.C.M. (1997). Exact ART: A complete implementation of an ART network. *Neural Networks, 10* (4), 649-669.
13. Raijmakers, M.E.J., van der Maas & Molenaar, P.C.M. (1997). Numerical Bifurcation Analysis of Distance-Dependent On-Center Off-Surround Shunting Neural networks. *Biological Cybernetics 75*, 495-507.
14. Van Ooyen, A., van Pelt, J. & Corner, M.A. (1995). Implications of activity dependent neurite outgrowth for neuronal morphology and network development. *Journal of Theoretical Biology, 172*, 63-82.

Development and Category Learning

Developing Knowledge about Living Things: A Connectionist Investigation

Samantha J. Hartley

Abstract

This paper describes empirical work and connectionist investigation of the featural basis of children's knowledge about living things. The empirical data shows differences in the rate at which children acquire subcategories of living things, differences in the timing of changes in knowledge organisation, and changes in the distribution of feature types children use to represent their knowledge. The connectionist model was developed to investigate the role of feature based representations in determining the organisation of knowledge during childhood, and to investigate the possible mechanisms involved in organisational change. The model was trained to associate living thing concepts with their associated features, using a data set derived from studies of children's featural representations. Analysis showed that a pattern of knowledge organisation strikingly similar to that derived from studies of children's category fluency and sorting behaviour developed in the model. The results inform on how the featural knowledge base may affect the structure of children's knowledge and performance on tasks drawing upon this knowledge.

1. Introduction

1.1 Featural Representations in Children's Semantic Memory

Support for the importance of featural representations of conceptual knowledge is well documented in the literature. Major theories of semantic memory processing rely heavily on featural representations [e.g. 3,20]. Research suggests that featural representations play an important role in children's conceptual knowledge development and processing [13,1,16]. Recent accounts have suggested that conceptual development may occur on two levels [15] - a feature based level, representing featural aspects of the concept, and a 'theory' level, representing theories about categories and concepts [2,14]. Theories are thought to act upon the featural level [15] allowing for example reasoning and categorisation tasks to be performed. Featural level representations may also play a role in the development of theories about concepts, which conceivably could result at least in part from statistical regularities in the featural knowledge base. In turn, theoretical level representations may direct attention to certain aspects of concepts to constrain or direct the acquisition of featural representations to those that are most useful or important [1]. Thus, the theory and featural levels of representation may interact in the development of conceptual knowledge.

Despite the overwhelming evidence for the importance of featural level representations there has been little if any work looking directly at the features children actually encode and how these determine the structure of their knowledge. The featural representations of adults have been assessed by asking them to list features of concepts they know (see [18] for a recent study using this technique). Possibly the major reason this approach has not been adopted in the developmental literature, is that it is difficult for young children to understand what is required in such a task. Such an assertion was supported by the results of a pilot study carried out prior to the first study discussed in this paper, illustrating that children are unable to provide responses to a traditional feature generation (listing) task, even when questioned about familiar objects such as animals and when simple language was used. Grounding the task in a context the children appreciated – describing an object to an alien who has no knowledge of it -enabled them to complete the task without particular difficulty.

An important issue tackled in this paper, is the relationship between feature based representations and the development of knowledge structures in memory. Recent studies [8,9,11] investigated knowledge structures of biological kind concepts (plants, animals and body parts) in primary school aged children. The results provide data to which comparisons can be made to determine how the features named by children relate to the development of knowledge structures. Consequently a primary aim of the work presented was to investigate the nature of representations at the featural level, how they change with age and how they may determine similarity relations between concepts and the thus contribute to the organisation of conceptual knowledge in memory.

1.2 Connectionism and Semantic Memory Development

Researchers have argued that natural kind categories are determined by correlated attributes of features [16,23]. The connectionist approach provides a method of investigating how these correlations may be extracted during learning and how they might affect generalisation and the development of knowledge structures. Furthermore, a working connectionist model may enable predictions to be made about the developmental process beyond the age ranges represented by the training data, which may be tested empirically. This is a particular benefit in the investigation of conceptual knowledge, since young children lack the language abilities to describe their knowledge, rendering in depth investigation of their knowledge representations difficult.

Hartley & Prescott, [10] argue that information about distributions of features in children's representations is necessary for effective connectionist investigation of the featural basis of semantic memory development. The first study presented describes the collection of a data set providing information about children's featural representations of fourteen living thing concepts. The second study describes a connectionist model trained on this data and investigation of the relationship between children's feature based representations, and the development of knowledge structures in memory using this model.

2. Feature Based Representations in Children

The first study describes the collection of a data set representing children's feature based representations using the feature generation paradigm, and discusses some of the observed changes in the distribution of features with age and experience. (Further details can be found in [9]).

2.1 Method

2.1.1 Participants

Sixty children attending a primary school in Sheffield, U.K. There were equal numbers from years 1, 3 and 5. Year 1 children had a mean age of 6 years (range 5:6 to 6:5), year 3 children had a mean age of 8 years (range 7:6 months to 8:5), Year 5 children had a mean age of 10 years (range 9: 6 months to 10:5). All of the children had English as a first language and none had any identified learning difficulties.

2.1.2 Stimuli

Fourteen living thing concepts were investigated, eight animals and six plants. All of the eight animal concepts chosen (lion, elephant, giraffe, sheep, pig, cow, cat and dog), and the plant concepts 'flower' and 'tree' had age of acquisition (AoA) scores below 44.5 months [19]. It was not possible to find AoA scores for the remaining plant concepts investigated (rose, sunflower, Christmas tree and apple tree). These were, however, concepts named at high frequency by children in all years and consistently grouped in the same subcategories in category fluency tasks [8,9,11].

2.1.3 Procedure

Familiarity with the fourteen target concepts was determined by asking children to identify them in a display of black and white line drawings containing a number of visually similar distracters from the living thing category (animals from [21], plants hand drawn). Only children who were able to correctly identify all 14 concepts were included in the study. Following picture identification the child was distracted for 5 minutes with general chat about the school day, interests etc. The child was introduced to a soft toy 'alien' and tape recorded instructions in an 'alien like' voice were played, explaining that the alien had not been to earth before, and wanted the child to tell the experimenter as much about the concepts named as possible. Children were given around one minute to describe each concept. Two practice runs were performed for the concepts 'frog' and 'daisy'. The 14 test items were presented verbally by the experimenter in a semi-random order (the items were randomised, but some ordering was changed so that similar items e.g. sunflower and rose, were presented at least two items apart). The experimenter transcribed the child's responses in production order.

2.2 Results and Discussion

2.2.1 Production Frequencies

Comparisons were made between the mean number of features produced by each child, the means are illustrated in Figure 1. Three children (one from year 5 and 2 from year 3) were excluded from this analysis to prevent their scores from biasing results, as their production frequencies were notably lower than those of other children tested. ANOVA revealed highly significant main effects for year group (age) for the mean number of features produced per concept ($F(2,54)=12.351$, $p<0.001$). Significant main effects for year group (age) were also found when the concepts were split into animal concepts ($F(2,54)=12.052$, $p<0.001$) and plant concepts ($F(2,54)=7.873$, $p<0.01$). Post Hoc Tukey's HSD revealed that year 3 and year 5 children generated significantly more features than year 1 children for both animal and plant concepts ($p<0.05$). There were no significant differences in production frequencies between years 3 and 5.

	Year Group	Mean Production Frequency (s.d)
Animals	1	7.98 (1.49)
	3	9.72 (1.11)
	5	10.04 (1.58)
Plants	1	4.94 (1.20)
	3	6.10 (1.18)
	5	6.38 (1.22)

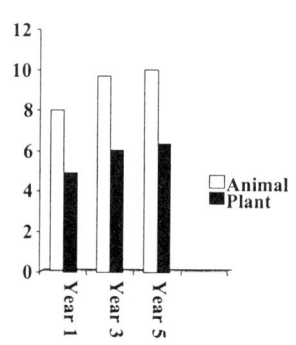

Figure 1: Mean production frequencies by year

2.2.2 Most Frequently Produced Features

The most frequently produced features appeared to be largely perceptual attributes. For most of the animal concepts, these were features which enabled distinctions to be made between animal concepts at the basic level i.e. features which are only shared (on the whole) with very close relatives, and are thus somewhat unique to the creature. For example, the giraffe's 'long neck', the elephant's 'trunk' and the lion's 'roar'. For many animals, these distinguishing features are noises that the animal makes. Animal noises are something children are likely to learn from an early age, both through experience, and through explicit teaching from parents and at nursery school. Of generally slightly lower frequency were those features shared by most members of the animal category, for example, ' has four legs', 'has eyes', 'has ears' etc. These features may be more likely to be used for superordinate level categorisation. Slightly lower in frequency still were those features which are characteristic of the animal, but perhaps less immediately useful for superordinate or basic level categorisation. Such features may be considered more general knowledge, which could be used in tasks which require reasoning or generalisation. Habitat features were within the top ten most frequently produced for wild animals (elephant, giraffe, lion) and within the top 5 most frequently produced for domesticated animals (farm animals - pig, sheep and cow; pets : dog, cat). This certainly suggests that children consider habitat to be a salient feature for animal concepts. This is consistent with the results of our work and that of other researchers using fluency fluency that show animals are grouped on the basis of typical habitat [see 6,8,9,11,22 for details].

It is possible that for domestic animals, habitat is even more salient than for wild animals. Two reasons may be suggested for this. Firstly, children may have more experience of domestic animals, and thus more knowledge of their habitat. Secondly, habitat may be a more useful feature in reasoning about or developing theories about domesticated animals, as their habitat likely explains something about their function. For example farm animals are often used for food, pets for company. An animal's habitat may also correlate with information about their behaviour or common interactions with humans. For example, a pet may be assumed to be friendly and something to be loved and cherished, whilst a farm animal may be considered to be less friendly, and of lower sentimental value to the human, but a producer of something useful.

Furthermore, for sheep and cow, 'has wool' and 'has milk' are within the top 5 features produced, and are produced by 65% and 60% of children respectively, suggesting that what they produce are highly salient features in the child's concepts of these animals. That sheep and cow are producers may be inferred from the fact that they live on a farm. For wild animals, habitat may be important in determining likely surroundings and which objects the animal might interact with, but is not perhaps as informative as for domestic animals and thus not as salient in the child's reasoning processes. Thus habitat appears to be a salient feature which may be useful in reasoning and generalisation. Thus it would appear that nature may have provided a useful basis for structuring animal knowledge.

As with the animals, the most frequently produced features for plant concepts were the most distinctive. For the flowers, this is generally the most typical colour of the flower (red for roses, yellow for sunflowers), for trees, the fruit born (e.g. apples for apple tree), or the shape of the leaf (spikes/prickles for Christmas tree). Next are those features which would be shared by members of the basic level (tree/flower) category such as 'has petals', 'has stem' and 'can grow' for rose and sunflower, and 'has leaves', 'has branches' and 'has trunk' for apple tree and Christmas tree. These shared features are those named most frequently for the basic level concepts tree and flower. Again, perceptual features far outweighed those of any other type in those most frequently named, suggesting that children may use primarily perceptual features to categorise plant concepts.

2.2.3 Categorisation of Features by Type

Figure 2 shows how features named by more than 3 participants in each year were categorised by type using a modified version of a system in [18]. The number of perceptual features produced far outweighed the number of functional features in all year groups, however, there was a trend towards the proportion of perceptual features in the representations decreasing with increasing age (8.1:1 year 1, 7.4:1 year 3, 5.4:1 year 5), consistent with the literature suggesting children's representations become more conceptual than perceptual as the knowledge base increases. The data suggests however, that there remains a strong reliance on perceptual features in representations of animal concepts [16] even in adults, for example, [7] found a ratio of 2.86 perceptual to functional features for animate entities.

Category (example features)	Feature type frequency (% of total)			
	year 1	year 3	year 5	overall
Perceptual (has fur) (is big)	32(76.2)	37 (74)	38(63.3)	54(62.7)
Type 1 Functional information: how used by man (used for climbing) (used for food)	2(4.8)	3(6)	3(5)	5 (5.8)
Type 2 Functional information: things the concept can do (can eat) (can fight)	2(4.8)	2(4)	4(6.7)	10 (11.6)
Habitat (lives in Africa)	3(7.1)	7(14)	6(10)	7 (8.1)
Nutritional (needs food) (eats leaves)	2(4.8)	0	6(10)	6 (7)
Interactions with other living things (visited by bees)	1(2.38)	1(2)	0	1(1.16)
Classification (is a flower)	0	0	3(5)	4 (4.7)

Figure 2: Classification of features by type

3. Connectionist Investigation

This second study describes an investigation of the relationship between the features children generated and the development of conceptual knowledge structures in memory using a simple connectionist model.

3.1 Deriving Training Data from Feature Generation Data

Input patterns consisted of 14 bit binary vectors, where each bit represented one of the fourteen concepts. Output training vectors were constructed from the year 5 data, and consisted of individual feature representations for each of the 14 concepts using similar methodology to previous studies in the adult literature [18]. Unit values in the output training vectors represented the number of participants naming feature j for concept i [$unit_{ij}$ (i=14, j=1-123)]. The resulting feature vectors were then log transformed. Thus, output target vectors maintained a scaled representation of the features named and their production frequency, which is considered to reflect the importance or salience of this feature in representing the concept [18].

3.2 Model Architecture and Training

A fully connected feedforward was batch trained using backpropagation of error to associate inputs representing each of the fourteen concepts with the appropriate feature based target output vectors. Training ceased when the output mean squared error fell below 0.001.

3.3 Results and Discussion

3.3.1 Development of Internal Concept Representations

The Euclidean distances between the hidden unit activation vectors was calculated and analysed using the 'gtree' algorithm [4] to produce an Additive Similarity Tree representation of the model's internal knowledge structure (see Figure 3). Related concepts are clustered on the tree. We have used this method of analysis to investigate knowledge structures derived from children's category fluency and sorting behaviour [e.g. 9,11,12] and it has been demonstrated over several studies in different labs that animal knowledge appears to be principally organised on the basis of habitat. The clustering seen in figure 3 (a) bears a striking resemblance to that seen in children (see [8,9,11] and figure 3 (b) for comparison). The fact that the model learns a similar similarity structure from children's featural representations to that derived from fluency studies supports the view that the representations underlying the processing and organisation observed in fluency studies are feature based. Studies suggest that children also organise plant knowledge on a taxonomic basis, however, this category appears to develop later than that of animals.

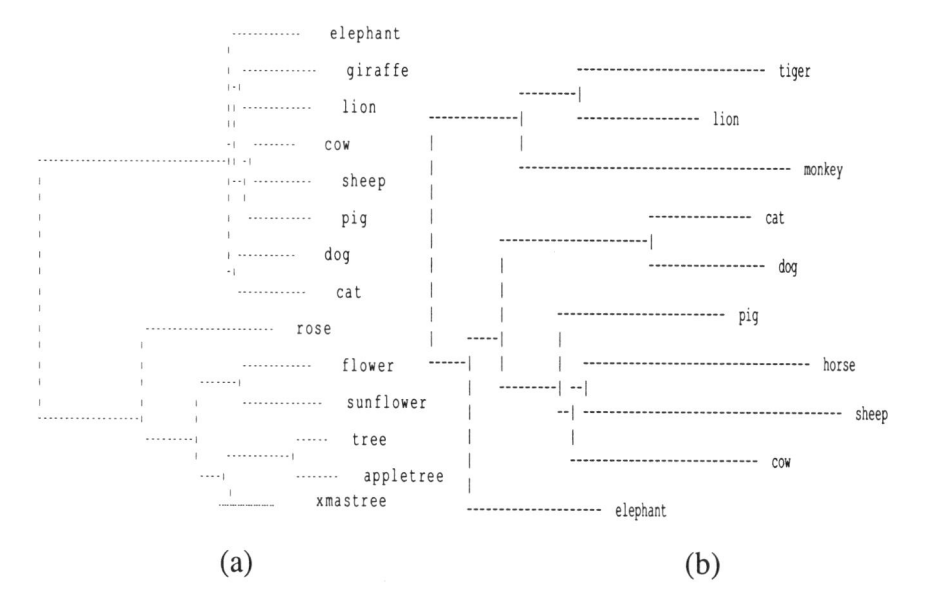

```
              ........   elephant                                      
           |  ........   giraffe                  ---------------- tiger
           |-|                                    ---------|           
           || ........   lion            -------|  -------- lion       
           ||                           |        |                     
          -| -...       cow             |        --------------- monkey
------------||-|        ......          |                              
|          |--| .......  sheep          |         ------- cat          
|          | |                          |   -----------|               
|          |  ........   pig            |              -------- dog    
|          |                            |                              
|          | ........    dog            |                              
|         -|                            |            ------------ pig  
|          ........      cat            |    -----|  |                 
|      ................. rose           |         |  ------------- horse
|      |                                |    ------|--|                
|      |   .........     flower    -----|             ------------ sheep
|      | -------                        |                              
|      | |     ........  sunflower      |                              
---------|                               |         ----------- cow     
          ---------      tree           |                              
                |                       |                              
           -----|        appletree      |                              
                |                        ------------ elephant         
          .......        xmastree                                      

              (a)                                  (b)
```

Figure 3: Internal knowledge structure (a) developed in the trained model (b) derived from fluency studies for animal concepts for children in year 5

3.3.2 Age of acquisition and developmental errors in the model

Figure 4 shows that in line with developmental data, the model learns the animal concepts very quickly, however the plant concepts are learned more slowly. This phenomenon may not be captured in models that use handcrafted data sets [17]. Children produced more features for animal than plant concepts, arguably reflecting their less developed knowledge of this category and possibly a lower variability in the featural information available about plants. The model's learning pattern reflects the relative sparsity of knowledge about plants compared to that of animals. It is important to capture developmental patterns such as this in models of semantic memory development, if reliable predictions are to be made about children's knowledge acquisition. Empirical data suggests that the use of habitat as a primary organising principle is robust and is seen in adulthood as well as in childhood [22]. Changes in the structure of animal knowledge are subtle, and new members of the category are integrated into a 'habitat-based' structure [see 8,9,11]. It is possible that early learning of animal concepts sets limits for later learning [5], promoting convergence to nearby local optima relative to the already established pattern of organisation. This assertion would suggest that other less well established categories might be more prone to organisational restructuring and provides an interesting avenue for further investigation.

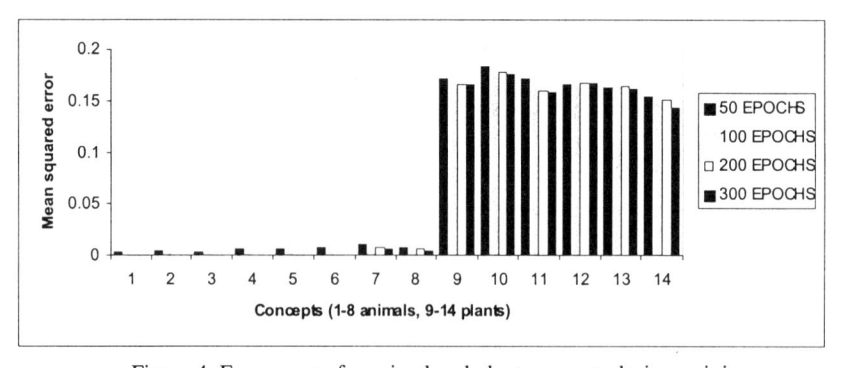

Figure 4: Error scores for animal and plant concepts during training

4. Conclusions

This paper has shown a high consistency between the trees derived from the model and those derived from fluency and sorting studies undertaken with children, suggesting that children's knowledge structures have a featural basis, and also that children do have access to the representations underlying their knowledge. The results provide support for the role of feature correlations in determining psychological proximity which is consistent with the assumptions of the family resemblance theories. The connectionist investigation presented in this paper is able to show a number of interesting phenomena reflected in empirical studies of children. This work is currently being extended to include a wider concept base, and use of incremental learning, interleaving new training patterns during learning to reflect age/experience related the changes in knowledge base is being explored.

References

1. Barrett, S., Abdi, H., Murphy, G.L. & McCarthy Gallagher, J. (1993). Theory-based correlations and their role in children's concepts. *Child Development, 64*, 1595-1616.
2. Carey, S. (1985). *Conceptual Change in Childhood*. Cambridge, MA, Bradford Books/MIT Press.
3. Collins, AM. and Loftus, EF. (1975). A spreading activation theory of semantic processing. *Psychological Review, 82,* 407-428
4. Corter, J.E. (1998). An efficient metric combinatorial algorithm for fitting additive trees. *Multivariate Behavioral Research, 33(2),* 249-272
5. Elman, JL. (1993). Learning and development in neural networks – the importance of starting small. *Cognition, 48,* 71-99.
6. Grube, D., & Hasselhorn, M. (1996). Children's freelisting of animal terms : Developmental changes in activation of categorical knowledge. *Zeitschrift fur Psychologie, 204,* 119-134.
7. Harley, T. A. (1998). The semantic deficit in dementia: connectionist approaches to what goes wrong in picture naming. *Aphasiology, 12,* 299-318.
8. Hartley SJ & Prescott TJ. (Submitted). Continuity and change in the development of knowledge structure: insights from category fluency. Submitted manuscript.
9. Hartley, S. J. (1999). *The Development of conceptual knowledge: Connectionist and experimental insights*. Unpublished PhD. Thesis submitted to The University of Sheffield, UK.
10. Hartley, S. J., Prescott, T. J. & Nicolson, R.I. (1999). Feature Distributions and Experimental Evaluation in Connectionist Models of Semantic Memory. In D. Heinke, G. W. Humphreys & A. Olson. (Eds.) *Connectionist models in cognitive neuroscience : the 5th Neural Computation and Psychology Workshop.* London, Springer Verlag.
11. Hartley, S.J., Prescott, T.J. & Nicolson, R. (1998). Experimental and Connectionist Perspectives on Semantic Memory Development. In *Proceedings of the Twentieth Annual Cognitive Science Conference*, M.A. Gernsbacher and S.J. Derry, Eds. Mahwah: Lawrence Erlbaum.
12. Jarrold, C., Hartley, S.J., Phillips, C. & Baddeley, A.D. (2000). Word fluency in Williams syndrome: Evidence for unusual semantic organisation? *Cognitive Neuropsychiatry, 5.*
13. Keil FC. (1994). Explanation, association and the acquisition of word meaning. *Lingua, 92,* 169-196 .
14. Keil, F. C. (1989). *Concepts, kinds and cognitive development.* Cambridge, MA, MIT Press.
15. Keil, F. C., Smith, W. C., Simons, D.J., & Levins, D.T. (1998). Two dogmas of conceptual empiricism: implications for hybrid models of the structure of knowledge. *Cognition, 65,* 103-135.
16. Mandler, J. M. (1997). Development of catgorisation: Perceptual and conceptual categories. In G. Bremner, A. Slater and G. Butterworth. (Eds) *Infant Development: Recent Advances.* Hove, Psychology Press.
17. McClelland, J. L., & McNaughton, B. L. (1995). Why there are complementary learning systems in the hippocampus and neocortex: Insights from the successes and failures of connectionist models of learning and memory. *Psychological Review, 102,* 419-457.

18. McRae, K., de Sa, V. R., & Seidenberg, M.S. (1997). On the nature and scope of featural representations of word meaning. *Journal of Experimental Psychology: General, 126,* 99-130.
19. Morrison, C. M., Chappell, T. D., & Ellis, A.W. (1997). Age of acquisition norms for a large set of object names and their relation to adult estimates and other variables. *Quarterly Journal of Experimental Psychology Section a-Human Experimental Psychology, 50,* 528-559
20. Rosch E., & Mervis, C.B. (1975). Family resemblances: studies in the internal structure of categories. *Cognitive Psychology, 7,* 573-605
21. Snodgrass, J. G., & Vanderwart, M. A. (1980). Standardised set of 260 pictures: Norms for name agreement, image agreement, familiarity and visual complexity. *Journal of Experimental Psychology : Human Learning and Memory, 6,* 174-215
22. Storm, C. (1980). The semantic structure of animal terms – a developmental study. *International Journal of Behavioural Development, 3,* 381-407.
23. Younger, B., & Mekos, D. (1992). Category construction in preschool-aged children: The use of correlated attributes. *Cognitive Development, 7,* 445-466

Paying Attention to Relevant Dimensions: A Localist Approach

Mike Page

Abstract

Localist models of, for example, the classification of multidimensional stimuli, can run into problems if generalization is attempted when many of the stimulus dimensions are irrelevant to the classification task in hand. A procedure is suggested by which a localist model can learn prototype representations that focus on the relevant dimensions only. These permit good generalization which would be lacking in a simple exemplar-based model.

1. Introduction

The research described here represents the first stages of development of a localist neural network for supervised learning that improves its classification performance by paying attention to input dimensions relevant to the task at hand. The work started very much as an applied problem which, as will be seen, benefits from a more theoretical analysis than was attempted at first.

1.1 The Problem

The problem involved the classification of 50-dimensional vectors of reals which had previously been derived from gray-scale images of faces. The faces had been preprocessed using "morphing" techniques so as to standardize the images to a common face-shape. The 110 faces, each comprising 10 000 pixel values, were then subjected to a principal components analysis (PCA), which allowed the faces to be represented as a compressed vector of 50 numbers. Each number represents a coordinate along an axis corresponding to one of those 50 principal components (PC) with the highest eigenvalues. The details of this preprocessing, and the motivation behind it, are given in more detail in [1].

The face set comprised a number of subjects, each posing 7 different expressions, namely anger, disgust, fear, happiness, neutral, sadness and surprise. The numbers of each emotion were approximately balanced and were 17, 15, 15, 18, 14, 17 and 14 respectively. Each 50-dimensional vector could therefore be labelled by the identity of the subject and by the emotion posed. The task was to design a localist network to learn to classify the faces into categories defined by their emotional expression.

As will be seen, and as is perhaps intuitively obvious, this task is more difficult than classifying the faces by identity.

The decision to use a localist network was motivated by earlier work (e.g., [7] [8]) that highlighted the advantages of such models. Of course, given that the task is one of classification, it would have been possible to train, using the back-propagation (BP) learning rule, a standard three-layer (of units) network, with 50 input units and 7 output units each one representing a localist coding of the correct expression-category. Nonetheless, given the reservations expressed by myself and others (see [8] and accompanying commentary) with regard to the plausibility of BP learning, an alternative model was sought. Similarly, a simple two-layer network trained by the delta-rule was avoided in favour of a network constrained such that each category was represented by an output unit whose activation would be maximal for a prototypical category member — not a natural consequence of applying delta-rule learning. This constraint encouraged the use of a radial-basis-function (RBF) network, as will be described below.

2. A Naive First Step

A naive first step, that helped to clarify the nature of the problem, was to attempt a simple nearest-neighbour classification of a given test face-pattern. To be specific, each face was classified according to the emotional label of its nearest neighbour in 50-dimensional space. Performance was extremely poor for the following reason: because each subject posed each expression (with a few exceptions) it is likely that the nearest face to that of subject A posing expression 1 is that of subject A posing a different expression. In these circumstances, in which distance between different expressions posed by the same model is smaller on average than distance between different models posing the same expression, performance of a nearest neighbour classifier is guaranteed to be poor in the expression classification task. One might say that similarity between vectors is dominated by identity at the expense of expression. This is intuitively plausible: It is not difficult to imagine that, even in the full 10 000-dimensional face space, an image of one person posing surprise is more similar to an image of the same person posing, say, happiness than it is to one of another person posing surprise. We can see, therefore, that the task faced by the network is in some sense to learn to pay attention to that subset of the 50 dimensions which defines a subspace in which expressions of the same type do indeed cluster, regardless of the identities of the models.

In order to formalize this idea, and taking an RBF-type approach, I next tried a simple two-layer classifier with 50 input nodes and 7 output nodes each representing a given expression. The weight, w_{ji}, incoming to the ith output node from the jth input node, represented the mean value along the jth dimension of patterns in class i. On presentation of a given pattern, p, for classification, the activation, A_i of the ith output node was given by

$$A_i = K - f(\sum_j g(\alpha_{ji} d_{ji})) \tag{1}$$

where K is a constant, f and g are functions to be defined, α_{ji} represents the attention paid by the ith output node to the jth input dimension and d_{ji} is a measure equal to

$|w_{ji} - p_{ji}|$, the absolute value of difference between the jth element of the input vector and the corresponding weight. This idea of attentional weighting is very similar to that found in Nosofsky's generalized context model ([4]) and Kruschke's ALCOVE model ([3])with one exception: in the model described here, each output node can allocate its attention differently. This seemed a useful development since the dimensions relevant to the classification of one emotion might well differ from those relevant to the classification of other emotions. In the work of Nosofsky and Kruschke, the attentional parameters have been envisaged as allowing the "stretching and "shrinking" of the input space to permit more appropriate classification. Here the aim is for each output node (i.e., each emotion classifier) to learn to stretch and shrink its input space in a manner such that patterns corresponding to that emotion class are well clustered.

The first network tested used the simplest version of (1), namely

$$A_i = K - \sum_j \alpha_{ji} d_{ji} \qquad (2)$$

There were seven output nodes, one for each expression category, and the bottom-up weight vector to a given node was set to the 50-D mean vector for the corresponding category. Classification of a test vector was implemented by clamping the test vector at the input and activating the output nodes according to (2). The category of the test vector was assumed to be that corresponding to the most active of the output nodes.

Attentional weights, α_{ji} in (2), were initialized to unity so that the classifier starts as a 1-nearest-neighbour classifier in an undistorted input space. We then attempted to ameliorate the poor performance of this classifier (described earlier) by adjusting only the values of the attentional weights. At first, this was effected by a learning rule which can be qualitatively summarized as:

1. present pattern and classify by the nearest weighted prototype.
2. if classification is correct reduce attention to badly matching dimensions (i.e., those with high d_{ji}), increase attention to well matching dimensions.
3. if classification is incorrect increase attention to badly matching dimensions, decrease to well matching dimensions.

The intuition underlying this procedure was that if a test pattern was classified correctly despite a large mismatch (i.e., d_{ji}) along a given dimension, then a sensitivity to values along that dimension is not likely to be critical in calculating the activation of the node representing that category. Conversely, for an erroneously responding category node, more attention should be paid to badly matching dimensions.

Beyond an intuitive appeal, it can also be seen that such a procedure minimises a measure of error with respect to the attentional weights using a gradient-descent-based rule. From (2),

$$\frac{\partial A_i}{\partial \alpha_{ji}} = -d_{ji} \qquad (3)$$

which indicates that if we wish to increase the activation of a given node in response to a particular test pattern, then we should subtract a value proportional to d_{ji} from the corresponding attentional weights. A large value of d_{ji} will lead to a large decrease in attention to that dimension, a small value of d_{ji} will lead to a small attentional decrease. The qualitative procedure enumerated above suggests an increase in attention to well-matched dimensions under these circumstances, rather than a small decrease. This can be achieved by renormalizing the attentional weights to a constant sum (or length) after each weight change.

The simple learning rule described above was tested extensively, using a leave-one-out crossvalidation regime. (This regime involves, for each pattern in the training set, training the network on all other patterns in the set and testing the resultant network's performance with the pattern itself – this gives a good test of generalization while maximizing the size of the training set in each case.) While it proved possible to increase performance on the training set using the attentional learning rule, generalization performance was poor. The performance never approached the 95% correct for training patterns and 78% correct for untrained (leave-one-out) test patterns that could be achieved using a standard linear discriminant analysis on this dataset.

3. Theoretical Considerations

It was decided to make a more detailed theoretical analysis of the problem. In particular, classification was conceived of as a Bayesian maximum-likelihood decision. For a multidimensional Normal distribution centred on mean vector m

$$p(x) = \frac{1}{(2\pi)^{d/2}|\Sigma|^{1/2}} \exp\{-\frac{1}{2}(x-m)^T \Sigma^{-1}(x-m)\} \quad (4)$$

where $p(x)$ is a probability density function and Σ is the covariance matrix. This implies that

$$\log p(x) = \log K - \frac{1}{2}\log(|\Sigma|) - \frac{1}{2}(x-m)^T \Sigma^{-1}(x-m) \quad (5)$$

where K is a constant. (It is often more convenient to deal with the logarithm of the probability rather than the probability itself, since the probability values can become very small. Looking for a class with the maximum log posterior probability is equivalent to seeking the class with the maximum posterior probability because the log function is monotonic increasing.) Various assumptions can be made about the covariance matrix for a given category. For example, if we assume that the off-diagonal elements are zero and that the on-diagonal elements (i.e., the variances of each dimension) are equal to $(1/\alpha_1, 1/\alpha_2, \ldots, 1/\alpha_n)$ (where the second subscripted index of α_{ji} has been dropped here because we are only talking about the distribution within a single category) then

$$\log p(x) = \log K + \frac{1}{2}\sum_{j=1}^{n} \log \alpha_j - \frac{1}{2}\sum_{j=1}^{n} \alpha_j (x_j - m_j)^2 \quad (6)$$

where n is the number of input dimensions. Thus the log probability density at a given vector x is given by a constant minus a linear combination (i.e., attentional weighting) of the dimensionwise distances squared, this time with an additional normalizing factor $\frac{1}{2}\sum_{j=1}^{n} \log \alpha_j$. If this function (with the constant $\log K$ term dropped for convenience) is allowed to replace the previous activation function for output nodes in our RBF network then, given a test vector as input, the output nodes will respond with activations equal to the posterior probability of each class given that test vector (assuming uniform priors) providing the our two assumptions are true, that is, that the covariance matrix is diagonal, and that the attentional weight on a given dimension is equal to the reciprocal of the variance along that dimension. Picking the most active output node corresponds, therefore, to a maximum likelihood decision process that assumes uniform priors.

Looking at the partial derivative of $\log p(x)$ with respect to α_j gives

$$\frac{\partial \log p(x)}{\partial \alpha_j} = \frac{1}{2}\left(\frac{1}{\alpha_j} - d_j^2\right) \tag{7}$$

which is similar to the earlier partial derivative but with the additional $1/\alpha_j$ term. A learning rule based on this partial derivative has the correct form such that α_j tends towards the reciprocal of the within-class variance along the jth input dimension. This is, of course, exactly as is required to satisfy the second of our assumptions above.

Unfortunately, the first assumption, that of diagonal within-class covariance matrices, is overly restrictive, even in cases when, as here, the fact that the 50-D vectors are themselves derived from a PCA ensures that the covariance matrix taken across the whole 110-pattern set is indeed diagonal. Of course it is possible to use a full covariance matrix and to perform the appropriate mathematical calculations to give the value of the posterior log probability of a given class but the neural network implementation becomes somewhat complicated due to the appearance of unwanted crossproduct terms. Another way around this problem, the one adopted here, is to "whiten" each of the classes, that is to preprocess the members of a given class such that their covariance matrix becomes diagonal. One way of doing this is to perform class-conditional PCAs, that is, for each class perform a PCA using only members of that class. For each class, this results in a number of linear components (at most one fewer than the number of patterns in the class) representing a rotated space of reduced dimensionality for which the class-conditional covariance matrix is indeed diagonal.

The structure of a network for implementing this class-conditional whitening is shown in Fig. 1. Each class has a group of preprocessing nodes which are dedicated, via PCA (or similar), to the representation of members in that class in a whitened space. There are a number of self-organizing networks which can perform PCA ([2, 5, 6, 9]), and these might be employed in the learning of the layer 1 to layer 2 connections for each class. This was not done for the preliminary tests presented here. A standard PCA algorithm was run separately for each class to produce directly the values of the corresponding network weights. Because of the restrictions of the PCA algorithm available and the fact that the smallest of the classes only contained 14 faces, the PCA was run only on the first 13 values of the original 50-dimensional pattern set (i.e., those 13 with the highest eigenvalues), to give 13-dimensional vectors at the output of each of the class-specific preprocessing modules. The layer 2 to Layer 3 weights then encode the mean vectors for each of the seven classes, taken across the preprocessed patterns for that class.

[Figure 1: The structure of the network — shows Layer 1: Input pattern at bottom, fully connected by adaptive preprocessing weights to Layer 2: Preprocessed input (seven class-specific modules), which feed Layer 3: Output nodes (one per category) with weights encoding category means.]

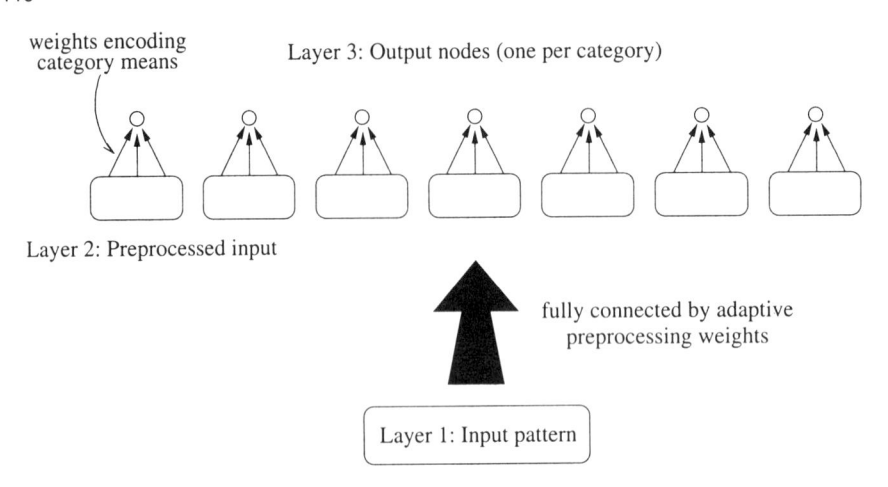

Figure 1: The structure of the network

The classification of a test face now involves the following: the reduced 13-dimensional pattern corresponding to the face is clamped at layer 1; each of the class-specific preprocessing modules then processes that vector, resulting in seven different 13-D output vectors, one for each class; these vectors are then processed further using attentional weights which are set to the reciprocal of the within-class variances in the whitened class-specific spaces. Again, in the preliminary testing presented here, these weights were calculated rather than learned. They should, however, be learnable via the rule described in (7). Each output node activates to an extent equal to the log posterior probability of membership of the corresponding class. Picking the most active of the output nodes implements a maximum likelihood decision. Prior probabilities can be incorporated into the model by adding the relevant bias to each of the output units.

4. Results

Preliminary results were encouraging. Using only the first 13 of the 50 dimensions in the original input pattern set, and the preprocessing strategy described above, training pattern performance of 95% correct was achieved. Cross-validation performance using a leave-one-out method, resulted in 93% correct. As a caveat regarding this latter figure, we note that all the training patterns were used in performing the class-specific PCAs from which the weights in the preprocessing stages were derived; likewise, all patterns were used in the setting of the class-specific attentional weights(i.e., reciprocal variances). The left-out pattern was not, however, used in the calculation of the class-conditional mean vectors for the preprocessed patterns. 93% correct is thus likely to be an overestimate of the crossvalidated performance of the network. Nonetheless, for the reduced 13-D input patterns, a linear discriminant analysis gives only 65% of training patterns correct and a crossvalidated (leave-one-out) performance of only

42% correct. This suggests that the ability of the two-stage network effectively to model fully general within-class covariance matrices, confers a considerable performance advantage, though at the cost of increased network complexity.

5. Conclusion

The ideas and results presented here are only preliminary, but they suggest a way in which standard unsupervised learning procedures can be combined to give a network whose generalization abilities are much improved over simple localist alternatives, such as unrefined nearest-neighbour techniques. The enhanced classification relies on using a subnetwork dedicated to each category, which produces as its output the posterior probability of that category given the test stimulus. In doing so, the classification network concentrates its "attention" on tranformed dimensions which show low within-class variance. Put another way, the subnetwork dedicated to a given class examines the test stimulus for evidence of invariant patterns that characterize that class.

In this preliminary work, many implementational shortcuts have been taken, such as the external calculation of class-specific PCs and subsequent within-category variances. Ideally, further work would build such functionality into the framework of a fully self-organizing network. One interesting point to note is that traditional PCA pulls out first those linear combinations that "soak up" the most variance in the input. Because of their high variance, these are the components to which the subsequent classification process pays least attention. It might be better to seek to extract first (nonzero) linear combinations with low variance, since these best characterize what is invariant about a given class, and are the dimensions to which most attention will subsequently be paid.

The network trained as described above, is able to perform classification of faces into emotional categories equivalent to a Bayesian maximum-likelihood decision rule. It assumes that all faces in a given category come from a single normal distribution centred on the category-mean vector. This assumption might not be appropriate – there may be more than one different "type" of happy face. This suggests that the network might usefully be extended by performing an early unsupervised clustering of faces from a given class, with whitening and calculation of the log probability performed separately for each cluster rather than just assuming that each class is equivalent to a single cluster. In the example presented here, however, this procedure does not seem necessary to permit good classification performance. Whichever procedure is used, it is clear that the preprocessing networks generated for the various emotional classes will not be appropriate for other classication tasks, of the faces into, say, identity or gender classes. (Compare the inappropriateness for a given task of hidden units in a BP net trained on another task.) If other classifications are required, then training can proceed as before, with preprocessing networks added accordingly. The resulting network might be described as modular, with separate subnetworks dedicated towards the recognition of emotions and identities. The double dissociations that have been found in patient populations, between emotion and identity recognition, and between the recognition of different emotions, might be seen as supportive of this modular structure.

References

1. Calder, A. J., Burton, A. M., Miller, P., Young, A. J., & Akamatsu, S. (submitted). A Principal Component Analysis of Facial Expressions. *Vision Research*.
2. Földiák, P. (1989). Adaptive Network for Optimal Linear Feature Extraction. In *International Joint Conference on Neural Networks (Washington)*. IEEE NewYork, Vol. 1401–405.
3. Kruschke, J. K. (1992). ALCOVE: An Exemplar-based Connectionist Model of Category Learning. *Psychological Review, 99*, 22–44.
4. Nosofsky, R. M. (1986). Attention, Similarity and the Identification-Categorization Relationship. *Journal of Experimental Psychology: Learning, Memory and Cognition, 115*, 39–57.
5. Oja, E. (1982). A Simplified Neuron as a Principal Component Analyzer. *Journal of Mathematical Biology, 15*, 267–273.
6. Oja, E. (1989). Neural Networks, Principal Components and Subspaces. *International Journal of Neural Systems, 1*, 61–68.
7. Page, M. P. A. (1998). Some Advantages of Localist over Distributed Representations. In J. A. Bullinaria, D. W. Glasspool, & G. Houghton, (Eds.), *Proceedings of the Fourth Neural Computation and Psychology Workshop: Connectionist Representations* (pp. 3-15). London: Springer-Verlag.
8. Page, M. P. A. (2000). Connectionist Modelling in Psychology. *Behavioral and Brain Sciences, 23*, 443–467.
9. Sanger, T. D. (1989). Optimal Unsupervised Learning in a Single-layer Linear Feedforward Neural Network. *Neural Networks, 2*, 459–473.

Coordinating Multiple Sensory Modalities While Learning to Reach

Matthew Schlesinger & Domenico Parisi

Abstract

By the onset of reaching, young infants are already able to coordinate vision of a target with the felt position of their arm [7]. How is this coordination achieved? In order to investigate the hypothesis that infants learn to link vision and proprioception via the sense of touch, we implemented a recent computational model of reaching [22]. The model employs a genetic algorithm as a proxy for sensorimotor development in young infants. The three principal findings of our simulations were that tactile perception: (1) facilitates learning to coordinate vision and proprioception, (2) promotes an efficient reaching strategy, and (3) accelerates the remapping of vision and proprioception after perturbation of the multimodal map. Follow-up analyses of the model provide additional support for our hypothesis, and suggest that touch helps to coordinate vision and proprioception by providing a third, correlated information channel.

1. Introduction

Until recently, the prevailing view among motor development researchers was that infants develop hand-eye coordination by visually guiding their hand as they learn to reach [5, 6, 18, 21, 24]. According to this view, infants learn to coordinate the felt position of their arm (via proprioception) with the seen position of a target object by watching their hand as it approaches the target. A number of recent studies, however, suggest that infants do not rely on visual feedback from the hand as a coordination strategy during early reaching. For example, 5-month-old infants rely on ballistic rather than visually-guided reaching strategies when targets are displaced during a reach [1]. In addition, 6-month-olds produce comparable kinematic reaching patterns regardless of whether or not visual feedback of the hand is available [8].

Perhaps the strongest evidence against the "visual-guidance" hypothesis is offered by Clifton, Muir, Ashmead, and Clarkson [7], who found that at the onset of reaching (around age 12 weeks), infants are just as likely to begin reaching for a target in a fully-lit room as they are to reach for a luminous target in a dark room. These findings are important in several respects. First, they suggest that infants do not learn to reach by visually guiding their hand toward a seen target. Second, they also reveal an early intermodal coordination between sight of the target and felt position of the arm, which is presumably not achieved through visual feedback from the hand during reaching. How then are vision and proprioception coordinated in young infants?

One possibility is that infants possess an innate multimodal map that computes a two-way transformation between the visual and proprioceptive spaces. Support for this hypothesis comes from studies of newborn infants, who show a tendency to generate reflex-like preparatory reaching movements ("prereaches"; see [9, 10, 12]). An alternative hypothesis is that infants learn to coordinate vision and proprioception during the development of reaching. According to this second account, learning to reach creates a task-specific context for infants to coordinate multiple information streams.

In this paper we investigate the second possibility, namely, that visual-proprioceptive coordination emerges as infants learn to reach. In particular, we focus on the role of tactile feedback from the target as a third information source that helps link vision and proprioception. The rest of the paper is divided into four sections. In the following section, we consider visual and tactile feedback as sources of sensory information available to infants as they learn to coordinate vision and proprioception. Next, we briefly describe a recent model of sensorimotor development in young infants [22] that we use to simulate early reaching. We then present a series of simulation studies that contrast the roles of visual and tactile feedback while learning to reach.

2. Linking Vision and Proprioception

If infants are not born with an intermodal map linking vision and proprioception, then how does this coordination develop? We propose that infants learn to "triangulate" the position of their hand by combining information from the target and the arm with a third information source. In principle, at least two sensory channels may provide this third source of information. One possibility is that infants use the sight of their hand to help determine its position relative to a target. The advantage of using visual feedback from the hand to coordinate eye and arm movements is that it is continuously available during a reach. However, vision of the hand is potentially ambiguous (e.g., the hand may occlude the target without touching it), and may require additional computation to determine the hand's position relative to the target. In addition, monitoring of the hand position also requires attentional resources above and beyond those available for keeping track of the target and controlling arm movements. These reasons may help explain why young infants do not appear to use visual feedback from the hand as an initial reaching strategy.

An alternative information source comes from tactile feedback during contact with the target. Unlike vision of the hand, tactile feedback is unambiguous. That is, once contact is made with the target, the hand's location can be equated with the seen position of the target. Thus, tactile feedback provides a unique sensory signal for associating seen locations in space (i.e., the target) with felt positions of the arm. However, compared to vision of the hand, which provides continuous feedback during a reach, tactile feedback is only available *after* a successful reach. In addition, because this coordination begins to develop *prior* to the onset of consistent reaching, only infrequent or sporadic sensory updates (after successful reaches) are available for coordinating sight of the target with felt position of the arm.

3. A Model of Reaching in Young Infants

The Schlesinger et al. model [22] is designed to represent two core features of infant sensorimotor development: (1) learning by trial-and-error exploratory movements [3, 11, 23], and (2) use of coarse feedback for producing subsequent movements (i.e., the success or failure of a reach). Accordingly, the model implements an unsupervised, variation-and-selection learning algorithm analogous to evolutionary learning [13] as a stochastic optimization procedure. Specifically, a population of artificial neural networks is used to represent the diverse set of reaching movements available to infants. Because the networks inhabit a simple ecology (including their bodies) we refer to them as "econets" [20]. Like the movements generated by infants, econets adapt under selective pressure to produce the most direct, efficient reaching patterns.

It is important to stress that the model simplifies many of the details of sensorimotor control systems, and should not be interpreted as a biologically plausible model of sensorimotor development in infants. Rather, our principal goal is to use the model as a framework for abstracting out the most relevant aspects of multimodal sensory control, and simulating the developmental relations among these factors.

The model replicates several key aspects of reaching development in infants (see [22] for a complete description), including (1) limited use of redundant joints [4]; (2) a predominance of muscle co-contractions during early reaching [4, 23]; (3) the early emergence of stereotyped reaching movements [4, 15]; and (4) an early suppression of prereaches [11]. We present here a brief overview of the Schlesinger et al. model. Interested readers may refer to [22] for a full description, including the details of the learning algorithm and a complete list of all major parameter values.

3.1 The econet

Figure 1a presents a schematic diagram of the econet in its two-dimensional reaching workspace. The econet remains at the center of the 100 x 100 unit workspace. The target is a small object (triangle), placed randomly in the workspace at the start of each trial. The econet has a monocular visual system with a 64° visual field (see Figure 1a). The econet also has a two-segment arm that spans the workspace. The trunk, head, and eye are positioned along the same axis, and so move as a single unit with the rotation of the eye. As Figure 1a illustrates, the eye

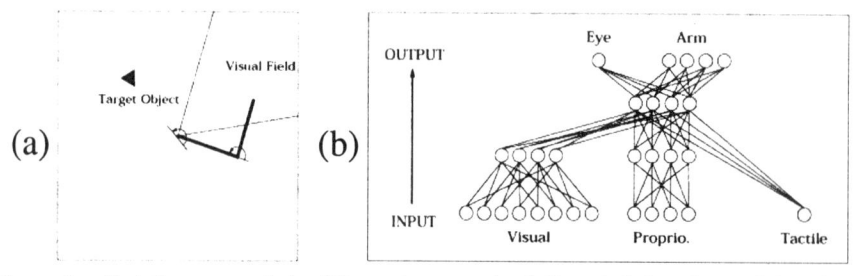

Figure 1: (1a) the econet it its 2D workspace; the left and right edges of the visual field are shown, as well as the rotational limits of the shoulder and elbow joints. (1b) the 4-layer, feedforward neural network used to control the movements of the econet (not all connections shown).

can rotate 360° in either direction while the arm's movement is limited to 180° rotation at the shoulder and elbow joints. A multi-layer feedforward neural network (see Figure 1b) uses visual input from the eye, proprioceptive input from the arm, and tactile input from the hand to control the movements of the eye and arm. Note that the visual and proprioceptive sensory inputs first pass through an intramodal hidden layer before combining at the second intermodal hidden layer. This partially modularized architecture helps minimize crosstalk or interference effects between sensory modalities, and results in improved performance [14, 22].

Each visual input unit spans 8° of the visual field, and encodes the presence of either the target or hand (scaled from 0 to 1 as a function of the distance of the object from the econet). The proprioceptive input units encode the amount of stretch in the arm muscles, represented by a linear mapping from the shoulder and elbow joint angles. Finally, the tactile input unit encodes contact of the hand with the target as a binary signal. Movements of the econet's eye and arm are produced by activation on its output units. A single unit moves the eye either left or right, while the remaining four units activate the flexor-extensor muscle pairs of the upper arm and forearm.

3.1 The learning algorithm

We employ a genetic algorithm as a proxy for trial-and-error sensorimotor development. An initial generation of econets is produced, with random connection weights. Each econet is given 20 trials (of 100 timesteps) to reach a nearby object; fitness is increased by 1 point for each timestep spent in contact with the target (for a theoretical maximum of 2000 fitness points). The top 20 econets of each generation are selected and used to produce 100 new econets (5 offspring per parent). During the reproduction process, the connection weights of the parent are copied with a 2% chance of random mutations. An 8-bit binary string is used to represent each connection weight; mutations are accomplished by switching the randomly selected bit from 1 to 0 or vice versa. Note that while the connection weights mutate from parent to offspring, all other characteristics (e.g., the neural network architecture, body parameters, etc.) remain constant across generations. Also, no other variation mechanisms (sexual reproduction, crossover) are employed.

4. Contrasting Visual and Tactile Feedback

We now present a series of simulation studies that allow us to contrast the relative roles of visual and tactile feedback while learning to reach. In the first set of simulations, we illustrate how tactile feedback facilitates learning to reach. The second set demonstrates the reaching strategies that emerge when either visual or tactile feedback is available. In the final set, we investigate how the model responds to a perturbation of the visual-proprioceptive coordination.

4.1 Touch facilitates learning to reach

Figure 2a presents the average fitness (i.e., time spent in contact with the target) in two populations of econets, over 10 simulation runs. (In all analyses and figures, we adopt the convention of presenting the results of the 20 most fit individuals from each generation.) Both populations receive visual and proprioceptive sensory inputs. In the "tactile-on" population, sensory input is also received from the tactile input unit (when the target is contacted); while no tactile inputs are received in the "tactile-off" population. Figure 2b presents the average number of trials (out of 20) in which the target was reached. As the figure illustrates, tactile feedback not only results in reaching the target more often, but also in maintaining contact with it for more time.

Can we conclude that tactile feedback leads to better coordination of vision and proprioception? One counterargument is that because the tactile-on econets receive both tactile *and* visual feedback about the location of their hand, it is the *combination* of the two forms of feedback that lead to better intermodal coordination. To test this possibility, we simulated a variant of the tactile-on population that could not see their hands (tactile-on-luminous). This condition corresponds to an infant that learns to reach for luminous objects in a dark room, and

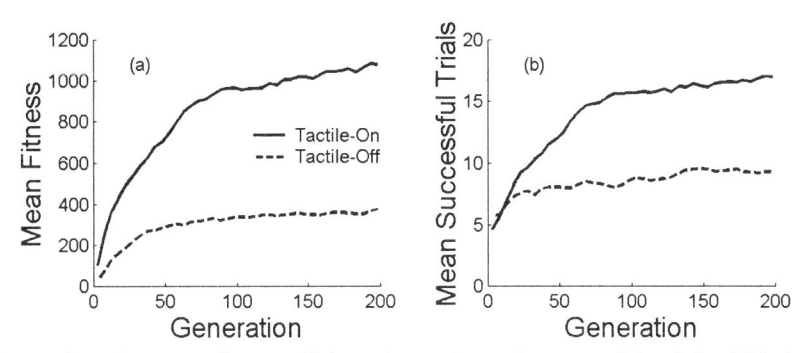

Figure 2: Average fitness (2a) and number of successful trials (2b) in two populations of econets. "Tactile-on" econets receive tactile feedback when they touch the target, while "tactile-off" econets receive no tactile feedback.

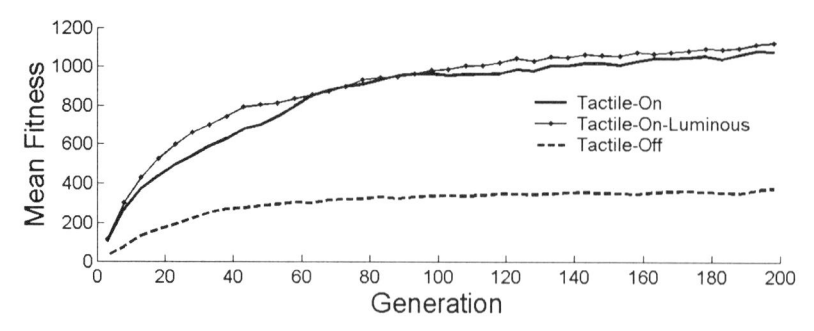

Figure 3: Average fitness in the tactile-on, tactile-off, and a third, tactile-on-luminous population. Tactile-on-luminous econets can see the target, but their hand is invisible.

allows us to compare reaching when only tactile feedback, or both tactile and visual feedback are available.

Figure 3 presents the results of the previous analysis, along with the average fitness in the tactile-on-luminous population. There are two important findings. First, in both of the tactile-on populations, performance and rate of learning are near equal. Thus, visibility of the hand does not appear to play a major role in coordinating vision and proprioception when tactile feedback is available. Second, this result also parallels the findings of Clifton et al. [7], who observed similar onset ages for reaching in normal and luminous-object conditions.

A second question concerns the fact that the tactile sensory signal not only indicates when the hand is touching the target, but also when the econet's fitness changes. (Recall that touching the target increases fitness.) Thus, touch is confounded with fitness. However, we can separate these two factors by asking what happens when touching the target is no longer rewarded, but remaining close to it is (i.e., increased fitness for "hovering" near the target). Figure 4 presents the average fitness for the tactile-on and tactile-off populations in the "hover" condition (i.e., 0 fitness when touching the target, +1 for remaining near). While the advantage of the

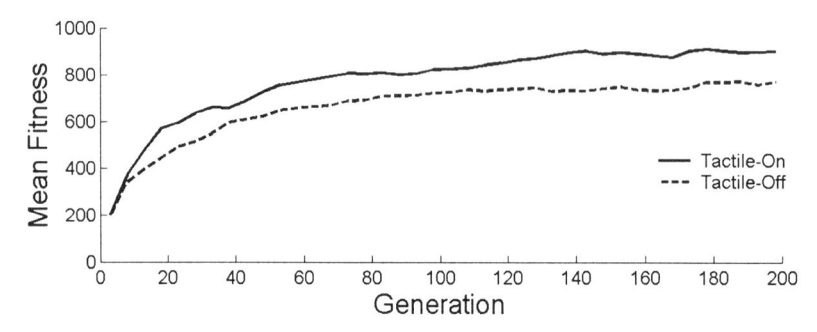

Figure 4: Average fitness in the "hover" condition, where econets are rewarded for keeping their hand near (but not on) the target.

tactile-on population is diminished (compare to Figure 2a), reaching is still significantly better with tactile feedback.

4.2 Touch promotes an efficient reaching strategy

We can explore the coordination between vision and proprioception more closely by analyzing the specific reaching strategies employed by the tactile-on and tactile-off populations. Reaching the target requires first determining the positions of the target and the hand, and then moving the hand toward the target. We might expect different sequencing strategies to emerge depending on how well vision and proprioception are coordinated. Specifically, if they are well-coordinated, econets should employ a "target-first" strategy in which the target is located, and then the hand is brought to it as the target is fixated. However, the target-first strategy depends on knowing the position of the hand, especially when it is outside the visual field. If vision and proprioception are poorly coordinated, we should instead expect a "hand-first" strategy, where econets first bring their hand into the visual field and then search for the target.

Figure 5 presents the relative fixation time of the hand and the target in the tactile-on and tactile-off populations. The tactile-on population employs a target-first reaching strategy: the majority of time is spent fixating the object (59%), while the hand is only fixated near the end of the reach (41%). In contrast, tactile-off econets find the hand, and then visually guide it toward the target: without tactile feedback, the hand is fixated more than the target (62% and 38%, respectively).

4.3 Touch accelerates intermodal recalibration

For the final set of analyses, we ask how econets adapt to a perturbation of the visual-proprioceptive coordination. In particular, we create an intermodal conflict by simulating a 10° prismatic displacement of the visual field. Thus, while touch and

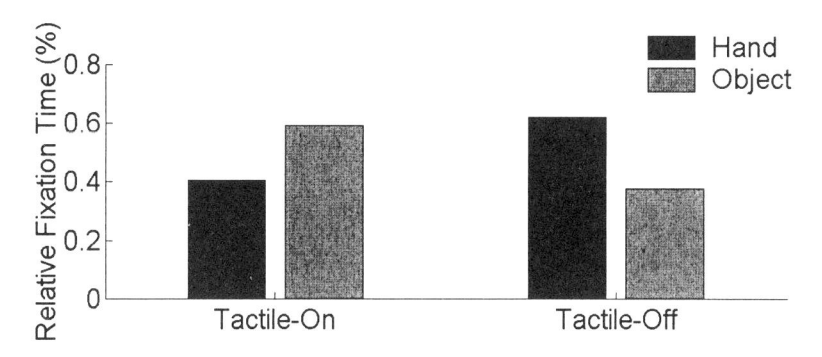

Figure 5: Relative time spent fixating the hand or the object, in the tactile-on and tactile-off populations. Tactile-on econets find the target first, and bring the hand toward it, while tactile-off econets find the target first, and then search for the target.

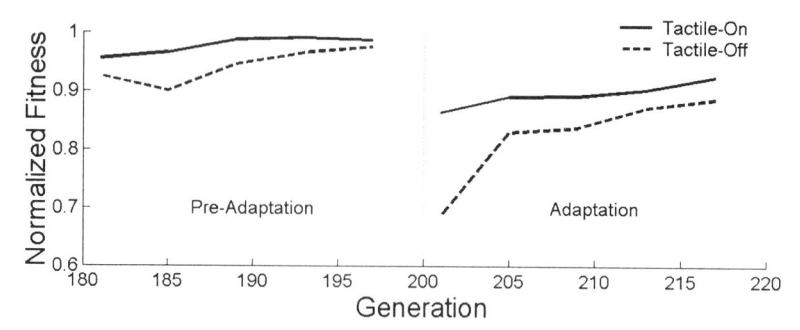

Figure 6: Normalized fitness in the tactile-on and tactile-off populations, during the final 20 generations of training ("Pre-Adaptation") and for 20 generations of simulated prismatic displacement of visual input ("Adaptation").

proprioception remain veridical, the visual sensory system now encodes objects in the visual field as 10° to the left of their actual position. This condition is analogous to prismatic adaptation studies with infants, who quickly learn to accommodate their reaching movements while wearing displacement prisms [18, 19].

Prismatic adaptation is simulated by first training the tactile-on and tactile-off econets as before (i.e., for 200 generations). After the initial training phase ("Pre-Adaptation"), prismatic displacement is simulated by shifting the apparent position of all objects in the visual field 10° to the left. Training during the adaptation phase then continues for an additional 20 generations ("Adaptation"). Figure 6 presents the mean normalized fitness in the tactile-on and tactile-off populations, during the pre-adaptation and adaptation phases. There are two major findings. First, prismatic displacement has a larger initial effect on the tactile-off population. Second, the tactile-on population recovers a higher level of performance during the adaptation phase.

5. Conclusion

Our simulation results provide support for the hypothesis that tactile feedback facilitates learning to reach. In particular, when both visual and tactile feedback are available, econets tend to rely on tactile feedback. Indeed, visually-guided reaching strategies only emerge when tactile feedback from the hand is unavailable (e.g., tactile-off econets). We also found that like human infants, econets do not need to see their hand in order to learn to coordinate vision of a target with felt position of their arm (e.g., the tactile-on-luminous condition). This result not only replicates the findings of Clifton et al. [7], but also helps us understand how infants may begin to coordinate vision and proprioception *before* the onset of consistent, successful reaching.

In considering these results, however, three important issues should be raised. First, the current findings may depend in part on the specific choice of network architecture and learning algorithm (however, see [22] for other a comparison with other architectures). It is critical to evaluate the hypothesis that tactile feedback facilitates coordination of vision and proprioception using other network architectures and unsupervised learning algorithms (e.g., reinforcement learning).

Second, the visual-proprioceptive coordination achieved by the econet model is task-specific. Rather than learning a general intermodal mapping that links vision and proprioception, a specific set of behavioral strategies are discovered and employed to solve a particular task. Other modeling approaches may shed light on how a general intermodal coordination might develop (e.g., see [16, 17] for a simulation based on infants' spontaneous arm movements).

A third question concerns how our simulation findings should be reconciled with later stages of reaching development during infancy, as well as with reaching behaviors in adults. While infants do not attend to their hand as they learn to reach, visual feedback will come to play a critical role as infants begin to attempt high-precision reaches (e.g., pincer grasp of a small object). In addition, visual feedback plays a central role for adults in recalibrating the visual-proprioceptive map after perturbation [2]. One way these findings may be integrated is by proposing a reliance on tactile feedback during early infancy, when visual perception (e.g., visual acuity, depth perception, visual tracking, etc.) may be no better at locating the hand than proprioception. As visual perception improves and stabilizes, however, infants should transition from the use of tactile feedback to visual feedback [7].

References

1. Ashmead, D.H., McCarty, M.E., Lucas, L.S., & Belvedere, M.C. (1993). Visual guidance in infants' reaching toward suddenly displaced targets. *Child Development, 64*, 1111-1127.
2. Beers, R.J. van, Sittig, A.C, & Gon J.J. (1999). Integration of proprioceptive and visual position-information: An experimentally supported model. *Journal of Neurophysiology, 81*, 1355-1364.
3. Berthier, N.E. (1996). Learning to reach: A mathematical model. *Developmental Psychology, 32*, 811-823.
4. Berthier, N.E., Clifton, R.K., McCall, D.D., & Robin, D.J. (1999). Proximodistal structure of initial reaching in human infants. *Experimental Brain Research, 127*, 259-269.
5. Bower, T.G.R. (1974). *Development in infancy.* San Francisco: Freeman.
6. Bushnell, E.W. (1985). The decline of visually guided reaching during infancy. *Infant Behavior and Development, 8*, 139-155.
7. Clifton, R.K., Muir, D.W., Ashmead, D.H., & Clarkson, M.G. (1993). Is visually guided reaching in early infancy a myth? *Child Development, 64*, 1099-1110.
8. Clifton, R.K., Rochat, P, Robin, D.J., & Berthier, N.E. (1994). Multimodal perception in the control of infant reaching. *Journal of Experimental Psychology: Human Perception and Performance, 20*, 876-886.
9. Ennouri, K., & Bloch, H. (1996). Visual control of hand approach movements in new-borns. *British Journal of Developmental Psychology, 14*, 327-338.

10. Hofsten, C. von. (1982). Eye-hand coordination in the newborn. *Developmental Psychology, 18*, 450-461.
11. Hofsten, C. von. (1984). Developmental changes in the organization of prereaching movements. *Developmental Psychology, 20*, 378-388.
12. Hofsten, C. von, & Ronnqvist, L. (1993). The structuring of neonatal arm movements. *Child Development, 64*, 1-46-1057.
13. Holland, J.H. (1975). *Adaptation in natural and artificial systems*. Ann Arbor, MI: University of Michigan Press.
14. Jacobs, R.A., Jordan, M.I., & Barto, A.G. (1991). Task decomposition through competition in a modular connectionist architecture: The what and where vision tasks. *Cognitive Science, 15*, 219-250.
15. Konczak, J., & Dichgans, J. (1997). The development toward stereotypic arm kinematics during reaching in the first 3 years of life. *Experimental Brain Research, 117*, 346-354.
16. Kuperstein, M. (1988). Neural model of adaptive hand-eye coordination for single postures. *Science, 239*, 1308-1311.
17. Kuperstein, M. (1991). INFANT neural controller for adaptive sensorimotor coordination. *Neural Networks, 4*, 131-145.
18. McDonnell, P.M. (1975). The development of visually guided reaching. *Perception and Psychophysics, 19*, 181-185.
19. McDonnell, P.M., & Abraham, W.C. (1979). Adaptation to displacing prisms in human infants. *Perception, 8*, 175-185.
20. Parisi, D., Cecconi, F., & Nolfi, S. (1990). Econets: Neural networks that learn in an environment. *Network, 1*, 149-168.
21. Piaget, J. (1952). *The origins of intelligence in children*. New York: Basic Books.
22. Schlesinger, M., Parisi, D., & Langer, J. (2000). Learning to reach by constraining the movement search space. *Developmental Science, 3*, 67-80.
23. Thelen, E., Corbetta, D., Kamm, K., Spencer, J.P., Schneider, K., & Zernicke, R.F. (1993). The transition to reaching: Mapping intention and intrinsic dynamics. *Child Development, 64*, 1058-1098.
24. White, B.L., Castle, P., & Held, R. (1964). Observations on the development of visually directed reaching. *Child Development, 35*, 349-364.
25. Zernicke, R.F., & Schneider, K. (1993). Biomechanics and developmental motor control. *Child Development, 64*, 982-1004.

Modelling Cognitive Development with Constructivist Neural Networks

Gert Westermann

Abstract

Based on recent evidence from cognitive developmental neuroscience, I argue for the importance of constructivist models of cognitive developmental phenomena. This point is empirically investigated with a constructivist neural network model of the acquisition of past tense/participle inflections. The model dynamically adapts its architecture to the learning task by growing units and connections in a task-specific way during learning. In contrast to other, fixed-architecture models, the constructivist network displays a realistic, U-shaped learning behaviour. In the trained network, realistic "adult" representations emerge that lead to aphasia-like dissociations between regular and irregular forms when the model is lesioned. These results show that constructivist neural networks form valid models of cognitive developmental processes and that they avoid many of the problems of fixed-architecture models.

1. Introduction

The computational modelling of psychological processes necessarily involves abstractions. Every modeller makes often implicit assumptions about which aspects of the original system can be abstracted away without compromising the model's value for explaining the observed phenomena. For example, much has been written about the relationship between biological and artificial neural networks, and it seems clear that they only share the most basic principles of operation, abstracting away most of the details of neural processing that are not considered essential. Nevertheless, the use of artificial networks in psychological modelling is valuable because they are thought to capture the characteristic principles of biological neurons, such as learning based on complex nonlinear associations.

However, in designing a model it is important that none of those aspects of the original system are abstracted away that are essential for the functioning of the modelled process. For example, a model of bird flight that does not take the properties of air into account will most likely not be useful.

In this paper, I argue that in the case of cognitive development, constructivist learning, that is, the dynamic adaptation of the learning system's architecture to the learning task, is such an essential property, and it should not be abstracted away in developmental models. This argument is based on recent research in brain and cognitive development. To illustrate the practical usefulness of constructivist models of cognitive development, I present a constructivist neural network model that acquires the English past tense in a realistic way, and that develops a final architecture which, when lesioned, displays a breakdown profile comparable to that of agrammatic aphasics.

The next two sections briefly review some neurobiological and cognitive developmental arguments for constructivism. Section 4 describes the neural network model of past tense/participle acquisition and impaired processing, and in section 5 I discuss the model's implications for cognitive modelling.

2. Activity-dependent Cortical Development

In recent years it has become clear that the architecture of the developing, but also of the adult cortex is shaped by neural activity. The activity-dependent loss or stabilization of neural connections have long been recognized as elements in shaping neural circuits [e.g. 5]. More recently, however, the activity-dependent directed *growth* of neural structures has been suggested as an equally important mechanism [19]. These claims have been supplemented with findings from neurobiology that show neural activity to have rapid and dramatic effects on neural function and morphology [reviewed in 24]. Often these effects are mediated by neurotrophins, a group of structurally related proteins that are released and taken up by neurons in the central nervous system in activity-dependent ways. For example, neurotrophins have been implicated in the branching complexity and target-finding of axons, and they play a central role in regulating the complexity of developing dendrites. These effects can be very specific, so that different neurotrophins have different effects on different neurons, and even on different locations of the same neuron. Experiments in which rats were trained on complex tasks showed that the newly generated structures were localized in those areas of the cortex that subserved the learned function [7].

From these findings emerges a constructivist view of cortical development in which activity, often dependent on sensory inputs, shapes the neural circuits that are responsible for the processing of these inputs, and thus learning itself changes the architecture of the learner which in turn determines what can be learned in subsequent steps.

While synaptic adjustments between neurons can be modelled in static networks by adjusting the connection strengths between units, morphological effects produce a change in the structure of the network that are not found in conventional artificial models.

3. Constructivism in Cognitive Development

The initial processing limitations caused by a gradual construction of the brain's neural architecture might be seen as a disadvantage because a young learner cannot function independently in the world. Several researchers have argued, however, that these restrictions are instead beneficial to learning because they allow for the gradual build-up of complex representations.

For example, it has been suggested that for normal development it is important that infants are able to focus visually only on a narrow range about 10 inches away from the eyes [22]. This limitation might serve to filter out a visually overly complex world and might be needed to learn that the perceived size of objects shrinks with increasing distance. Most significantly, because young children can only focus on things close enough to be touched or grasped, this limitation might be a prerequisite for the integration of different sensory modalities. These ideas were backed by experiments with rat pups whose eyes were prematurely opened directly after birth [8]: these rats developed an abnormal homing behavior.

Similarly, the *Less is More* hypothesis [13] argues that late learners of language do not reach native speaker competence because they are too advanced in their cognitive abilities and store linguistic units as a whole. By contrast, early learners, due to their limited processing capabilities, break down the units into their constituents and can later generalize them better to new situations.

Other results show how development can make use of processing resources as they are needed: in people that have been blind from an early age, primary visual cortex can be activated by Braille reading and other tactile tasks [21], and this additional area used for tactile processing is functionally relevant [3].

Taken together, these results suggest an important role for initially limited processing capabilities which lead to greater flexibility, and the absence of such early limitations would lead to a different, often worse outcome.

4. A Case Study: Verb Inflections in the Brain

The domain of verb inflection has attracted considerable attention in modelling but has mainly been used to contrast rule-based and associative (connectionist) approaches [see e.g., 4, 10, 11, 15–17, 20, 25]. Here, this domain is used to assess the viability of constructivist modelling in comparison with fixed-architecture neural network models [4, 10, 17, 20, and an equivalent, but non-constructivist version of the present model].

4.1 The Model

The constructivist neural network model (CNN) used in the simulation experiments is shown in fig. 1. It consists of an input layer and an output layer that are fully interconnected, and a hidden layer with Gaussian receptive fields that is constructed during learning. The input layer receives the verb stem in a templated phonological form that was developed by MacWhinney and Leinbach [10] for their past tense neural network model.

In contrast to other models of past tense learning, in the CNN the formation of the past tense is considered a classification task: instead of producing the phonological form of the past tense, 23 classes were defined according to the way in which the past tense form is generated from the stem. For example, the class $/x/ \to /6/$ (where $/x/$ stands for and arbitrary phoneme and $/6/$ is the notation of the vowel in e.g., *stuck*), contains the verbs *string, strike, swing, stick, fling, cling, spin, hang,* and *dig*. Viewing past tense formation as a classification task eliminates several confounding

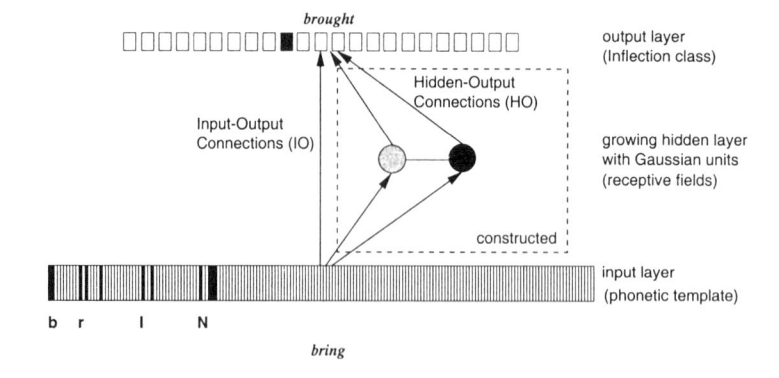

Figure 1: The architecture of the CNN model.

factors such as different phonological realizations of the regular ending in *wanted*, *looked*, and *loved*.

The CNN model starts out with direct connections between the input and output layer. The weights are trained with the quickprop algorithm [6]. When this weight adaptation fails to further improve the network's performance, a new hidden unit is inserted and fully connected to the input and output layer. The hidden units have Gaussian activation functions, and they are inserted in those areas of the input space that generate high error. After the insertion of a unit, weight adaptation proceeds as before, until the next unit is inserted. The algorithm stops when all verb mappings have been learned.

The principle of this algorithm is that resources in the network are attributed when and where they are needed, resulting in an architecture that is adapted to the specific learning task at hand. As such it corresponds conceptually to notions of constructivist learning in developmental psychology.

4.2 Data

The training corpus for the network consisted of 8,000 verb tokens, corresponding to 1,066 types, and was a modified version of one used in previous past tense models [9, 10]. It accurately reflects the type and token frequencies of regular and irregular verbs in language. While other models have simulated development by manipulating the training corpus during training [17, 20], here the focus is on the development of the system itself, and therefore the verbs were presented in a non-incremental fashion and each token was presented at each epoch.

The simulation was run with six networks, all differing in the initial setting of the weights.

4.3 Results

The model learned to classify all verbs correctly after on average 1,672 epochs. By contrast, comparable fixed-architecture models had difficulties with learning all irregular verbs (95% [20], 90.7% [10], and 92% [4]). An analysis of the developed network architecture showed that hidden units were preferentially inserted for irregular forms that are harder to learn, and this suggests that the success of the CNN is a direct consequence of its ability to allocate resources where they are needed.

The most striking feature of past tense acquisition in children is the U-shaped learning curve: an unlearning of previously correct irregular past tense forms and their subsequent re-learning (e.g., saw—seed/sawed—saw), and a plausible model should follow this well-documented [12] course. However, existing models of past tense learning have either modelled this phenomenon through unrealistic manipulations of the training data [20], have failed to make a distinction between irregulars that are newly introduced into the corpus and that have been produced correctly before [17], or have altogether failed to show a decrease in performance like in the left half of the U-shaped learning curve [10].

By contrast, the CNN model displayed a U-shaped learning curve for many of the irregular verbs in the training corpus, where a period of overregularization (i.e., a classification of the verb as belonging to the regular class) was preceded by a phase of correct classification. Corresponding to psycholinguistic evidence [12], the irregulars generally displayed so-called *micro* U-shaped learning, i.e., a phase of correct production followed by overregularizations at individual times for different verb (as opposed to *across-the-board* U-shaped learning affecting all verbs simultaneously, which does not occur in children). This learning behaviour was a direct consequence of the constructivist learning process: initially, without a hidden layer, all forms were produced in the direct IO-connections of the network. These were most of the regulars together with the most frequent irregulars. When subsequently the hidden layer is constructed, the Gaussian units initially cover both regular and irregular verbs, and this stage can lead to a temporary unlearning of the correct form for the irregulars.

It might be possible that the success of the CNN in comparison with previous neural network models of past tense acquisition is due to the fact that here this task is treated as a classification problem, whereas other models require the production of the phonological past tense form at the output layer. To exclude this possibility and to establish that the U-shaped learning curve was a consequence of the constructivist learning, a non-constructivist version of the CNN was trained on the same task. This model started out with a full hidden layer and did not add units during training, but was otherwise identical to the CNN. Indeed, this model failed to display a U-shaped learning curve and instead overgeneralized irregulars from the onset of training, quickly learning the correct forms.

To further investigate whether the difference between classification and actual production of the past tense form is a fundamental one, the CNN model was enhanced to produce the phonemic past tense form of the verbs. For this, a new output layer was added to the network that received inputs from the input layer and the previous "classification" output layer. The task was therefore to generate the phonemic past tense form based on the phonemic input together with the inflection class information. This model learned the correct phonological form of the past tense of all verbs after 210 epochs, showing that the mapping from phonological stem plus inflection class to phonological past tense form is a linearly separable problem. This result further strengthens the argument that the difference between the tasks of previous models and the CNN is not sufficient to explain the better performance of the CNN model.

In summary, the CNN displayed a realistic course of acquisition of the English past tense, and it performed better than other, non-constructivist models in accounting for the human data, both in the acquisition profile and in the final success of the learning task.

4.4 Generalization to Novel Verbs

The trained CNN was tested on its generalization to a set of 60 pseudo-verbs that had also been tested on human subjects [18]. These verbs consisted of blocks of ten which were prototypical, intermediate and distant in their similarity to existing regular and irregular verbs.

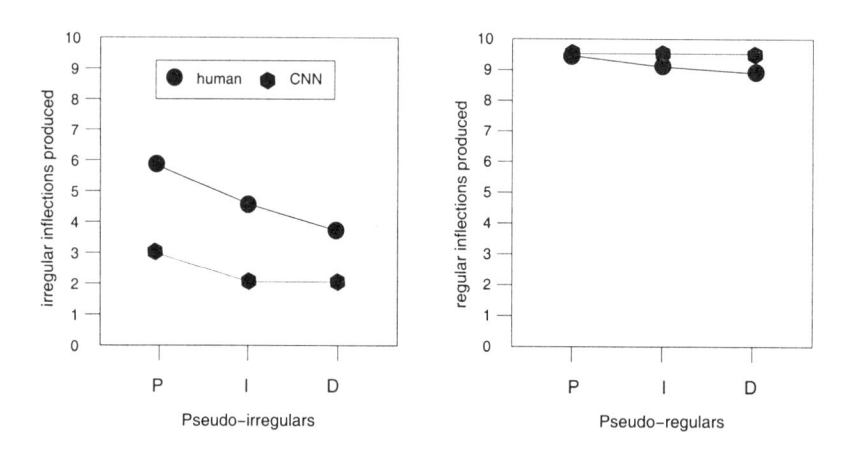

Figure 2: Generalization of the CNN to different classes of pseudo-verbs, in comparison with humans, the SPA, and M&L's network. P = Prototypical, I = Intermediate, D = Distant.

The results of the generalization experiments are shown in fig. 2. The CNN had a stronger tendency to regularize novel "irregular" pseudo-verbs than human subjects, but like humans it showed a decreasing tendency for irregular inflection with decreasing similarity to existing irregulars. The test case for realistic generalization behaviour are those verbs that are dissimilar to any existing verbs, i.e., the "distant" regulars. A regularization of such verbs indicates that the regular case has been learned as the default: it is applied in cases where a similarity-based inflection is not possible. Such behaviour is often taken as evidence that regular forms are produced by a symbolic rule instead of by association with existing verbs. However, the CNN model performed very much like the human subjects and regularized 29 of the 30 pseudo-regulars without recourse to an implemented rule. This result indicates that in the associative CNN model, the regular case was learned as the default.

4.5 Agrammatic Aphasia in the Adult Network

The previous section described the realistic acquisition and generalization behaviour of the CNN model. In this section, these studies are extended by investigating the representations developed in the "adult", fully trained model. These studies were done by training the model on the German participle, and then lesioning the final architecture, comparing the resulting performance breakdown with that of German agrammatic aphasics.

The German participle is functionally very similar to the English past tense and equally comprises regular (no stem change, inflectional ending *-t*) and irregular (possible stem change, inflectional ending *-en*) verbs. The study of the German participle inflection paradigm has become popular because, unlike in the English past tense, regular verbs do not form the majority of verb tokens. For the present model of agrammatic aphasic processing, I was able to collaborate with Martina Penke who investigated the performance of eleven German agrammatic aphasics in detail [14].

Agrammatic Broca's aphasia is a language disorder that is generally caused by a stroke affecting anterior parts of the left hemisphere. One of the characteristic symptoms of Broca's aphasia is the tendency to omit or confuse inflections, and the majority of the investigated German aphasics displayed a selective impairment of irregular with better preserved regular inflections [14], while some showed an across-the board breakdown of both regulars and irregulars.

For the network simulations, a training set of 20,018 tokens, corresponding to 664 types, were extracted from the CELEX database [1]. Like in the English past tense experiments, the verbs were assigned to classes according to the way in which the participle was formed from the verb stem. This resulted in 22 classes: one for the "stem + *-t*" (regular) class, and 21 for all other participles. The training proceeded identically to the English simulations for five networks that differed only in their initial weight settings. Again, all verbs were learned, but this time the goal was to investigate the "adult" architecture that was developed by the model through the constructivist learning process.

The CNN model has two sets of connections (fig. 1): the initial direct input-output (IO) connections, and the hidden-output (HO) connections that developed together with the hidden layer during training. To assess the functional role that these connections took on over the course of training, they were selectively lesioned.

In four of the five trained networks, the lesioning of the pathways produced double dissociations between regular and irregular verbs. The reported results are for these four networks and they are shown in fig. 3. Lesioning the HO connections resulted in a marked decrease of the performance for irregular verbs, with regular inflections remaining nearly fully intact. By contrast, lesioning the IO connections resulted in the opposite profile: the performance of regulars was significantly more impaired than that of irregular verbs. It is important to note that this double dissociation emerged as a result of the structure of the training data together with the constructivist development of the model and was in no way pre-specified. The hidden layer resources were allocated mainly for the harder to learn irregular verbs and thus the HO- lesioning affected these verbs more.

Removing the HO connections in the network thus modelled the basic deficit in the

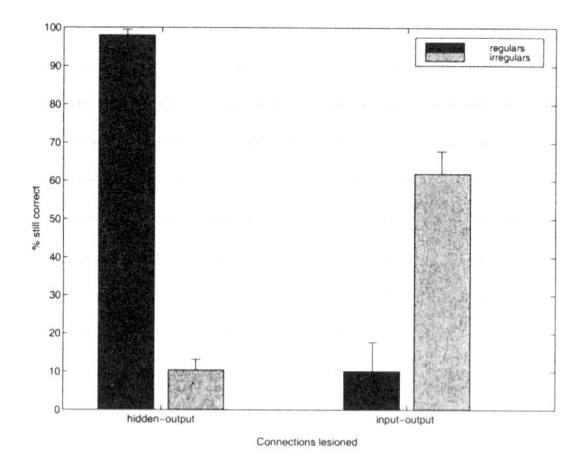

Figure 3: Double dissociation between regular and irregular verbs after lesioning the two pathways in the networks.

inflection of agrammatic aphasics, namely, the breakdown of irregular and selective sparing of regular participles. More detailed results reported in [14] could also be modelled: more of the wrong irregular participles were regularized (aphasics: 63.3%; CNN: 73.7%) than wrong regulars were irregularized (aphasics: 14.3%, CNN: 6.5%). Further, like aphasics, the model showed a frequency effect for irregulars (i.e., the more frequent irregulars were better preserved than the less frequent ones), but not for regulars (the regular errors made were independent from the frequency of these verbs).

To establish in how much this result relied on the constructivist learning process, again an equivalent non-constructivist model was trained on the German participle and then lesioned in the same way. Although in this model a double dissociation also occured, it was much less pronounced than in the constructivist model.

5. Discussion

Combining the results from the English past tense acquisition model and the German participle lesioning experiments, a comprehensive picture of the processing of verb inflection emerges: the model displays a realistic acquisition profile, and it develops an "adult" architecture that is prone to the same deficits as agrammatic aphasics. The regular case is learned as a default and applied to novel verbs that are dissimilar to any existing verbs (fig. 2), and it can be selectively spared when the system is lesioned.

The dissociations between regular and irregular verbs that are observed in human processing are often taken as evidence for two qualitatively distinct, encapsulated mechanisms underlying the production of regulars and irregulars, respectively [e.g. 2, 16]. Regular forms are assumed to be generated by a rule, whereas irregulars are claimed to be stored in an associative lexicon. The CNN model offers an alternative view: dissociations emerge through the structure of the environment (here, the training set) together with a constructivist learning system that adapts its architecture to the specific learning task. In this way, the CNN shows how the strong assumption of qualitatively distinct processing mechanisms and rule processing can be abandoned when the constructivist nature of development is taken into account. These result also suggest a replacement of the binary dichotomy *regular–irregular* with a graded distinction between *easy* and *hard* forms [23], where the "easiness" of a form is determined by factors such as high frequency, many similar verbs that share the same stem–participle transformation, few similar verbs with different transformation, and unambiguity of the inflectional morpheme. Easy verbs correspond largely to regulars, but the match is not perfect, and indeed "hard regulars" seem to behave more like irregulars in aphasic subjects [23].

Equivalent, but non-constructivist models were trained both on the past tense acquisition task and on the German participles, but they did not display the U-shaped learning curve found in children, and the resulting functional dissociation was less pronounced than in the constructivist case. The CNN model therefore suggests an important role for constructivist models in the simulation of developmental processes.

Acknowledgments

This research was conducted while the author was at the Centre for Cognitive Science at the University of Edinburgh, Scotland UK.

References

1. Baayen, H., Piepenbrock, R., van Rijn, H. (1993). *The CELEX Lexical Database*. CD-ROM. Linguistic Data Consortium. University of Pennsylvania, PA.
2. Clahsen, H. (1999). Lexical entries and rules of language: A multidisciplinary study of German inflection. *Behavioral and Brain Sciences*, 22(6), 991–1013.
3. Cohen, L., Celnik, P., Pascual-Leone, A., Corwell, B., Falz, L., Dambrosia, J., Honda, M., Sadato, N., Gerloff, C., Catalá, M., Hallett, M. (1997). Functional relevance of cross-modal plasticity in blind humans. *Nature*, 389, 180–183.
4. Daugherty, K. Seidenberg, M. S. (1992). Rules or connections? The past tense revisited. *Proceedings of the 14th Annual Conference of the Cognitive Science Society*. Erlbaum, Hillsdale, NJ.
5. Edelman, G. (1987). *Neural Darwinism: The Theory of Neuronal Group Selection*. Basic Books, New York.
6. Fahlman, S. E. (1989). Faster-learning variations on back-propagation: An empirical study. *Proceedings of the 1988 Connectionist Models Summer School*. Morgan Kaufmann, San Mateo, CA.
7. Greenough, W., Larson, J., Withers, G. (1985). Effects of unilateral and bilateral training in a reaching task on dendritic branching of neurons in the rat motor-sensory forelimb cortex. *Behavioral and Neural Biology*, 44, 301–314.
8. Kenny, P. Turkewitz, G. (1986). Effects of unusually early visual stimulation on

the development of homing behavior in the rat pup. *Developmental Psychobiology, 19(1),* 57–66.
9. Ling, C. X. Marinov, M. (1993). Answering the connectionist challenge: A symbolic model of learning the past tenses of English verbs. *Cognition, 49,* 235–290.
10. MacWhinney, B. Leinbach, J. (1991). Implementations are not conceptualizations: Revising the verb learning model. *Cognition, 40,* 121–157.
11. Marcus, G., Brinkmann, U., Clahsen, H., Wiese, R., Pinker, S. (1995). German inflection: The exception that proves the rule. *Cognitive Psychology, 29,* 189–256.
12. Marcus, G. F., Pinker, S., Ullman, M., Hollander, M., Rosen, T. J., Xu, F. (1992). Overregularization in language acquisition. Monographs of the Society for Research in Child Development, Serial No. 228, Vol. 57, No. 4.
13. Newport, E. L. (1990). Maturational constraints on language learning. *Cognitive Science, 14,* 11–28.
14. Penke, M., Janssen, U., Krause, M. (1999). The representation of inflectional morphology: Evidence from Broca's aphasia. *Brain and Language, 68,* 225–232.
15. Pinker, S. (1991). Rules of language. *Science, 253,* 530–535.
16. Pinker, S. (1999). *Words and Rules: The Ingredients of Language.* Basic Books, New York.
17. Plunkett, K. Marchman, V. (1993). From rote learning to system building: Acquiring verb morphology in children and connectionist nets. *Cognition, 48,* 21–69.
18. Prasada, S. Pinker, S. (1993). Generalization of regular and irregular morphological patterns. *Language and Cognitive Processes, 8(1),* 1–56.
19. Quartz, S. R. Sejnowski, T. J. (1997). The neural basis of cognitive development: A constructivist manifesto. *Behavioral and Brain Sciences, 20,* 537–596.
20. Rumelhart, D. E. McClelland, J. L. (1986). On learning past tenses of English Verbs. In: David E. Rumelhart and James L. McClelland (eds.) *Parallel Distributed Processing, Vol. 2.* MIT Press, Cambridge, MA, 216–271.
21. Sadato, N., Pascual-Leone, A., Grafman, J., Ibañez, V., Deiber, M., Dold, G., Hallett, M. (1996). Activation of the primary visual cortex by Braille reading in blind subjects. *Nature, 380,* 526–528.
22. Turkewitz, G. Kenny, P. A. (1982). Limitations on input as a basis for neural organization and perceptual development: A preliminary theoretical statement. *Developmental Psychobiology, 15,* 357–368.
23. Westermann, G. (2000). A constructivist dual-representation model of verb inflection. *Proceedings of the 22nd Annual Conference of the Cognitive Science Society.* Erlbaum, Hillsdale, NJ.
24. Westermann, G. (2000). *Constructivist Neural Network Models of Cognitive Development,* PhD thesis, Division of Informatics, University of Edinburgh, Edinburgh, UK.
25. Westermann, G. Goebel, R. (1995). Connectionist rules of language. *Proceedings of the 17th Annual Conference of the Cognitive Science Society.* Erlbaum, Hillsdale, NJ.

Learning Action Affordances and Action Schemas

Richard Cooper & David Glasspool

Abstract

Several theories of action selection assume that the environment both suggests and constrains possible actions at each moment. We present an interactive activation model (based on an existing model of routine sequential action selection) in which actions are organised into partially ordered schemas for simple task elements, and the current state of the environment contributes to selecting actions and schemas. Previous versions of the model have not accounted for learning. We show that a simple reinforcement learning paradigm allows environment-action associations to be acquired through unguided exploration of the environment. The basic model is limited in the types of environment/action associations that it acquires. We explore ways in which greater diversity of learning (and behaviour) may be achieved, and suggest that, in order to acquire a broad range of environment/action associations, mechanisms for boredom avoidance or novelty seeking are required.

1. Introduction

The task of action selection is commonly argued to be mediated by constraints imposed by the environment. Thus, Thorndike [13] argued that behaviour was structured into habit families, in which any situation (or stimulus) was associated with a hierarchically structured set of responses. Thorndike held that the response or behaviour selected at any point in time was that at the top of the currently applicable habit family, and that habit families were modified (i.e., learnt) through the laws of exercise (promote or strengthen responses that are attempted) and effect (promote or strengthen responses that are successful but demote or weaken responses that fail or are inappropriate).

More recently, Gibson [4] proposed that objects and/or situations might "afford" certain responses. Thus, a light switch might afford being flicked on or off, and a knife might afford cutting. While the mechanisms underlying affordances (notably so-called direct perception) are often portrayed as mysterious and anti-cognitivist, affordance-like notions have arisen in the cognitive literature. In the theory of automatic and willed action control proposed by Norman & Shallice [9, 10], for example, action schemas (corresponding to abstractions over individual actions or action sequences) participate in an interactive activation network, with action being controlled by the most highly active schema nodes. Environmental triggering is held to be a significant

---

*This work was supported in part by grant RSRG 20546 from the Royal Society to Richard Cooper and grant #R01 NS31824-05 from the National Institutes for Health to Myrna Schwartz.

source of excitation within the network. Indeed, such triggering is assumed to underly some common forms of action lapse (e.g., capture errors: cf. [12]) and behaviours of certain neurological patients (such as utilisation behaviour [7], where patients appear to be unable to inhibit environmentally appropriate actions).

Similar requirements are apparent in approaches to action selection developed within Artificial Intelligence. Thus, excitation of nodes within Maes [8] interactive activation network for sequential action control is partially dependent upon the satisfaction (by the current state of the environment) of action preconditions.

The common elements of the above theories are that, at any point in time, the environment is held to make a set of possible actions available, and that behaviour involves selecting and performing one action from this set. That is, the theories all assume environment/action associations (affordances, triggering conditions or preconditions). However, none can provide an account of the acquisition of these associations.

This incompleteness is particularly well illustrated in the case of Norman & Shallice's "Contention Scheduling" account of automatic action control. An implementation of the account has recently been developed by Cooper & Shallice [2] (see also [3]). The model is able to produce complex well-formed action sequences, and is also able to account for a range of errors (including those seen in normals when distracted or fatigued and those characteristic of a number of classes of neurological patient). An important part of the model is the environmental triggering of schemas: following Norman & Shallice [9, 10], each schema has a triggering function that determines the extent to which it is excited (or inhibited) by the state of the environment at any point in time. A significant weakness of the model, however, is that it is unable to acquire these triggering functions — they must be specified by hand for all schemas within the model's repertoire. This raises both practical difficulties (specifying such triggering functions by hand is time-consuming and error-prone) and methodological difficulties (countless degrees of freedom are introduced into the model because the theory fails to specify the degree to which a given state of the environment triggers each action schema). Although the severity of these difficulties may be reduced by the incorporation of general principles governing the form of triggering functions, the question of acquisition remains.

There is thus a need for a fully explicit computational account of the acquisition of environment/action associations. Thorndike's laws of exercise and effect provide a starting point for such an account. The remainder of this paper presents a model that uses reinforcement learning (effectively implementing Thorndike's law of effect) to acquire environment/action associations within the context of the Cooper & Shallice model of routine action selection. The basic premise of the model is that environment/action associations are learnt through trial and error with feedback, with the conditions under which successful actions are performed being reinforced beyond the conditions under which unsuccessful actions are performed. The basic model is only able to acquire a small set of environment/action associations. We therefore explore some simple modifications and extensions to the basic model. These explorations suggest that, in order to acquire appropriate associations relating to a broad range of actions, mechanisms for "boredom avoidance" or "novelty seeking" are required. We conclude by relating our findings to more general problems in learning to act, especially those concerning the learning of action sequences.

2. The Basic Model

An essential feature of this account of the acquisition of environment/action associations is that the shape of such associations is moulded by what is possible within the environment. Actions should only be associated with, or triggered by, states of the environment in which they may be successfully realised. It is therefore essential to develop a sufficiently realistic model of the environment in tandem with the model of the agent. The COGENT modelling environment [1] was used to implement and evaluate such a system, with the agent model and the environment model encapsulated in separate, communicating, functional modules.

2.1 The Model of the Environment

The environment model is based on the coffee preparation domain used by Cooper & Shallice [2, 3]). It consists of a set of objects, two hands, and a set of functions for manipulating objects with the hands. Each object has a set of features including size, shape, position, and relevant aspects of state. Thus, the world includes several containers and implements (an empty mug, a jug of water, a closed bowl of sugar, a teaspoon and a coffee stirrer). The manipulation functions correspond to basic actions (pickup, discard, open, close, pour, dip spoon, empty spoon and stir). Each includes a set of preconditions that specify the physical conditions under which the action may be successfully performed. The functions are called at various times by the agent, as described below. If a manipulation function's preconditions are satisfied by the state of the world when the function is called, then the world is modified according to that function and positive feedback is returned to the agent. If the preconditions are not satisfied, the world is not modified and negative feedback is returned.

2.2 The Model of the Agent

We assume that the agent is capable of a variety of basic actions, corresponding to (a subset of) the manipulation functions available within the environment. The agent's task is to learn to select actions through time that are compatible with the changing state of the environment. Essentially the agent must internalise action preconditions. Minimally, the agent comprises three processes: perceptual processes that transduce the state of the environment into internal representations; selection processes that select one action from those available to the agent on the basis of its internal representation of the environment, and a learning process that uses feedback from the environment to shape the selection process.

The perceptual processes are finessed by a set of rules that map each object in the environment to a feature vector (of width eight in the implementation reported here). Essentially the environment is modelled in symbolic terms, and the perceptual processes re-represent each object as a vector of features. The internal representation of the complete environment is therefore a set of such vectors. For present purposes such vectors include both perceptual and functional features, although the functional features employed (e.g., *implement*) may be taken as short-hand for simple combinations of perceptual features (e.g., *one primary axis* and *between 150 mm and 300 mm in length*). The approach is similar in spirit to that of Plaut & Shallice [11].

Following Norman & Shallice [9, 10] (see also [2, 3]), we use an interactive activation network to mediate the selection process. There is one node in this network for each action. Processes of lateral inhibition and self activation operate over the nodes to ensure that at most one node is highly active at any time. Nodes may also receive excitation (or inhibition) from the representation of the environment (as described below). Selection occurs when a node's activation exceeds a threshold.

Once a node has been selected, its corresponding action is performed. Performance of an action entails selecting objects on which to act and effectors with which to act. Thus, once a basic action such as pickup has been selected, it is necessary also to select an object to pick up and an effector to do the picking up. Not all actions require one object and one effector, and different actions have different argument requirements (e.g., pickup requires an effector that is not full, but discard requires an effector that is holding an object). Each action therefore has an associated set of object and effector selection criteria.

In the model of Cooper & Shallice [2, 3], selection criteria were functions that mapped object representations to numeric values. The numeric value essentially indicated the appropriateness of an object or effector for the given argument role. In the current model these selection criteria must be learnt. We thus associate with each action a list of argument roles. The number and type of arguments for an action (e.g., two objects and one effector) is assumed to be intrinsic to the action (and hence hardwired), but the mapping of objects and effectors to these argument roles is assumed to be subject to learning. Each argument role is therefore assumed to correspond to a vector of weights, with the appropriateness of an object for an argument role being determined by the dot product of the object's featural representation and the argument role's weight vector. Learning alters weight vectors, leading to more accurate measures of appropriateness.

While the specification of argument roles in terms of weight vectors addresses the issue of argument selection, the issue of action triggering by the representation of the environment remains. Cooper & Shallice [2, 3] used separate triggering functions for this purpose, but such functions introduce additional degrees of freedom and, in fact, are not required. The current model instead assumes that action nodes are triggered to the extent that their argument selection criteria are met by objects in the model's representation of the environment. Thus, selection criteria do double duty: once an action has been selected they are used in the selection of object and effector representations to fill the action's argument roles, but, prior to selection of an action, they mediate excitation and inhibition of the action's node within the interactive activation network.

After the agent model selects and attempts an action, it receives feedback from the environment model. If the action succeeds the feedback is positive, and each argument and effector weight vector of the action is adjusted through delta rule learning so that its dot product with the representation of the actual object or effector used is nearer $+1$. If the action fails the feedback is negative, and each argument and effector weight vector of the action is adjusted through delta rule learning so that its dot product with the representation of the actual object or effector used is nearer -1. In all cases the node within the interactive activation network corresponding to the attempted action is subsequently inhibited, allowing other nodes to become active and hence other actions to be performed.

2.3 Functioning of the Model

In order to establish that appropriate selection restrictions would yield a model capable of generating random sequences of appropriate actions, the model's behaviour was first assessed with hand-coded selection restrictions. The selection restrictions in this "expert" model embodied all action preconditions. For example, the featural encoding of objects included several features indicating an object's position. One of these features was set (by perceptual processes) to +1 for objects that were held and −1 for objects that were not held. In the hand-coded selection restriction for the object argument role of pickup the corresponding feature was set to −1. Consequently, objects that were not held would tend to match the selection restrictions of pickup better than objects that were held. This matching would also lead to greater excitation of the pickup action by objects that were not held than by objects that were held.

This assessment was successful in that, with appropriate parameter settings (persistence = 0.85; self activation = 0.50; lateral inhibition = 0.50; environmental triggering = 0.20; selection threshold = 0.60; noise variance = 0.001), the hand-coded selection restrictions yielded well-formed, but somewhat aimless, sequences of actions. Figure 1 (left) shows one such sequence, and figure 1 (right) shows the activation profiles of node throughout this sequence.

To investigate learning, selection restrictions for all actions were initialised to random vectors near zero. (Specifically, the components of each vector were initialised to random values normally distributed around 0 with variance 0.01.) The same parameters were used in the learning model as in the expert model, with the exception of self activation and lateral inhibition, which were both increased to 0.60, and the addition of the learning rate parameter, which was set to 0.05. The increase in self activation and lateral inhibition was necessary to ensure robust functioning of the learning model during the early stages of learning. Prior to learning, little excitation is provided by the environment. Additional excitation within the action node network is needed to ensure that action nodes (including those corresponding to inappropriate actions) may become sufficiently active to allow action selection.

```
 38: pickup stirrer with left hand
 88: stir jug with stirrer
133: discard contents of left hand
144: pickup teaspoon with left hand
188: stir jug with teaspoon
211: dip teaspoon into jug
252: discard contents of left hand
272: pickup teaspoon with right hand
```

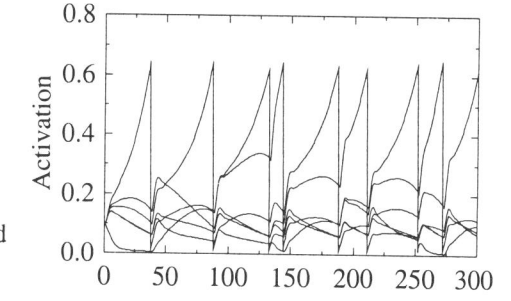

Figure 1: Left: A well-formed sequence of actions generated by the model when given hand-coded selection restrictions. The left column shows the time of action initiation, in terms of processing cycles. Right: Activation profiles of all action nodes throughout the sequence shown on the left. Time, in terms of processing cycles, is shown on the horizontal axis. Peaks in activation correspond to selected actions.

Figure 2 (left) shows the behaviour of the model prior to learning. Actions are attempted apparently at random, and generally with inappropriate arguments. In this example, only one of the first ten actions is well-formed (that of picking up the mug with the left hand). However, each time an action is attempted its selection restrictions are adjusted as described above. Consequently, the model rapidly discovers appropriate and inappropriate actions, internalising an approximation to each action's preconditions. In general, 100 to 150 action attempts are required before the model's approximate selection restrictions are sufficient for error-free action performance. Figure 2 (right) shows a fragment of such performance.

3. Towards Diversity

Although the action sequence of figure 2 (right) is error-free, it is also highly stereotyped. Two complementary actions are repeated *ad nauseum*. The model has not learnt selection restrictions for other actions, and examination of the selection restrictions it has learnt reveals that they are highly specialised (applying in the case shown in figure 2 (right) to the stirrer and the left hand, but not to other objects or to the right hand). Different runs of the model do yield different long-term behaviour (with different acquired selection restrictions), but that behaviour typically comprises performance of two complementary actions (such as pickup and discard, or open and close) with only a limited subset of objects and effectors.

In order to acquire general selection restrictions for the full range of actions it is necessary to successfully perform the full range of actions with a variety of arguments. Two factors prevent the model as described above from acquiring full generality. First, the only exogenous source of activation to the action network is due to environmental triggering by appropriate actions. Because reinforcement strengthens such triggering, actions whose selection restrictions are acquired first will dominate and be performed in favour of other actions, preventing attempts at such other actions and hence preventing the acquisition of triggering conditions for other actions. Second, argument selection is determined entirely through selection restrictions, so, once selected, an action will always be applied to the object that best matches its selection restrictions.

32: empty mug into stirrer*	
50: stir teaspoon with mug*	2978: pickup stirrer with left hand
58: pour mug into jug*	2989: discard contents of left hand
105: close jug with left hand*	3001: pickup stirrer with left hand
126: open bowl with right hand*	3010: discard contents of left hand
160: discard contents of right hand*	3022: pickup stirrer with left hand
187: pour stirrer into stirrer*	3031: discard contents of left hand
200: pickup mug with left hand	3043: pickup stirrer with left hand
217: open stirrer with left hand*	3053: discard contents of left hand
243: pickup mug with left hand*	

Figure 2: Left: A fully random action sequence, containing mostly errors, generated by the model prior to learning selection restrictions. Erroneous actions are marked with an asterix. Right: An error-free action sequence following acquisition of approximate selection restrictions.

```
34868: discard contents of left hand
34889: pickup jug with left hand
34908: close jug using right hand
34924: open jug using right hand
34930: discard contents of left hand
34947: pickup jug with left hand
34972: close jug using right hand
34992: open jug using right hand
34995: discard contents of left hand
```

Figure 3: Action sequences generated when extreme inhibition (500 units) is applied to attempted actions.

The first difficulty appears to require further exogenous sources of excitation and inhibition to the action network. In particular, it appears that some mechanism is needed to inhibit or suppress actions whose triggering conditions have been acquired. We consider one such mechanism below.

The second difficulty is solved in the closely related model of routine action selection (without learning) of Cooper & Shallice [2] through the incorporation of interactive activation networks for object representations and effectors. The networks, analogous to the action network, include one node for each represented object, and one node for each effector. Nodes receive excitation and inhibition from action nodes (mediated by selection restrictions) and (potentially) from attentional processes. In essence, the use of object representation and effector networks allows greater control over the arguments that are selected. Below we therefore consider augmenting the existing model with such networks.

3.1 Incorporating Action Inhibition

When the activation of an action schema node exceeds a threshold (0.6 in the above simulations) the action schema is selected, its argument roles are filled, and the resulting action is performed. The action schema node is then inhibited, allowing other action schema nodes to become active. Moderate inhibition (5 units in the above simulations) returns the selected action schema node's activation to approximately rest (a magnitude comparable to that of its competitors), allowing competition between all action schema nodes to proceed afresh.

However, by dramatically increasing the inhibition on action schema nodes once they have been performed, such schema nodes can be effectively removed from competition for an extended period, giving greater opportunity for the selection of other schemas, and thus encouraging greater diversity of action. Figure 3 shows action sequences produced by the model when inhibition was increased to 500 units.

Although high levels of inhibition result in the model taking significantly longer to acquire selection restrictions that reliably yield appropriate action selection, greater diversity is apparent in the selection restrictions thus acquired. With inhibition at 500 units, reasonable approximations to selection restrictions for four distinct actions are acquired. This approach does not, however, lead to greater diversity in the objects to which actions are directed. This is reflected in the acquired selection restrictions, which still converge to features of specific objects. Thus, the object role of pickup converges to match the favoured object (the open jug, in the case shown in figure 3), and not to match just the features of that object that are relevant to the action (e.g., that the object is not already in hand). Nevertheless, the approach does demonstrate that learning can be improved by manipulating the actions that are performed.

3.2 Incorporating Additional Networks

Incorporation of object representation and effector networks as described above into the model pose no difficulties. Activation flow within the additional networks is governed by the same principles (including self activation and lateral inhibition) as that within the action network. In addition, object representation nodes are excited or inhibited by action nodes in proportion to the product of the activation of the action node and the degree to which the corresponding object representation matches the action's argument selection restriction, and *vice versa* (i.e., action nodes are equivalently excited or inhibited by object nodes). Nodes in the effector network interact with the other networks in an analogous fashion.

On selection of an action, the action's arguments are filled by those objects and effectors that are both highly active and a good match for the action's selection restrictions. Arguments are selected to maximise the product of these two factors. Figure 4 shows an action sequence generated (after learning) by the augmented model. The model includes moderate inhibition of actions, object representations and effectors on each action attempt. The figure may thus be compared with figure 2 (right).

The introduction of object representation and effector networks results in considerable diversity of argument selection. Although only two action schemas are selected (cf. figure 4) both hands are employed to manipulate a variety of objects. The model has therefore acquired less restrictive, and more appropriate, selection restrictions for these two actions.

4. General Discussion

We have shown that a schema-based model of environmentally triggered action selection may acquire appropriate environment/action associations through trial and error. Diversity of behaviour arises when the basic model is augmented with mechanisms to

> 4131: pickup bowl with left hand
> 4139: pickup teaspoon with right hand
> 4144: discard contents of right hand
> 4150: pickup stirrer with right hand
> 4158: discard contents of left hand
> 4164: pickup jug with left hand
> 4167: discard contents of right hand
> 4173: discard contents of left hand
> 4178: pickup stirrer with right hand

Figure 4: Action sequences generated when object representation and effector networks are employed.

inhibit action reselection and to differentiate objects on the basis of activation. However, the use of extreme inhibition to encourage variety of action results in a system that is constantly switching between actions. As is apparent from figure 3, this dramatically slows learning. A more sophisticated mechanism might monitor learning and/or behaviour and manipulate excitation of both action schemas and their arguments in order to optimise learning. Such a mechanism might, for example, embody some notion of boredom detection, repeating behaviours until they are well learnt and then inhibiting them in favour of less well learnt behaviours. (Curiously, infants show just this kind of behaviour.)

The necessary distinction between novel and routine behaviour is already present in the Norman & Shallice [9, 10] framework. The former is handled by a supervisory attentional system (SAS) that responds to a novel situation by generating a strategy for dealing with it. If an initially novel situation is repeated the required strategies eventually become well learnt and may be automatically executed by contention scheduling (CS), the routine action selection system, without the intervention of SAS. At this point the situation has effectively become routine. Mechanisms thus already exist for (i) detecting whether a well-learnt schema exists in CS to deal with the current situation and (ii) supervising the learning of a new action schema if required. In its normal mode of operation the CS/SAS system employs existing schemas wherever possible, and only uses the limited resources of the SAS when necessary. We can however envisage a different mode of operation ("exploration" or "play") in which this precedence is reversed so that situations that stimulate the SAS are deliberately sought, and activity is abandoned when it becomes routine. We hypothesise that such an operational mode is essential to the acquisition of general selection restrictions for a broad range of actions.

While the selection restrictions acquired by the model may be viewed as serving the function of Gibsonian affordances, they are not true affordances in the Gibsonian sense. To Gibson, the concept of an affordance went hand in hand with that of direct perception. Affordances were held to be directly available from the environment, without the integration of perceived features into object representations. Because selection restrictions within the model are object-based, the model requires that features are first bound together as objects before they can affect an action schema's activation. Within the framework adopted here, the binding of features into objects is a necessary precondition for action triggering. The model thus sides against direct perception.

The model is also intended only as a fragment of a model of intelligent action selection. While the model does acquire action preconditions, it lacks goal directedness. The mechanisms of action selection embodied within the model are compatible, however, with those required by CS/SAS theory, and consistent with those of Cooper & Shallice's implementation of CS [2]. In fact, the model has implications for the CS implementation, suggesting that triggering functions and selection restrictions, currently separate elements of CS, might be merged.

Contention scheduling is a general mechanism for schema selection that may operate at multiple levels over a hierarchically structured sets of schemas. The model developed here illustrates learning only at the lowest level. Two additional mechanisms are required if the current learning mechanisms are to generalise to higher-order action schemas. First, the model must be instructable via higher-order supervisory processes. This is relatively straightforward: such a process may, through selective excitation and inhibition of nodes in the various networks, exert detailed control over behaviour. Second, the model must be able to acquire and represent higher-order action schemas, including appropriate sequential relations within those schemas. Two factors drive sequential behaviour in the current model: The changing state of the environment, and the inhibition of previous actions which prevents their immediate repetition. An existing class of models of sequential behaviour, Competitive Queueing models (e.g., [5, 6]), have successfully accounted for many aspects of serial behaviour in similar terms. In these models actions are given a *gradient* of activations, such that the earlier an action is to be performed the more active it is. Sequential behaviour is achieved by repeatedly performing and inhibiting the most active action. Since appropriate selection and inhibition mechanisms already exist in the current model, the learning of simple sequential behaviour might involve nothing more than the acquisition of an appropriate activation gradient over actions. This may again be achieved by reinforcement learning, with actions that occur too early receiving negative feedback and actions that occur too late receiving positive feedback. Further work will explore this possibility.

References

1. Cooper, R., & Fox, J. (1998). COGENT: A visual design environment for cognitive modeling. *Behavior Research Methods, Instruments, and Computers, 30*, 553–564.
2. Cooper, R., & Shallice, T. (2000). Contention Scheduling and the control of routine activities. *Cognitive Neuropsychology, 17*, 297–338.
3. Cooper, R., Shallice, T., & Farringdon, J. (1995). Symbolic and continuous processes in the automatic selection of actions. In Hallam, J. (Ed.), *Hybrid Problems, Hybrid Solutions*, pp. 27–37. IOS Press, Amsterdam.
4. Gibson, J. J. (1979). *The ecological approach to visual perception*. Houghton Miflin, Boston, MA.
5. Glasspool, D. W. (1998). *Modelling Serial Order in Behaviour: Studies of Spelling*. Ph.D. thesis, Department of Psychology, University College, London, UK.
6. Houghton, G. (1990). The problem of serial order: A neural network model of sequence learning and recall. In Dale, R., Mellish, C., & Zock, M. (Eds.), *Current Research in Natural Language Generation*, chap. 11, pp. 287–319. Academic Press, London, UK.
7. Lhermitte, F. (1983). Utilisation behaviour and its relation to lesions of the frontal lobes. *Brain, 106*, 237–255.
8. Maes, P. (1989). How to do the right thing. *Connection Science, 1*, 291–323.
9. Norman, D. A., & Shallice, T. (1980). Attention to action: Willed and automatic control of behavior. Chip report 99, University of California, San Diego.
10. Norman, D. A., & Shallice, T. (1986). Attention to action: Willed and automatic control of behavior. In Davidson, R., Schwartz, G., & Shapiro, D. (Eds.), *Consciousness and Self Regulation: Advances in Research and Theory, Volume 4*, pp. 1–18. Plenum, New York, NY.
11. Plaut, D. C., & Shallice, T. (1993). Perseverative and semantic influences on visual object naming errors in optic aphasia. *Journal of Cognitive Neuroscience, 5*, 89–117.
12. Reason, J. T. (1984). Lapses of attention in everyday life. In Parasuraman, W., & Davies, R. (Eds.), *Varieties of Attention*, chap. 14, pp. 515–549. Academic Press, Orlando, FL.
13. Thorndike, E. L. (1898). Animal intelligence: An experimental study of associative processes in animals. *Psychological Monographs, 2*.

A Three-Layer Configural Cue Model of Category Learning Rates

Paul Bartos and Martin Le Voi

Abstract

The relative difficulty of six different category structures revealed in Shepard, Hovland, and Jenkins' 1961 classic category learning paradigm [18] is predicted using simple channel capacity calculations. This approach is subsequently used to inform the design of a three-layer connectionist network based on Gluck and Bower's configural cue model of category learning [3]. The extra layer of nodes in this model consists of intermediate or bottleneck nodes which lie between each spatial group of nodes, representing particular cues and cue configurations, and each category label node. The weights in these nodes learn to approximate the correlation between the output of the 'space' and the target output. The model, using two free parameters, shows a superior fit to the human data than the configural cue model and its variants evaluated by Nosofsky, Gluck, Palmeri, McKinley, and Glauthier [16] in their replication of the Shepard et al. experiment [18]. The reason for this and the applicability of the model to other category learning paradigms is discussed.

1. A Category Learning Task

In their classic research on category learning rates Shepard, Hovland, and Jenkins [18] required participants to categorise eight, three-dimensional patterns or objects into two groups of four according to one of six experimenter-defined category types. The dimensions used were separable, for example big-small, black-white, and triangle-square. The abstract structures of these types are shown in figure 1. Dimensions are labelled for the purposes of this article as x, y, and z, with individual objects indexed by a binary number between 000 and 111.

Type I simply requires knowledge of one dimension to successfully predict the category label. Type II is a condensation or XOR task in which knowledge of two dimensions is required. Types III to V are rule-plus-exception structures where all three dimensions are relevant but to different extents. Finally, for type VI there are no 'compressions' possible and knowledge of all three dimensions is the minimum requirement.

The experiment, replicated in 1994 [16], employed a repeated presentation-guess-feedback cycle. Each block consists of 16 trials, each object presented twice with order randomised.

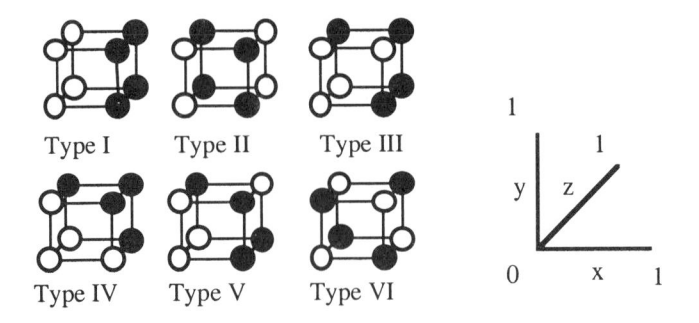

Figure 1: Abstract structure of the six category types used by Shepard, Hovland, and Jenkins [18]. Category assignments are shown by black or white circles.

The results of these experiments, shown in figure 2 for the Nosofsky *et al.* replication [16], revealed systematic differences in the rates at which the different category structures were learnt. The type I being the easiest, followed by the type II. Types III to V were next, being of approximately the same difficulty, with the type VI hardest of all.

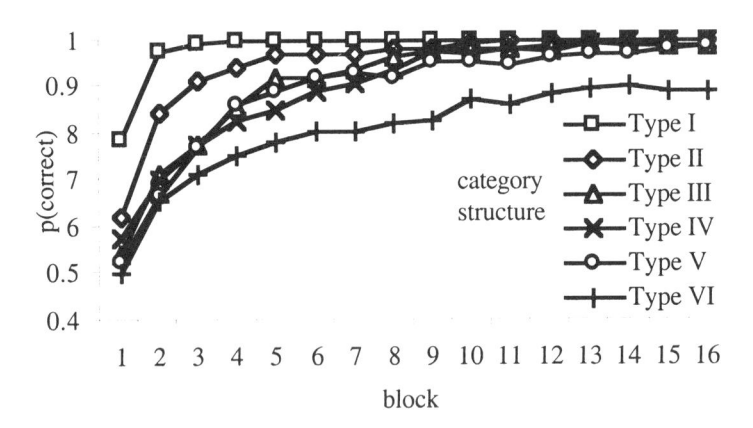

Figure 2: Human category learning rates, with mean p(correct) per block represented; from Nosofsky, Gluck, Palmeri, McKinley, and Glauthier (1994) [16].

1.1 Modelling Issues

The results are important for category learning models, as they cannot be predicted by models based purely on cue conditioning or stimulus generalisation [18,17], or prototypes ([2] p.49-54). Numerous theorists, including Shepard *et al.* [18], have proposed that the reason for these systematic differences in difficulty is that selective attention to diagnostic (predictive) dimensions may enhance the rate of task learning for structures where knowledge of less than all three dimensions is required. The principle difficulty for many models is that they fail to account for the faster learning of the type II structure in relation to the types III to V. The attention to dimensions account provides some explanation of this in that for the rule plus exception structures some attention must be allocated to all three dimensions, whereas for the type II attention to only two dimensions is required. In their partial replication of the experiment Nosofsky et al. [16] tested a number of connectionist models which had shown some success at predicting other features of category learning.

The most successful model in these tests was the attention learning covering map or ALCOVE [7, 8], which combines connectionist learning rules with the exemplar representations of the context model [13] and general context model [14,15].

This model uses a selective attention algorithm based on back-propagation of error, which controls the sensitivity to differences in input along certain dimensions. A capacity limitation is implemented by a trial-by-trial normalisation of attention weights to each dimension. The result is a system that maximises intra-category similarity and inter-category differences.

Another of the models tested was the configural cue model [3]. This model tended to learn the type III task more quickly than the type II. The model displays very similar performance on types II, III, IV, and V.

The reason for this is that there are more perfectly diagnostic cues or cue configurations for problem types III to V than for the type II. While these extra cues are in separate, semi-diagnostic 'spaces', there is nothing in the model that means that these cues acquire associative strength at any lower a rate than those in a fully diagnostic space.

Nosofsky et al. [16] introduced a form of selective attention algorithm to the model, using a variant of the dimensionalized, adaptive learning rate, or DALR, model [4]. While showing slight improvements, the augmented model failed to show the significant advantage of the type II over the type III tasks. It was suggested [16] that the reason for this was that the system still gave too much weight to the individual cues in the partially diagnostic two-dimensional spaces.

2. A Channel Capacities Approach

If the dimensional attention is implemented in a way which more accurately reflects the ongoing differences in diagnosticity between spaces, the fit to the data may be improved. The concept of channel capacity provides a ready index of the diagnosticity of a set of orthogonal inputs (such as those in the seven sub-networks within the configural cue model).

In this approach each of the spatial sub-networks would be regarded as an *encoder*. It is possible to describe the rate at which capacity accumulates for the encoder in terms of the capacity of a channel, with inputs consisting of the cues or cue configurations occurrent within each encoder and outputs in the form of the category labels. These indices of correlation will describe both the maximum capacity of such a channel and also be a measure of the rate at which discrepancy based error signals may have an effect on the capacity of the channel. Combining the maximum capacities for each channel, for each category structure, should provide an index of the relative difficulty of each task for this particular architecture. The higher the combined capacity, the easier the task will be to learn.

The capacity of such a channel c, M_c, consists of the uncertainty regarding the category label, H(label), minus the *conditional* uncertainty regarding the label given the input (for a channel c), H_c(label|input). For the Shepard *et al.* [18] task where there are two equiprobable category labels A and B the label uncertainty H(label) is given by

$$H(label) = -\left(\left(p(A)\log_2 p(A)\right) + \left(p(B)\log_2 p(B)\right)\right) = 1 \qquad (1)$$

The conditional uncertainty for a given channel, H_c(label|input) is evaluated as follows;

$$H_c(label|input) = \sum_{input}\sum_{label} -p(input)p(label|input)\log_2 p(label|input) \qquad (2)$$

Table 1 gives the capacities of each of the seven channels for each of the six category structures used. The table shows the simple summed capacities and the summed squared capacities in the bottom two rows. As can be seen the summed capacities show an ordering which suggests that the type II structure will be harder to learn than the types III to V. Squaring the capacities prior to summing reveals figures which qualitatively fit the order of difficulty observed in the experiments.

This squaring may be analogous to the gating of encoder output via a weight whose value reflects the average capacity of the encoder.

		category structure				
channel	I	II	III	IV	V	VI
X	1	0	0.18872	0.18872	0.18872	0
Y	0	0	0	0.18872	0	0
Z	0	0	0.18872	0.18872	0	0
XY	1	0	0.5	0.5	0.5	0
XZ	1	1	0.5	0.5	0.5	0
YZ	0	0	0.5	0.5	0	0
XYZ	1	1	1	1	1	1
ΣM_c	4	2	2.87744	3.06616	2.18872	1
ΣM_c^2	4	2	1.82123	1.85685	1.53562	1

Table 1: Maximum capacities (M_c) for each of the seven spatial sub-networks or encoders for the six category structures used by Shepard et al. [18].

2.1 A Simple Learning Model

The model assumes that these channels begin with zero capacity and then approach their maximum at a rate proportional to that maximum. The variables being altered are the conditional probabilities p(outputs label A| input is a member of category A), known as p(correct), and p(outputs label B|input is a member of category A), known as p(incorrect). This involves a further simplification in that each channel is conceptualised as a binary symmetrical channel such that p(incorrect) = 1-p(correct). Representation of the B channel is not necessary, as development of this channel will be symmetrical and subject to identical constraints.

The probabilities begin at 0.5 with p(correct) being modified, each iteration, upwards, where $M_c>0$, according to the update rule shown below,

$$\Delta p(correct) = \left(1 - \sum_c R_c^2\right)(M_c - R_c)\frac{1}{n_c}k \quad (3)$$

where n_c is the number of sources for channel c (2, 4, or 8) and k is constant (0.1 used). R_c is the ongoing capacity of the channel, evaluated as follows.

$$R_c = 1 + \left((p(correct)\log_2 p(correct)) + (p(incorrect)\log_2 p(incorrect))\right) \quad (4)$$

The learning rule does not relate directly to the conditional probabilities of the individual sources. It is assumed that these will move towards whatever maxima are appropriate to them given M_c. In any event, the capacity of the whole channel cannot exceed M_c and it should move towards it at a rate basically proportional to M_c and $1/n_c$. The p(correct) approached is the p(correct) for a channel with two equiprobable inputs and a maximum capacity M_c.

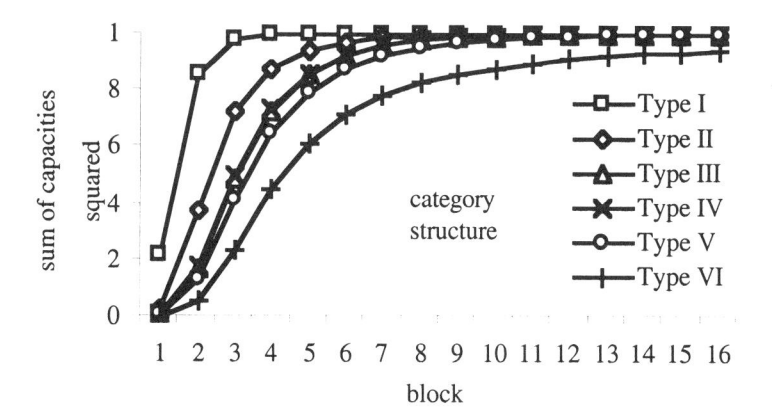

Figure 3: Mean sum of R_c^2 for the simple channel capacities learning model across the 16 trials of each block, for the six category types.

Each iteration is used to represent a single, average, trial. The results are presented in figure 3 with the values graphed being the mean value of the sum of R_c^2 across the 16 trials of each block. As can be seen the qualitative trends shown in the human data in figure 2 are reasonably well captured using this model.

3. A Connectionist Implementation

The facts that the sum of M_c^2 provides a qualitative fit to the observed difficulty, and the ongoing sum of R_c^2 enables a qualitative fit to the learning curves, (where sums of M_c and R_c do not), has a particular bearing on the design suggested. The two layer networks that make up each spatial channel are bound to acquire capacity at a rate proportional to $M_c \times 1/n_c$ towards a maximum M_c. The channel capacity model implies that acquisition of capacity and feedforward output is gated by a value that is, itself, a function of the capacity of the sub-network.

The simplest way to implement this is to locate a 'bottleneck' node or weight between each sub-network and the point at which the output is delivered. This is a similar approach taken to that used in modular 'mixtures of experts' networks [6, 5, 1]. The difference here is that there is no requirement for a normalised deterministic or stochastic gating mechanism for module output and weight increments.

If the weight in this bottleneck or intermediate node in some way tracks the capacity of the sub-network, then using it to gate feedforward and feedback error signals should allow some approximation of the behaviour of the channel capacity model. The approach taken was to develop a three-layer configural cue network, part of which is shown in figure 4.

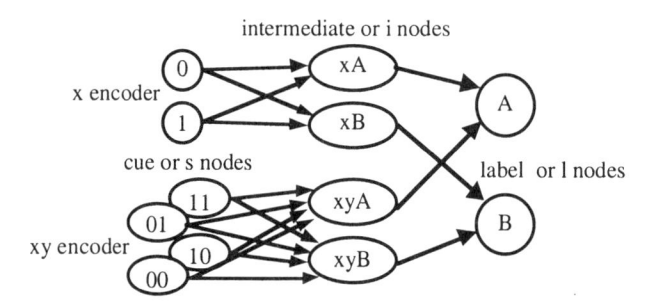

Figure 4: A section of the three-layer configural cue model. This shows only a one and two-dimensional encoder or sub-network and their associated bottleneck or intermediate nodes, all connections shown have modifiable weights.

3.1 Activation Propagation

Output from a cue node s to an intermediate node i, o_{si}, is the product of the s nodes activation and the association weight on the connection between the two, w_{si}.

$$o_{si} = a_s w_{si} \qquad (5)$$

The activation, a_s is 1 if the cue is present and zero if absent, $-1 < w_{si} < 1$. Activation of the an intermediate node i is the sum of the outputs from its s nodes as follows;

$$a_i = \sum_s o_{si} \qquad (6)$$

These nodes output to one label or l node, o_{il}, via their weights θ_i. This weight may be positive or negative and any size dependent on previous weight updates,

$$o_{il} = a_i \theta_i \quad (7)$$

The figure used to represent the probability of the network producing a particular output, in this case p(A) is based on the Luce choice function [11];

$$p(A) = \frac{e^{\sum_i o_{iA}}}{e^{\sum_i o_{iA}} + e^{\sum_i o_{iB}}} \quad (8)$$

3.2 Weight Updates

The weights between s nodes and their i node are initialised at zero and updated after each trial. The change is added to the weight before the next trial.

$$\Delta w_{si} = ((a_l \Theta_i) - o_{si}) a_s |\theta_i| (1 - w_{si}) k_w \quad (9)$$

Here Θ_i is the sign of θ_i, either 1 or -1. The parameter k_w is a constant set to 0.35 for the results produced here.

The θ weights are meant to track the mean product of the activation, a_i, of the node and the target or category label activation a_l. The label activation is -1 if absent and +1 if the label is present. The update rule, applied at the end of each trial, is as follows;

$$\Delta \theta_i = ((a_l a_i) - \tanh(\theta_i)) |\theta_i| k_\theta \quad (10)$$

The use of the hyperbolic tangent provides an advantage for fully diagnostic channels with values on these weights generally exceeding 1 as learning progresses. The parameter k_θ is a constant set, for these results, to 0.2. These weights were initialised at 0.5. Lower initial weights severely slow early learning due to the dependence of the update rules on the value of θ.

3.3 The Simulation

Object-label pairings were presented to the network in 16 blocks of 16 trials. As in Shepard et al. [18], each of the eight objects are presented twice in a random order. The first block is divided into 2 sub-blocks of eight trials during which each object is presented once. This was repeated for all six category structures, ten times each with input orders re-randomised on each run.

4. Results and Discussion

Figure 6 shows the performance of the model on the six category structures. The values plotted were produced by averaging the mean p(correct) per block (of 16 trials) for 10 runs through the experiment. This was repeated for each problem type to produce the six curves shown.

The results show a reasonably good qualitative fit in terms of the order of difficulty to the human data shown to the far left. There is a clear advantage for the type II task over the types III to V indicating that the θ weights in the extra layer of nodes makes the difference for the configural cue organisation. An examination of these weights indicates that they are restricted for semi-diagnostic channels to a value that approximates M_c. For this model this isn't actually M_c but a value reflecting the mean product of input to the i node and the target signal. This works out at approximately 0.5 for the 2d semi-diagnostic channels and 0.25 for the 1d semi-diagnostic channels. The θ weights for non-diagnostic channels head towards zero. This not surprising as a closer examination of equation 10 suggests that the same thing may be implemented using a delta rule with the sub-network output as target and the θ weight and label activation used for output and activation parameters. The label is learning the encoder output at the same time as the encoder is learning the label.

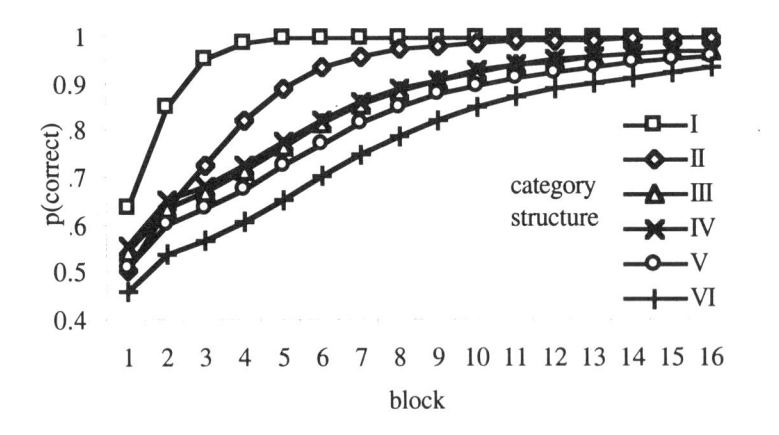

Figure 5: Mean p(correct) per block for the model on each of the six category structures.

The fact that the superiority of the type II does not emerge until about block three is being addressed by ongoing research. This is a problem with initialising the θ weights at 0.5. When the capacity of a sub-network is zero the θ weight falls towards zero relatively slowly.

This is a problem for the type II structure because it has a large proportion of non-diagnostic channels. The w weights in these channels will continue to track the target and thus produce, effectively, noise output. Eliminating this noise relates to theories of learned inattention (e.g. [12,10]). This basic model makes no commitment to mechanisms that might facilitate learned inattention or blocking of conditioning.

It may be noted that there are no global error signals in the learning equations for the connectionist model. The model does not seem to need any form of capacity limitation (either in terms of error signals or dimensional attention) but the state of weights at asymptote suggest that it may be a psychologically implausible solution. For the type I task, for example, high weights will accrue to the fully diagnostic (see table 1) two and three-dimensional channels. Again, this relates to learned inattention, this time due to the different reinforcement schedules which apply to cues in higher dimensionality encoders.

In addition the model does not have any mechanism by which attention (or inattention) to *dimensions* may perseverate between tasks, as seen in dimensional relevance shifts (e.g. [9]). For this model high capacity for the (say) xy encoder confers no advantage to the x and y encoders relative to the z encoder. Extra processes and weights which use relative encoder effectiveness to allocate attention to the dimensions upon which they are dependent may mitigate this problem. The implementation of such processes would have to take into account all of the various circumstances under which blocking or learned inattention may occur.

References

1. Erickson, M. A., & Kruschke, J. K. (1998). Rules and exemplars in category learning. *Journal of Experimental Psychology: General , 127,* 107-140.
2. Estes, W. K. (1994). Classification and Cognition. New York: Oxford University Press.
3. Gluck, M. A., & Bower, G. H. (1988). Evaluating an adaptive network model of human learning. *Journal of Memory and Language, 27,* 166-195.
4. Gluck, M. A., Glauthier, P. T., & Sutton, R. S. (1992). Adaptation of cue-specific learning rates in network models of human category learning. *Proceedings of the Fourteenth Annual Meeting of the Cognitive Science Society.* (pp 540-545) Hillsdale, NJ: Erlbaum.
5. Jacobs, R. A. (1997). Nature, nurture, and the development of functional specializations: A computational approach. *Psychonomic Bulletin and Review, 4,* 299-309.
6. Jacobs, R. A., Jordan, M. I., Nowlan, S. J., & Hinton, G. E. (1991). Adaptive mixtures of local experts. *Neural Computation, 3,* 79-87.
7. Kruschke, J. K. (1992). ALCOVE: An exemplar-based connectionist model of category learning. *Psychological Review, 99,* 22-44.
8. Kruschke, J. K. (1993). Human category learning: implications for backpropagation models. *Connection Science, 5,* 3-36.
9. Kruschke, J. K. (1996). Dimensional relevance shifts in category learning. *Connection Science, 8,* 201-223.
10. Kruschke, J. K., & Blair, N. J. (In Press). Blocking and backward blocking involve learned inattention. *Psychonomic Bulletin & Review.*
11. Luce, R. D. (1963). Detection and recognition. In R. D. Luce, R. R. Bush, & E. Galanter (Eds.), *Handbook of mathematical psychology.* (pp. 103-189). New York: Wiley.
12. Mackintosh, N. J. (1975). A Theory of attention: Variations in the associability of stimuli with reinforcement. *Psychological Review, 82,* 276-298.
13. Medin, D. L., & Schaffer, M. M. (1978). Context theory of classification learning. *Psychological Review, 85,* 207-238.

14. Nosofsky, R. M. (1984). Choice, similarity, and the context theory of classification. *Journal of Experimental Psychology: Learning, Memory and Cognition, 10,* 104-114.
15. Nosofsky, R. M. (1986). Attention, similarity, and the identification-categorization relationship. *Journal of Experimental Psychology: General, 115,* 39-57.
16. Nosofsky, R. M., Gluck, M. A., Palmeri, T. J., McKinley, S. C., & Glauthier, P. (1994). Comparing models of rule-based classification learning: A replication and extension of Shepard, Hovland, and Jenkins (1961). *Memory and Cognition, 22,* 352-369.
17. Nosofsky, R. M., & Palmeri, T. J. (1996). Learning to classify integral-dimension stimuli. *Psychonomic Bulletin and Review, 3,* 222-226.
18. Shepard, R. N., Hovland, C. I., & Jenkins, H. M. (1961). Learning and memorization of classifications. *Psychological Monographs, 75,* 13.

A Revival of Turing's Forgotten Connectionist Ideas: Exploring Unorganized Machines

Christof Teuscher and Eduardo Sanchez

Abstract

Turing had already investigated connectionist networks at the end of the forties and was probably the first person to consider building machines out of very simple, neuron-like elements connected together in a mostly random manner. The present paper aims to revive and shed more light on Turing's ideas. Turing's unorganized machines are analyzed and new types of machines are proposed by the authors. An example of a pattern classification task is presented using a Turing net and genetic algorithms for building and training the networks.

1. Introduction

In a little-known paper entitled "Intelligent Machinery" [7,15], Turing had already investigated connectionist networks at the end of the forties. His employer at the National Physical Laboratory in London, Sir Charles Darwin, dismissed the manuscript as a "schoolboy essay". Turing never had great interest in publicizing his ideas, so the paper went unpublished until 1968, 14 years after his death. Copeland and Proudfoot, directors of the *Turing Project*[1], a large ongoing project focused on Turing's life-work at the University of Canterbury, revived Turing's connectionist ideas in a Scientific American [5] publication.

In describing networks of artificial neurons connected in a random manner, Turing had written the first manifesto of the field of artificial intelligence. Many of the concepts that became later important to AI were introduced in this work. One of the questions he was always interested in was whether it is possible for machinery to show intelligent behavior or not. There is no doubt that Turing's work on neural networks goes far beyond the earlier work of McCulloch and Pitts [9]. However, it is interesting that Turing makes no reference in the 1948 report to their work. Turing's idea of using supervised interference to train an initially random arrangement of units to compute a specified function was new and revolutionary at that time and it is in fact not astonishing that his paper has been first dismissed as a "schoolboy essay".

Random boolean networks were seriously investigated many years later by Weisbuch [17] and Kauffman [8], who had probably never heard of Turing's networks that are in fact a subset of random boolean networks. Igor Aleksander is probably one of the few researchers that consider weightless neural architectures [1] as a serious

---
[1] http://www.alanturing.net

alternative to modern-day connectionist architectures. Other interesting work related to Turing's connectionism and random connected networks came from Allanson in 1956 [2], from Smith and Davidson in 1962 [13], from Rozonoér [11] in 1968, and in 1971 from Amari [3].

The present paper aims to revive and shed more light on Turing's ideas, to study, to simulate, and to analyze them. To the best of our knowledge, nobody, with the exception of the Turing project team and its former member Craig Webster[2], has ever analyzed in detail, simulated, implemented, and applied Turing's neural machines to pattern classification and other problems. Turing was not really interested in practical applications of his models. In a letter to Ashby, Turing remarked: "[...] I am more interested in the possibility of producing models of the action of the brain than in the practical applications to computing".

Our research about Turing nets goes far beyond Turing's proposals. Turing's unorganized machines, namely A-type and B-type networks, are analyzed and described in detail in section 2.. This section is a historical summary and provides the background for the rest of the paper. All the machines in section 2. were proposed by Turing. Section 3. introduces new machines, namely TB-type networks, proposed by the authors. Section 4. presents an example of a pattern classification problem on TB-type networks, and section 5. concludes the paper.

2. Unorganized Machines: Turing's Anticipation on Neural Networks

"Someday, perhaps soon, we will build a machine that will be able to perform the functions of a human mind, a thinking machine" [6]. This is the first sentence in Hillis' book on the *Connection Machine*, a legendary computing machine that provided a large number of tiny processors and memory cells connected by a programmable communications network. Much earlier in the 20th century, Alan M. Turing described different types of machines—some well-known today, others almost forgotten. Turing's vision was probably very comparable to Hillis'. However, he only had pencil and paper at his disposal to verify the functioning of his machines. Throughout his entire life, Turing was interested in creating intelligent machines. One of the best-known, yet abstract, machines invented by Turing is certainly the classical Turing machine. Turing's lifelong interest in thinking machines and his concern with modelling the human mind using machines were probably at the origin of his rather little known paper on mechanical intelligence, a report for the National Physical Laboratory [15]. In the remainder of this section, we shall present the neuron-like machines Turing described in this paper.

2.1 Unorganized A-type and B-type Machines

The term *unorganized machines* was defined by Turing in a rather inaccurate and informal way. Turing's *unorganized machine* is a machine that is built up "[...] in

---
[2] http://home.clear.net.nz/pages/cw/

a comparatively unsystematic and random way from some kind of standard components" [15, p. 9]. Turing admitted that the same machine might be regarded by one man as organized and by another as unorganized. He gave an example of a typical unorganized machine:

> "The machine is made up from a rather large number N of similar units. Each unit has two input terminals, and has an output terminal which can be connected to input terminals of (0 or more) other units. We may imagine that for each integer $r, 1 \leq r \leq N$, two numbers $i(r)$ and $j(r)$ are chosen at random from $1...N$ and that we connect the inputs of unit r to the outputs of units $i(r)$ and $j(r)$. All of the units are connected to a central synchronizing unit from which synchronizing pulses are emitted at more or less equal intervals of time. The times when these pulses arrive will be called "moments". Each unit is capable of having two states at each moment. These states may be called 0 and 1".

The complete system is thus updated synchronously, in discrete time steps. The state of each unit is determined by the previous value of the two inputs and the transmission of information among neurons requires a unit time delay (a "moment"). In terms of a digital system, the basic unit which Turing describes can be straightforwardly defined as an edge-triggered D flip-flop preceded by a two-input NAND gate (Figure 1a).

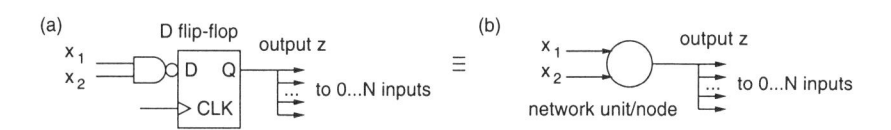

Figure 1: A basic unit regarded as a digital system: (a) D flip-flop preceded by a two-input NAND gate ([14]); (b) Symbolized network unit/node of an unorganized machine. Only this notation was used by Turing [15].

A positive-edge-triggered D flip-flop samples its D input and changes its Q output only at the rising edge of the controlling clock (CLK) signal [16]. In our case, the central synchronizing unit that emits the pulses is the global clock generator of the digital system.

In more modern terms, the unorganized machine Turing has proposed is a boolean neural network. However, modern networks are often organized in layers. For unorganized machines, the layered structure does not really exist since recurrent connections are allowed with no constraints. Furthermore, unorganized machines are normally constructed randomly and thus freely interconnected. The machines Turing proposed are in fact a subset of the random boolean networks described and investigated in detail by Stuart Kauffman [8].

Let us start with the simplest unorganized machine proposed by Turing. An A-*type unorganized machine* is a machine built up from units and connections as described in the example above. Figure 2 shows a five-unit A-type unorganized machine and a possible state sequence. Turing did not give any indication of how the inputs and outputs of his unorganized machines might interface with a given environment.

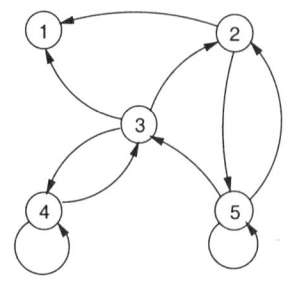

node	t	t+1	t+2	t+3	t+4	t+5
1	1	1	0	0	1	0
2	1	1	1	0	1	0
3	0	1	1	1	1	1
4	0	1	0	1	0	1
5	1	0	1	0	1	0

Figure 2: Example of an A-type unorganized machine built up from 5 units and a possible state sequence.

Turing suggested that an A-type unorganized machine is the simplest model of a nervous system with a random arrangement of neurons. Unfortunately, this statement came without any justification or proof.

A second type of unorganized machine described in his 1948 paper is called B-*type unorganized machine*. A B-type machine is an A-type machine where each connection between two nodes has been replaced by a small A-type machine called an *introverted pair*, as shown in Figure 3(a). This small A-type machine functions as a sort of switch (modifier) that is either opened or closed. In the next section, we will see how to operate this switch with an external stimulus. For now, this is not possible and the state of the switch (open or closed) only depends on the initial internal state of the small A-type machine that constitutes the interconnection. Turing simply wanted to create a switch based on the same basic element as the rest of the machine. He could have taken another element, i.e., a multiplexer, but he really wanted to construct the entire networks using the same basic element.

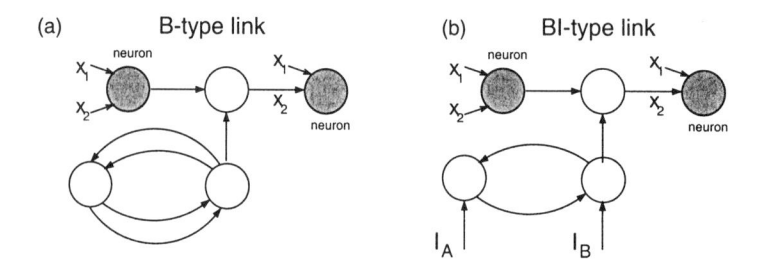

Figure 3: Turing's B-type links. Each B-type link is a small 3-node A-type machine also called an *introverted pair*. Part (b) shows a link with two interfering inputs.

The A-type machine that forms the B-type link can be in three different states of operation:

1. it may invert all signals (closed connection),
2. it may convert all signals into 1 (opened connection), or
3. it may act as in (1) and (2) alternately.

It can easily be seen that each B-type machine might be regarded as an A-type machine. The inverse statement does not hold since it is extremely unlikely that a B-type machine could be obtained by randomly assembling an A-type machine from a given number of units.

2.2 Turing's "Education": Organizing Unorganized Machines

So far, the unorganized machines we have described are, once initialized, no longer modifiable. It would clearly be interesting to modify the machine's interconnection switches (synapses) at runtime. Such a possibility would allow the use of an online learning algorithm that modifies the network's interconnections in order to train a net. Learning is undoubtedly the most difficult problem related to neural networks and much discussion is focused on it. Turing wrote: "It would be quite unfair to expect a machine straight from the factory to compete on equal terms with a university graduate" [15]. Thus, a machine, initially unorganized, must be organized: "Then by applying appropriate interference, mimicking education, we should hope to modify the machine until it could be relied on to produce definite reactions to certain commands" [15, p. 14]. This key idea is fascinating and can be found again in classical neural network learning algorithms. Turing went even further and wrote:

> "[...] with suitable initial conditions they [i.e., B-type machines] will do any required job, given sufficient time and provided the number of units is sufficient. In particular with a B-type unorganized machine with sufficient units one can find initial conditions which will make it into a universal machine with a given storage capacity" [15, p. 15].

However, Turing did not give a formal proof of this hypothesis because "[...] it lies rather too far outside the main argument". For a number of modern connectionist architectures such proofs have been given (see for example J. Pollack [10] and H. Siegelmann [12]).

In section 2.3 we will see that Turing was wrong and that it is impossible to compute every function with a B-type unorganized machine. Nevertheless, let us consider Turing's way of modifying a B-type interconnection. He simply proposed to replace a B-type link by a connection as shown in Figure 3(b). Each interconnection has two additional *interfering inputs* I_A and I_B, i.e., inputs that affect the internal state of the link. By supplying appropriate signals at I_A and I_B we can set the interconnection into state 1 or 2 (as explained in section 2.1). We will call this type of link a BI-type link (the I stands for "interference"). By means of this type of link, an external or internal agent can organize an initially random BI-type machine by disabling and enabling connections (switches). Neural networks usually have weighted connections with weights having real values. In contrast, there is nothing smooth in switching introverted pairs of a BI-type network: the connection is either enabled or disabled—a savage all-or-nothing shift.

2.3 The B-type Pitfall

It is easy to see that the principal computing element of an A-type network is the NAND operation since each node contains one 2-input NAND gate. Other logical functions thus have to be realized using NAND functions—which, as is a generally well known fact, is always possible since NAND units form a logical basis [16]. It thus directly follows that every logical function can be computed by an A-type network. The implementation of logical functions with lots of variables, however, can be very complex and tricky: the greatest difficulty lies in respecting the timing constraints due to the inherent delay of the D flip-flops.

As with A-type networks, the basic function of a B-type node is still a NAND operation. However, the two input signals of a node now pass by a B-type link that either inverts the signal (state 1) or always holds it to a 1 (state 2). One can see that a B-type node together with its associated input links is nothing more that a simple OR gate (when abstracting from the inherent delay)!

Since simple OR gates do not form a logical basis, it is now obvious that not all boolean functions can be computed with a B-type machine as defined by Turing. To understand the difficulty, one may attempt to design a B-type network that computes the XOR function or the NOT function. Turing's hypothesis about B-type machines stating that "[...] with suitable initial conditions they [i.e., B-type machines] will do any required job, given sufficient time and provided the number of units is sufficient" [15, p. 15] is wrong.

3. New Types of Unorganized Machines

In general, it is desirable to work with networks that offer universal computability, i.e., that are based on elements that form a logical basis. In this section, the authors present new types of machines that offer universal computability.

In order to obtain a computationally universal network, Copeland and Proudfoot [4] propose a remedy in form of the following B-type link modification: "[...] every unit-to-unit connection within an A-type machine is replaced by two of the devices [i.e., the introverted pairs, Figure 3(a)] [...] linked in series." Having two introverted pairs in each node-to-node connection adds useless nodes and connections in a network. In order to reduce the number of elements in the network, we propose to use another type of link—the TB-type link shown in Figure 4(a)—which is functionally equivalent to Copeland's and Proudfoot's proposal, but simpler. This simplification will be very useful for hardware implementations. The additional node simply inverts the input signal. One can see that a simple node together with its two associated TB-type links now forms a NAND operator. Since a NAND operator forms a logical basis, it is possible to realize any logical function with TB-type networks.

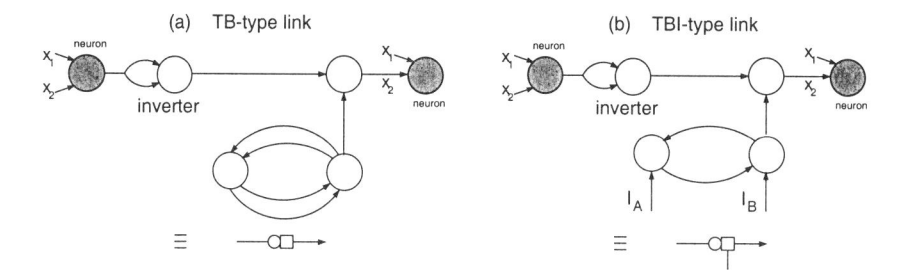

Figure 4: The TB-type (a) and TBI-type (b) link and notation proposed by Teuscher [14]. A simple node is used as an inverter in series with a normal introverted pair.

So far, we have seen that with a simple B-type or TB-type machine, the internal state of the links could not be modified due to the lack of interfering inputs. However, in section 2.2 we presented the BI-type machine that allows the interconnection switches to be modified by an external stimulus. It is easy to add this possibility also for TB-type links. Figure 4(b) shows how two interfering inputs may be added to the introverted pair of the TB-type link. The resulting link is called TBI-*type link*.

4. Pattern Classification using Turing Neural Networks and Genetic Algorithms

In order to simulate and test Turing neural networks, we implemented an object-oriented experimental framework [14]—called TNNSIM— that allows a detailed simulation of all Turing neural networks. In this section, we will present an experiment using genetic algorithms we performed using the TNNSIM simulator. Note that it was not the goal of this rather simple experiment to enter into competition with today's well-known and certainly better-performing pattern classification techniques.

The experiment consists in classifying into two classes (X and O) the twenty 16×16 dot patterns of Figure 5 without doing any explicit pre-processing. In order to determine the fitness of a network (individual of a GA population), all patterns are presented to the network. The network is set to its initial state (reset of all DFFs) before presenting a new pattern. The fitness of a network is defined as the number of correctly classified patterns. A steady-state genetic algorithm with a strong network encoding scheme has been used for this experience.

In this context, we call *learning* or *training* the process of modifying the interconnection switches since this process is very similar to the modification of synaptic weights in classical neural networks. We call *evolution* the process of building a suitable interconnection topology. During this phase, the switches are, in general, all closed.

A combination of topology evolution (with all connections enabled) and learning (connection modification only) was successful with a TBI-type network. A suitable interconnection topology was evolved during the first 200 generations. The topology was fixed, and only the switches were modified (enabled or disabled) during the learning phase. The network was built from 100 nodes. Figure 6 shows the fitness graph obtained with a population of 50 individuals (networks). This experiment showed that the initial interconnection topology is as least as important as the later learning phase. Learning is generally not successful if the underlying network structure is badly chosen.

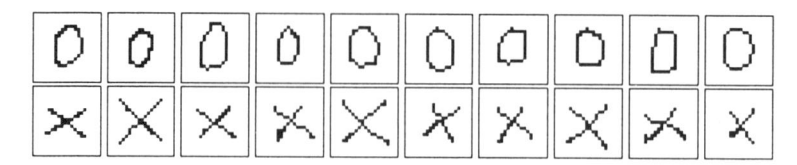

Figure 5: 16 * 16 dot patterns.

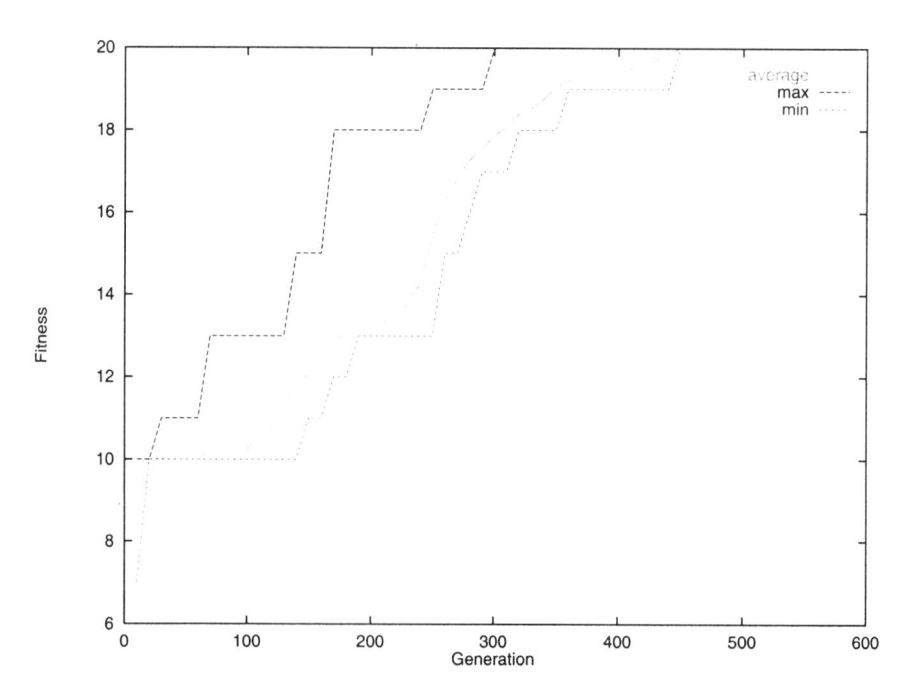

Figure 6: 16*16 dot pattern classification fitness with a TBI-type network. A network topology where each connection was enabled was first evolved. In a second step, only the connections were enabled or disabled. The population size was set to 50 individuals (networks).

5. Conclusion

Turing's idea of neural-like structures was new and promising in the forties. A universal and very simple basic element (neuron with associated input connections) in A-type and TB/TBI-type nets guarantees that, at least from a theoretical point of view, any function or system can be implemented using Turing's networks. The philosophy followed by Turing was to use the same type of node everywhere in the network, the reason why the modifiable interconnections are also built up from the same basic nodes.

We presented one example of a pattern classification experiment out of a series of experiments we performed with different characteristics and different levels of difficulty. It turned out that only the simpler pattern classification problems could be correctly solved by evolved Turing networks in a reasonable amount of time. However, due to the vast search spaces, it is often just a question of time to find a solution. We have shown that Turing's networks, like random boolean networks, "[...] inevitably classify and have internal models of their worlds" [8].

Turing's networks are probably more of theoretical than practical interest. Kauffman wrote: "[...] the classes formed by arbitrary Boolean networks are not necessarily natural, useful, or evolvable"[8]. However, randomly generated networks are very interesting to study the behavior of dynamic and complex systems, as Kauffman demonstrated impressively by his research.

Are Turing's nets of any interest for understanding biological systems? Yes, with limits. The neurons are far too simple to be compared with biological neurons. Turing himself, however, suggested that an A-type unorganized machine might be the simplest model of a nervous system with a random arrangement of neurons [15, p. 10]. The concept of using a large amount of identical and highly interconnected elements is powerful and could be the source of some insights on how biological systems might work and self-organize. Kauffman wrote [8]: "Most biological systems confront us with a vast number of connected elements. Typically we are ignorant of the details of the structure and logic by which the elements of such systems are coupled."

Turing might have asked if the networks give rise to higher-level cognitive functions, e.g., planning, abstraction, etc., since he was very interested in machines that think and demonstrate intelligent behavior. We have not focussed on that question so far.

Acknowledgements

The authors are grateful to Gianluca Tempesti for his careful reading of this work and his helpful comments.

References

1. Aleksander, I., & Morton, H. (1995). *An Introduction to Neural Computing*. I Boston, MA: International Thomson Computer Press,
2. Allanson, J. T. (1956). Some properties of randomly connected neural nets. In C. Cherry, (Ed.), *Proceedings of the 3rd London Symposium on Information Theory*, pp. 303–313, London: Butterworths.
3. Amari, S.-I. (1971). Characteristics of Randomly Connected Threshold-Element Networks and Network Systems. *Proceedings of the IEEE, 59,* 35–47.
4. Copeland, B. J., & Proudfoot, D. (1996). On Alan Turing's anticipation of connectionsim. *Synthese: An International Journal for Epistemology, Methodology and Philosophy of Science, 108,* 361–377.
5. Copeland, B. J., and Proudfoot, D. (1999). Alan Turing's Forgotten Ideas in Computer Science. *Scientific American*, April, 77–81.
6. Hillis, D. W. (1985). *The Connection Machine*. Cambridge, MA: MIT Press.
7. Ince, D. C., (1992). Mechanical Intelligence: Collected Works of A. M. Turing, North-Holland.
8. Kauffman, S. A. (1993). *The Origins of Order: Self–Organization and Selection in Evolution*. Oxford: Oxford University Press.
9. McCulloch, W. S., & Pitts, W. H. (1943). A logical calculus of the ideas immanent in neural nets. *Bulletin of Mathematical Biophysics, 5,* 115–133.
10. Pollack, J. B. (1987). *On Connectionist Models of Neural Language Process-ing*. Unpublished PhD thesis, Computer Science Department, University of Illinois, Urbana.
11. Rozonoer, L. I. (1969). Random logical nets I. *Automation and Remote Control, 5,* 773–781.
12. Siegelmann, H. T., & Sontag, E. D. (1992). On the Computational Power of Neural Nets. In *Proceedings of the 5th Annual ACM Workshop on Compu-tational Learning Theory*, pp. 440–449.
13. Smith, D. R., & Davidson, C. H. (1962). Maintened activity in neural nets. *JACM, 9.*
14. Teuscher, C. (2000). Study, Implementation, and Evolution of the Artificial Neural Networks Proposed by Alan M. Turing. A Revival of his "Schoolboy" Ideas. Master's thesis, Swiss Federal Institute of Technology Lausanne, Logic Systems Laboratory, EPFL-DI-LSL, CH-1015 Lausanne. http://lslwww.epfl.ch/turing_nets.
15. Turing, A. M. (1969). Intelligent Machinery. In B. Meltzer, & D. Michie, (Eds.) *Machine Intelligence*, volume 5 of *National Physical Laboratory Report*, pp 3–23. Edinburgh: Edinburgh University Press.
16. Wakerly, J. F. (2000). *Digital Design: Principles & Practices*. Prentice Hall International Inc.
17. Weisbuch, G. (1991). *Complex Systems Dynamics: An Introduction to Au-tomata Networks*, Redwood City, CA.: Addison-Wesley,

Visual Crowding and Category-Specific Deficits: a Neural Network Model.

John Done, Tim M. Gale, Ray J. Frank

Abstract

Various experimental approaches, using static 2D canonical views of living and non-living entities indicate that knowledge representations of these categories are distinct. In a series of experiments a Kohonen self organizing feature map was trained to recognise 2D digitised images. As a result, images of animals and musical instruments were represented within a shared set of processing units, which suggests that they are visually crowded categories, unlike clothing and furniture. These results are in keeping with those from other experimental approaches. Thus, it would appear that the simple interplay between a SOM and 2D images provides a valuable model of one route in visual object recognition.

1. Introduction

A number of different approaches have been adopted to understand how humans, and other primates, are able to recognise individual objects and classes of objects when presented visually. These approaches include neuropsychological studies of patients with object recognition problems - i.e. visual agnosias [1], studies of categorisation in infants [2,3], studies of object recognition and categorisation in adults [4] and electrophysiological studies in primates [5,6]. Together these provide a multidimensional approach to understanding the brain systems responsible for visual object recognition (VOR).

In the neuropsychological literature, patients with visual agnosias can exhibit category-specific recognition deficits (CSRDs). In such cases particular classes of pictured object, usually living things, are more prone to misidentification after brain damage than other classes, usually man-made objects, although the reverse dissociation has also been documented [1]. Such impairments have been attributed to different semantic representations for living and non-living entities. The sensory functional theory (SFT) regards knowledge of living entities to be predominantly sensory as opposed to functional with the reverse being true for non-living entities [7,1]. In support of this theory, living things have a particularly high level of correlation between their body parts [8]. Discrimination learning in monkeys produces a remarkably similar pattern of errors to visual agnosic patients afflicted by CSRD for living things [9].

Thus monkeys [9], infants [3], and adult humans with CSRD may well share similar processing limitations in the ventral visual pathway [6], which runs, in humans, from the primary visual cortex (V1-V2-V4), through the posterior inferotemporal cortex (TEO) and then the anterior inferotemporal cortex (TE) [5]. TE cells respond to either: a) configurations of features, rather than specific features [5] or, b) extract principal components where there are correlations between features [10]. Damage to the temporal lobes is implicated in many, although not all, cases of CSRD for living things [11].

In this paper we take an approach similar that of Schyns [12] over a series of experiments. However, a critical difference is that we present our model with real images depicting objects from a range of taxonomies. There is nothing 'semantic' about these pictures (i.e. there are no cues other than the topography of visual features to provide categorical information).

2. Method

In this section, we describe those methodologies that are common to all experiments detailed in this paper. Firstly, we describe the way in which our training data was assembled and, secondly, we discuss the modular architecture that is used to realise our model of VOR

2.1 Training Patterns: Image Choice and Preparation

Names (i.e. base level labels) of all objects were taken from commonly used sources [13,14] and colour pictures/high quality coloured drawings then found. To minimise experimenter bias, the task of choosing images was given to someone who was not part of the investigative team and who was also naive to the rationale of the study. The guidelines given to this person were to select either colour photographs or very high quality coloured drawings which depicted items in a *typical* orientation, *preferably* with no foreshortening of the principal axis [15]. Selected images were required to depict different but not atypical examples of each basic level category. 140 images were selected, 35 in each superordinate category of animals, musical instruments, clothing and furniture (7 base level with 5 subordinates per base level). The images were then reduced in size and centred such that each object's maximal dimension fitted exactly within a 50 by 50 pixel grid. These were then scanned into an Apple Macintosh AV microcomputer at a resolution of 100dpi. Two additional control measures were undertaken which resulted in an increased image set size: Firstly, a left-right inversion of each image was created to overcome the fact that some asymmetrical objects (e.g. fish) tended always to be depicted in the same orientation (e.g. with head to the left and tail to the right), whereas others (e.g. chairs) had a more variable orientation. Secondly, for each left-right inversion, a novel image was created by changing the image contrast by ±20%.

2.2 The Modular Artificial Neural Network (MANN)

The full model comprises an unsupervised visual processing module and a supervised categorisation module which are trained independently of each other. The visual processing module is based on Kohonen's self-organising feature map (SOM) [16,17]. Similar to Kohonen's later model [17], classification is distributed across all units in the SOM, providing a contoured internal representation of each image in the training set. Our SOM is characterised by an n-dimensional input vector, that is fully connected to a 2-dimensional map of output units (usually 10 units square). A detailed description of the model can be found elsewhere [18]. In our model the activation value of each SOM unit varies between 0 and 255, allowing us to view the SOM surface as a contour map with regions of high and low activity. The supervised module was included so that we could model the processes of both object recognition and acess to semantic knowledge.

Two different supervised modules, each being a 32 bit output layer, were used. To simulate object recognition, each object was represented by supra threshold activation on 2 units only, one representing the superordinate category and the other the basic level category i.e. localist representation. To simulate distributed, semantic representations, objects were represented by a pattern of activation across all 32 units (16 units representing superordinate features and 16 base level features). Using a standard multi-layered perceptron, a 100 unit input vector (i.e. the 10 by 10 output units of the SOM) fed into 15 hidden units that, in turn, connect to a 32 unit output layer. This output layer has bi-directional connections to a layer of 10 clean-up units.

3. The Experimental Investigations

3.1 Representation of Categories in the SOM

During training of a SOM the weight vector gradually evolves towards a state that approximates a particular input vector. Thus in our model, a weight vector generated image can be depicted as a 50 by 50 pixel greyscale image.

A single SOM of size 10 by 10 units was trained for 1500 epochs with the full training set of 560 greyscale images. The weight vectors for each trained SOM unit were recorded and converted to a weight vector generated image in order to provide a depiction of the type of input pattern for which each SOM unit was maximally responsive.

Mean frequencies of output unit activity for 10 SOMs are plotted in Figure 1 for each superordinate category. Here it can be seen that furniture and clothing images tend to generate feature maps in which the majority of units are only activated to low levels. The winning unit, in these cases, tends to 'win' by a considerable amount and may well, in itself, be diagnostic of the object's category (ie localist coding). Conversely, for animals and musical instruments, the majority of units on the SOM surface are activated to a higher level such that the winning unit exceeds its competitor's activation levels by a much smaller margin (ie sparse coding).

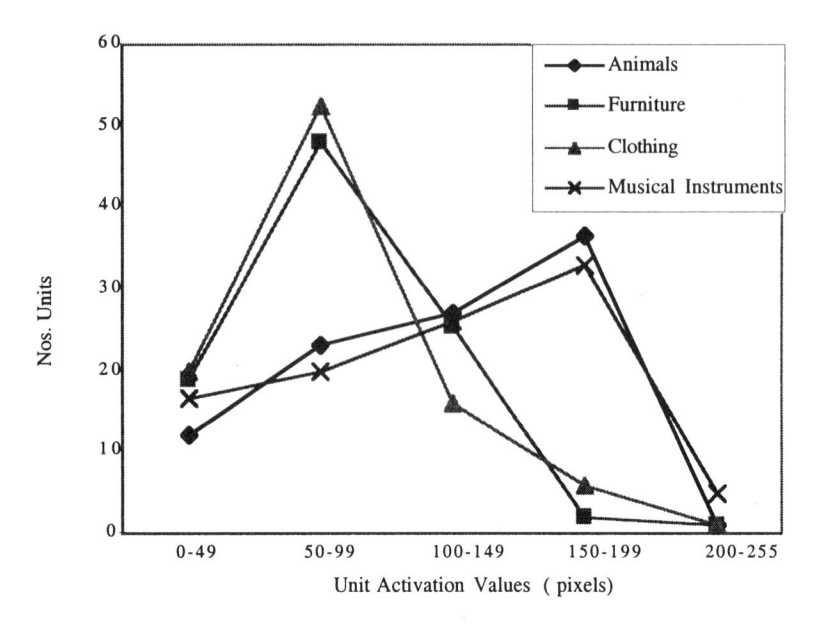

Figure 1. Activation Values for 4 categories in a 100 unit trained SOM.

Using an alternative method, we asked 10 naive subjects to try and identify each of the 100 weight vector images generated by each SOM (1000 images in total). If the subject was unsure or unable to identify the object within 5 seconds then the image was deemed 'unidentifiable'. The results showed that subjects found it easier to recognise furniture and clothing exemplars than animals and musical instruments, $F[4, 49] = 390$, $p < 0.0001$ (post-hoc Scheffe F tests: all pairwise comparisons significant at $p < 0.005$ except animals vs. musical instruments).

Summarising these results, it can be concluded that different classes of object produce different representations in a SOM. The basis for such differences must be inherent within the stimuli, since no explicit categorical information was encoded in the training images. Whilst some images generate contour maps with a sharp peak and a residual surface of relatively inactive units, other images generate contour maps with plateaux of high activity and less tendency for localised activation. The former are identifiable as representations of unique exemplars, whilst the latter tend to be represented by abstractions or 'prototypes'.

3.2 Modelling Visual Object Recognition: Learning

Training the supervised module to recognise the patterns of activation of the SOM , took substantially longer for the sparsely coded categories of animals and musical instruments. For each 20 epoch interval up to 300 epochs, the average RMS error was significantly greater (p<.05) for animals and musical instruments when compared with either furniture or clothing. After 520 epochs there was no statistically significant difference between any categories. The results are consistent with visual discrimination learning in monkeys [9].

3.3 Modelling Visual Object Recognition: Categorisation

Categorisation was operationalised by 2 tests of generalisating to novel images. In one test (Method 1), the supervised module was trained with only 80% of the SOM patterns. Generalisation was tested by analysing the classification performance with the remaining 20%. To generalise the supervised module would need to identify a SOM representation of a clock, not used in the training set, as both a clock and an item of furniture. Category effects in generalisation accuracy were observed at both the superordinate ($F[3, 87] = 20.99$, $p < 0.0001$) and base level ($F[3, 87] = 67.62$, $p < 0.0001$). At the superordinate level, accuracy for recognising furniture items was consistently lower compared with other categories, whilst accuracy for animals exceeded that of all other categories except musical instruments (Bonferroni corrected Scheffe F-tests, $p < 0.008$). At the base level, accuracy for recognising furniture was consistently lower than in all 3 other categories (Bonferroni corrected Scheffe F-tests, $p < 0.008$).

In a separate test (Method 2), a new test set of images comprising 2 examples from each animal and musical instrument category (28 images in total) was obtained from the 1995 Grolier-Dorling-Kindersley CD-ROM encyclopaedia. No images of furniture or clothing were contained in this volume. These new images were very high quality coloured line drawings depicting each object, in a canonical view [15], against a plain white background. Each image was converted to an 8-bit greyscale image, centred within a 50 by 50 pixel grid. These images were then presented to each of 6 SOMs that had already been trained with the original 560 image training set. Classification of the new sample of stimuli was compared with performance for the original stimuli (just animals and musical instruments) . For animals, there was no significant difference between performance on the new and original image samples for either superordinate recognition ($F[1, 47] = 2.74$, $p > 0.1$) or for base level identification ($F[1, 47] = 3.55$, $p > 0.05$). For musical instruments, base level performance was not significantly different between the 2 image samples ($F[1, 47] < 1$), but there was a significant difference at the superordinate level ($F[1, 47] = 4.3$, $p = 0.045$).

3.4 Modelling Visual Object Recognition: Individual Differences

One of the limitations of all other experimental methods of examining the organization of knowledge representations is that they have not explored individual differences, especially differences in processing efficiency or capacity. Modeling offers great scope for doing this. Here we explore the implications for restricting the capacity of the SOM and examine the implications on representation and performance.

Six SOMs were trained according to Method 1 for testing generalistion. These comprised 2 SOMs of size 100 units (10 x 10 - large), 2 of size 49 units (7 x 7 - medium) and 2 of size 25 units (5 x 5 - small). Training parameters were identical for each SOM . The criteria for successful classification were identical to those used in the previous experiments ie RMS error asymptotes at about 0.004 for the training set. The detrimental effect, on recognition accuracy, of reducing the size of SOM was significant for both superordinate and base level recognition for visually crowded categories only. However significant decline in performance was found for the smallest SOM only (ie only when capacity was reduced by 75%). Despite the non-significant reduction in base level categorisation for uncrowded categories, there were qualitative changes in the types of errors. Thus 'within category' confusions doubled (F [2, 179] = 11.5, p < 0.0001), whereas this type of error remained stable for the crowded categories (F [2, 179] = 1.25, p > 0.25). This result combined with observation of patterns of activity on the SOM surface suggest that by constraining the capacity of the SOM there was a shift toward coarse-coding in the perceptually heterogeneous categories.

4. Effects of Lesioning

All of the previous experiments used the supervised module with localist representations. For the lesioning experiments we used the supervised module with units representing semantic features i.e. distributed representations. Lesioning was carried out by removing 5%,10%,20% connections between either input and hidden (IH) , hidden and output (HO) or between output units (OO).

For IH lesions, (i) visually crowded categories were more vulnerable to damage (F [1, 138] = 37.7, p < 0.0001). and (ii) superordinate information was less prone to disruption than basic level information (F [1, 138] = 58.1, p < 0.0001). For HO lesions a different pattern emerged. The visually crowded categories were no more vulnerable to lesions that the uncrowded categories (F [1, 138] < 1), and this pattern hardly changed with lesion severity (F [2, 137] = 2.99, p = 0.05 for the interaction).

Using cluster analysis it appeared as though the hidden layer was, in general, performing different functions for visually crowded and visually uncrowded categories. For visually crowded categories, the hidden layer tended to 'pull apart' the patterns, rendering their representations less easily 'push together' diverse patterns such that representations which did not cohere well on a 'visual' basis, were now more similar to each other.

For OO lesions visually crowded categories were more vulnerable than uncrowded ones, but there was a significant 2 way interaction between visual crowdedness and level of abstraction (F [1, 138] = 35.5, p < 0.0001). Visually crowded categories were more vulnerable for base level attributes, but less vulnerable for superordinate attributes.

5. General Discussion

The traditional view of internal representations as static local states is losing ground to the view of the brain as a self organising system which arrives at stable states in the complex interplay between brain, body and local environment [19]. Visual object recognition (VOR) in the natural environment is typically such a complex interplay . However various experimental approaches using static 2D canonical views of living things and artefacts have produced a significant body of knowledge about one VOR route, namely living and non-living things seem to be represented differently.

In this series of experiments we used a Kohonen self organising feature map (SOM) presented with grey scale images from a diverse range of animals, musical instruments, clothing and furniture. Our aim was to simulate the development of object representations in the ventral visual pathway and explore the effects of lesioning the model. The model is, of course, a crude simplification of the underlying neural processing that we are trying to capture. Nevertheless, by using lines of converging evidence from electrophysiology and neuropsychology it is possible to demonstrate that this is a heuristically valuable model of one brain system responsible for VOR, and not merely a simulation of the outputs of such a system. We have shown how, when a large stimulus set is presented, units in a SOM, perhaps like cells in IT, behave somewhat paradoxically in that they may respond maximally to only one specific image yet, at lower levels of activation, can be involved in the coarse coding of a great many other images. Of fundamental interest is the implication that this has for recognising objects and classes of objects. We demonstrated that the contour map in the SOM depended on the class of object, with living things and musical instruments (visually crowded categories) tending towards a more distributed representation with overlapping contour maps. These qualitatively different types of representation generate different performance characteristics in learning , generalisation to novel items and post-lesion classification access to semantic attributes. Thus visually crowded categories may not simply require finer visual differentiation for stored knowledge to be retrieved [20,21] but, during learning, may actually give rise to different types of representation. Thus, when learning was complete within our model, the SOM was organised, to a certain extent, along categorical lines. This questions the merits of the debate as to whether category specific deficits indicate either categorical organisation of semantic knowledge, or the greater susceptibility of living things arising from their visual similarity [22,23,24]. It would appear that in a self-organising system, visual similarity could actually lead to some degree of categorical organisation.

Like the simulation of Humphreys et al. [25], the findings from our lesion experiments demonstrate how lesioning within the 'semantic' system can result in an impairment for living things even though the categorical effect derives from a pre-semantic factor (i.e. visual crowding). However, we also found that the hidden layer appeared to be exaggerating subtle inter-item differences (i.e. a pulling apart) for visually crowded categories, and reducing the amount of discrepancy between visually dissimilar items (i.e. a pushing together). Thus, semantic proximity can exert a strong modulatory effect upon visual similarity, a finding which concurs with the work of Dixon et al. [26].

As we have already pointed out, a relative strength of our model is its ability to account for a co-occurrence of recognition problems for both living things *and* musical instruments. That such an apparently bizarre pattern of patient deficit should be so consistently reported has been the focus of much interest [11,27]. Meanwhile, purported category specific deficits for non-living things tend to be restricted to small manipulable objects rather than large, outdoor objects [7] and there is frequently a co-occurrence with impaired knowledge for body parts. One explanation is that knowledge critical for recognising small manipulable objects is somato-sensory, or motor memory [11]. Thus, deficits in recognising non-living entities would arise from difficulties in accessing somato-sensory or motor representations, rather than visual-perceptual representations. If this is the case, the influence of perceptual crowding should hold true for somato-sensory relatedness as much as it does for visual relatedness in our simulations. Thus it is possible that if the somato-sensory representation system is also self-organising, the converse pattern of representation to that reported here will occur (i.e. greater coarse-coding for non-living things and little representational overlap for living things).

References

1. Caramazza A. (1998) The interpretation of semantic category-specific deficits: what do they reveal about the organization of conceptual knowledge in the brain ? *Neurocase, 4,* 265-272.
2. Behl-Chadha, G. (1996) Basic-level and superordinate-like categorical representations in early infancy. *Cognition, 60,* 105-141.
3. French, R. M. & Mareschal, D. (1998) Could category specific semantic deficits reflect differences in the distributions of features within a unified semantic memory ? In *Proceedings of the Twentieth Annual Cognitive Science Society Conference.* NJ:LEA. 374-379
4. Rosch, E. (1975) Cognitive representations of semantic categories. *Journal of Experimental Psychology: General, 104,* 192-233.
5. Tanaka, K. (1996) Inferotemporal cortex and object vision. *Annual Review of Neuroscience, 19,* 109-139.
6. Logothetis, N. K. & Sheinberg, D. L. (1996) Visual Object Recognition. *Annual Review of Neuroscience, 19,* 577-621.
7. Warrington, E. K. & McCarthy, R. A. (1987) Categories of knowledge: Further fractionations and an attempted integration. *Brain, 110,* 1273-1296.
8. McRae, K., de Sa, V. R., and Seidenberg, M.S. (1997) On the nature and scope of featural representations of word meaning. *Journal of Experimental Psychology: General,126,* 99-130.
9. Gaffan, D., and Heywood, C. A. (1993) A spurious category-specific visual agnosia for living things in normal human and non human primates. *Journal of Cognitive Neuroscience, 5,* 118-128.
10. Yamane, S., Kaji, S., Kawano, K. (1988) What facial features activate face neurons in the inferotemporal cortex of the monkey? *Exp. Brain Res., 139,* 209-214.
11. Gainotti, G., Silveri M. C., Daniele, A., Giustolisi, L (1995) Neuroanatomical correlates of category-specific semantic disorders: a critical survey. *Memory, 3,* 247-264.

12. Schyns, P. G. (1991) A modular neural network model of concept acquisition. *Cognitive Science, 15,* 461-508.
13. Battig, W. F., & Montague, W. E. (1969) Category norms for verbal items in 56 categories: A replication and extension of the Connecticut category norms. *Journal of Experimental Psychology, 80,* 1-146.
14. Snodgrass, J. G., & Vanderwart, M. (1980) A standardised set of 260 pictures: Norms for name agreement, image agreement, familiarity, and visual complexity. *Journal of Experimental Psychology: Human Learning & Memory, 6,* 174-215.
15. Palmer,S. E., Rosch,E., &Chase, P. 91981) Canonical perspective and the perception of objects. In Long,J. & Baddeley,A. (Eds.) *Attention and Performance Vol IX,* 131-151, Erlbaum.
16. Kohonen, T. (1988) *Self-Organisation and Associative Memory.* Berlin: Springer-Verlag.
17. Kohonen,T and Hari, R. (1999) Where the abstract feature maps of the brain might come from. *TINS, 22,* 135-139.
18. Gale,T. M., Done, D. J. and Frank , R. J. Visual crowding and category specific deficits for pictorial stimuli: A neural network model. *Cognitive Neuropsychology (in press).*
19. Clark, A . (1997) The Dynamical Challenge. *Cognitive Science, 21,* 461-481.
20. Forde, E. M. E., Francis, D., Riddoch, M. J., Rumiati, R. I. and Humphreys, G. W. (1997) On the links between visual knowledge and naming: A single case study of a patient with a category-specific impairment for living things. *Cognitive Neuropsychology, 14 ,* 403-458.
21. Humphreys, G. W., Riddoch, M. J., and Quinlan, P. T. (1988) Cascade processes in picture identification. *Cognitive Neuropsychology, 5,,* 67-103.
22. Farah, M.J ., Meyer, M. M., and McMullen, P. A. (1996) The living/non-living dissociation in not an artefact: Giving an a-priori implausible hypothesis a strong test. *Cognitive Neuropsychology, 13,* 137-154.
23. Kurbat M. A. (1997) Can the recognition of living things really be impaired? *Neuropsychologia, 35,* 813-827.
24. Kurbat, M. A., & Farah, M. J. (1998) Is the category-specific deficit for living things spurious? *Journal of Cognitive Neuroscience, 10,* 355-361.
25. Humphreys, G. W., Lamote, C., & Lloyd-Jones, T. J. (1995) An interactive activation approach to object processing: Effects of structural similarity, name frequency, and task in normality and pathology. *Memory, 3,* 535-586.
26. Dixon, M., Bub, D. N., & Arguin, M. (1997) The interaction of object form and object meaning in the identification performance of a patient with category-specific visual agnosia. *Cognitive Neuropsychology, 14,* 1085-1130.
27. Parkin, A.J ., and Stewart, F. (1993) Category-specific impairments? No. A critique of Sartori et al. *Quarterly Journal of Experimental Psychology, 46A,* 505-509.

Implicit Learning

Implicit Learning of Regularities in Western Tonal Music by Self-Organization

Barbara Tillmann, Jamshed J. Bharucha & Emmanuel Bigand

Abstract

Western tonal music is a highly structured system whose regularities are implicitly learned in everyday life. A hierarchical self-organizing network simulates learning of tonal regularities by mere exposure to musical material. The trained network provides a parsimonious account of empirical findings on perceived tone, chord and key relationships and suggests activation as a unifying mechanism underlying a range of cognitive tasks.

1. Introduction

Implicit learning is the acquisition of knowledge in an incidental manner without complete verbalizable knowledge of what is learned [39]. It has been investigated in the laboratory with artificial material based on statistical regularities. For example, letter strings have been generated by finite state grammars on the basis of restricted sets of letters. After passive exposure to grammatical strings, participants differentiated better than chance new grammatical strings from nongrammatical ones. Most were unable to explain the rules underlying the grammar in verbal free reports [1, 16, 35]. These results have been extended to more complex material attempting to bridge implicit learning of artificial grammars and environmental event sequences. In [37], for example, participants became sensitive to regularities in artificial language-like auditory sequences based on the transition probabilities of syllables.

Western tonal music is an example of a highly structured system in our environment. It constitutes a constrained system of regularities (i.e., regularities of co-occurrence, frequency of occurrence and psychoacoustic regularities) based on a limited number of elements. Experimental studies in music cognition have provided indirect evidence that listeners acquire implicit knowledge about these regularities through mere exposure. As the number of opportunities to listen to musical pieces obeying this system of regularities is so great in everyday life it is plausibly learned by the same implicit processes as investigated in the laboratory. The present paper reviews a neural network that simulates implicit learning of tonal regularities via mere exposure to musical material through self-organization. Once learning has occurred, the network in combination with a spreading activation mechanism accounts for a range of empirical findings on music perception. Before presenting the self-organizing network, basic regularities of the Western tonal system and some empirical findings on music perception are reviewed.

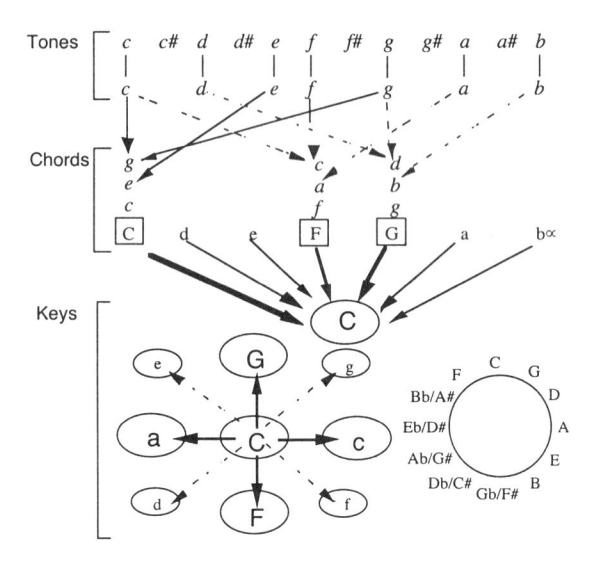

Figure 1: Schematic representation of the three levels in the tonal system using the example of C Major. Top) the 12 chromatic tones, followed by the diatonic scale in C Major. Middle) construction of three major chords, followed by the chord set forming the C Major key. Bottom) relations of C Major with close major and minor keys (left) and with all major keys forming the circle of fifths (right). (Tones are represented in italics, minor and major chords/keys in lower and upper case respectively)

2. Regularities in Western Tonal Music

The Western tonal system may be conceived of as a three-level hierarchical grammar that generates strong regularities in musical pieces (Fig. 1). It is based on a set of 12 pitch classes, which are organized in subsets of seven (defining major and minor diatonic scales). For each note of a scale, chords (e.g., major or minor) are constructed by adding two notes ≠ creating a second order of musical units. In the C major key for example, the chord C major is constructed by combining the notes E and G to the root note C. Based on tones and chords, keys define a third order of musical units. Keys sharing numerous tones and chords are said to be harmonically related. The strength of harmonic relationship depends on the number of shared events. For example, the *C* major key shares six tones and four chords with the *G* major key, but only one tone with the *F#* major key. In music theory, major keys are conceived spatially as a circle (i.e., the circle of fifths), with harmonic distance represented by the number of steps on the circle. Inter-key distances are also defined between major and minor keys. A major key (*C* major) is harmonically related to both its relative minor key (*a* minor) and its parallel minor key (*c* minor). The three levels of musical units occur with strong regularities of co-occurrence. Tones and chords belonging to the same key are more likely to co-occur in a musical piece than tones and chords belonging to different keys. Changes between keys are more likely to occur between closely related keys (e.g., *C* and *F* or *G* major) than between less related ones (e.g., *C* and *F#* major).

Within each key, tones and chords have different tonal functions creating tonal and harmonic hierarchies. For example, chords built on the first, fifth, and fourth scale degrees (referred to as tonic, dominant and subdominant respectively) usually have a more central function than chords built on other scale degrees. From a psychological point of view, the hierarchically important events of a key act as stable cognitive reference points [24] to which others are anchored [2]. Interestingly, these within-key hierarchies are strongly correlated with the frequency of occurrence of tones and chords in Western musical pieces. Tones and chords used with higher frequency (and longer duration) correspond to events that are defined by music theory as having more important functions in a given key [12, 19, 25]. Finally, as based on a restricted set of events, the Western tonal system has the characteristic that functions of tones and chords depend on the established key context. For example, the *C* major chord functions as a stable tonic chord in a *C* major context, as a less stable dominant or subdominant chord in an *F* or *G* major contexts respectively and as an out-of-key chord in a D major context.

3. Perception of Tonal Relations by Musicians and Nonmusicians

The multilevel relations between tones and chords, chords and keys, and between keys define a complex set of possible relations between musical events [25, 30]. Despite its complexity, numerous experimental studies in music cognition have provided evidence that even nonmusician listeners without explicit musical training or theoretical background are sensitive to musical structures and functions. While musicians usually exhibit better performance than nonmusicians, the overall responses show the same pattern. The sensitivity to tonal hierarchy was found to be similar in groups with different musical expertise [15, 22] as was the sensitivity to harmonic relations of chords [4, 8] or the ability to detect modulation [13, 14, 42]. Harmonic priming studies showed that independently of listenersπ musical expertise, the processing of a target chord was influenced by its harmonic relatedness with the previous context. This finding suggests that factors governing musical expectation are based on cognitive processes that do not require explicit knowledge of musical structure [6, 10, 43].

The overall pattern of results in music cognition research suggests that mere exposure to Western musical pieces suffices to develop implicit, but nevertheless sophisticated, knowledge of the tonal system. Just by listening to music in everyday life, listeners become sensitive to the structures of the tonal system without being necessarily able to verbalize them [19]. This implicit learning of tonal regularities may be viewed as an ecological validation of implicit learning analyzed in the laboratory [9].

Connectionist models allow the simulation of how knowledge may be learned through passive exposure and how this knowledge influences perception. The next section presents a connectionist model that learns tonal regularities via mere exposure and simulates experimental results on music perception via activation spreading through the learned representation.

4. Learning and Representing Tonal Knowledge

In the music domain, neural networks have been developed for selective aspects of music perception, e.g. pitch perception [5, 38, 40], melodic sequence learning [7, 20, 32] or the extraction of tonal centers [28, 29]. A model of tonal knowledge representation (named *MUSACT*, standing for *MUS*ic and *ACT*ivation) is proposed in [3], it is interesting for two features: a) the inclusion of the three organizational levels of the tonal system (tones, chords, keys) and b) the simulation of top-down influences through spreading activation. The general architecture of this network mirrors interactive-activation models of word recognition [31] and speech recognition [18] containing three levels of units (features, letters/phonemes, words). The simulation of top-down influences is an important characteristic of a knowledge model, for both language and music. For example, once the key of a musical context is recognized, the tones belonging to that key are perceived as more stable than other tones, even if they were not been present in the context [19, 25].

In *MUSACT*, tonal relations are not stored explicitly but emerge from activation reverberating between tone, chord and key layers. After reverberation, activation of chord units, for example, reflects the harmonic hierarchy of the key context, with higher activation for stable than unstable chords. *MUSACT* provides a framework for understanding how musical knowledge may be mentally represented and how this knowledge, once activated by a musical context, may influence the processing of tonal structure [6, 10]. However, it was conceived with music theoretic constraints and thus represents an idealized end-state of an implicit learning process.

Recent simulations modelled this implicit learning process via neural self-organization [44]. Through mere exposure to musical stimuli, the connection weights of a hierarchical self-organizing map adapt and internalize tonal regularities. The trained network in combination with a spreading activation mechanism simulates empirical data on perceived relations between and among tones, chords and keys. In the following, we briefly review the learning simulations and the simulations of empirical data (detailed presentation in [44]).

Principles of Self-Organization. As learning of tonal regularities presumably occurs without supervision, unsupervised learning algorithms that extract statistical regularities and encode events that occur often together [21, 23, 36] seem to be well suited to simulate implicit learning in music. The Self-Organizing Map (SOM) proposed in [23] is one unsupervised algorithm that creates topological mappings between input data and units of a map. Before learning, no particular organization exists among map units. With repeated presentation of the input data, the specialization of the map units takes place by competition among units. The unit that is best able to represent an input wins the competition and learns the representation even better by adapting its connection weights. The learning algorithm reinforces links coming from active input units and weakens links coming from inactive ones. The unit's response will be subsequently stronger for this same input and weaker for other patterns. In a similar way, other units specialize to respond to other input patterns. In SOM, learning is not restricted to the winning unit, but extended to its neighbor units. As neighbor units gradually specialize for similar inputs, the representation becomes topographically ordered on the map.

A Hierarchical Self-Organizing Map Learns Tonal Regularities. For the simulation of implicit learning of tonal regularities, a hierarchical self-organizing map was defined in order to allow the extraction of different levels of tonal organization. The input layer consisted of 12 units tuned to the 12 chromatic tones. The input layer was connected to a second layer map which in turn was connected to a third layer map (Fig. 2). Before training, all units between two layers were fully interconnected with weights initialized to random values. Four training simulations were realized with this network architecture. The network was trained with either simple harmonic material (i.e., set of chords belonging to a key) or more realistic short chord sequences. The input was defined by either a sparse coding (i.e., coding the presence of tones) or a psychoacoustically richer coding (i.e., including subharmonics of tones [33]). Training consisted of two phases, during which chords and groups of chords were presented repeatedly to the input layer.

In all four networks, units of the second layer became specialized in the detection of chords, and units of the third layer in the detection of sets of chords (keys). The maps were calibrated by naming each winning unit after the stimulus for which it won. For both maps, a topographic organization was observed (Fig. 2). In the second layer, neighbor units represented chords that share component tones[1]. In the third layer, neighbor units represented keys sharing tones and chords. The organization of key units reproduced the topology of the circle of fifth (cf. Fig. 1). During training, the initially randomized connections changed to reflect the regularities of occurrence between tones and between chords. With simple input coding, for example, each tone unit was strongly linked to six winning units in the chord layer, and each chord unit was strongly linked to three winning units in the key layer. The self-organization leads to a hierarchical encoding in which tones occurring together are represented by chord units, and similarly, chords occurring together are represented by key units.

The trained network was used as a feedforward system and as a reverberation system.[2] The feedforward activation consisted of bottom-up

---

[1] After training with rich coding, the chord layer showed a more global organization than just the differentiation based on shared pitches. The specialized units on one half of the map represent chords from one side of the circle of fifths, units on the other half the other side of the circle.

[2] As in [3], reverberation was defined by phasic activation spreading between units until equilibrium was reached. Equilibrium was defined to be reached when phasic activation was less than a threshold of .005 for each unit. For event sequences, activation due to each event is accumulated and weighted according to recency. The total activation of a unit is

information only: chord units π activation reflected the number of component tones shared with the input. After reverberation, the activation pattern changed qualitatively due to top-down influences of learned, schematic structures and reflected tonal relations. Across major chord units, for example, activation decreased monotonically with distance around the circle of fifths.

The different training material (sets of chords vs. chord sequences) and input coding (sparse vs. rich) impressed on the learned weights in the connection matrices. For example, for matrices learned with rich coding feedforward profiles had a shape closer to top-down profiles than for matrices learned with sparse coding. However, with reverberation top-down processes imposed a pattern of activation that was analogous for sparse and rich coding and overwrote influences of coding richness.

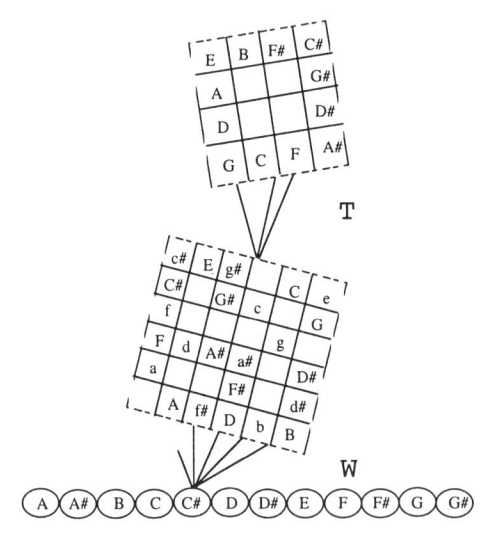

Figure 2: The hierarchical SOM: Input units code the 12 chromatic tones, the first map specialized in the detection of chords, the second map in the detection of keys. The represented maps are the result of a network trained with simple input coding and sets of chords. W and T refer to matrices with the learned connection weights between units of two layers (cf. text), with only some links represented for convenience.

In sum, the hierarchical self-organizing map extracted underlying regularities of musical material through mere exposure. Specialized representational units were formed for combinations of musical events (tones, chords) that occur with great regularity. Further analyses of activation patterns after reverberation indicated strong correlations between the four trained networks and the *MUSACT* model ($.984 < \underline{r} < .999$; $\underline{p} < .01$; $\underline{df} = 10$). This outcome suggests that the activation patterns of *MUSACT* that had been based on theoretical constraints can emerge automatically through self-organization.

---

thus the sum of three terms: the stimulus activation, the phasic activation accumulated during reverberation and the decayed activation due to previous events.

5. Simulation of Tonal Perception by Activation Spreading in Tonal Knowledge Representation

A crucial test for a neural net model of knowledge representation is to simulate the performance of human participants, in our case of Western tonal listeners in music perception studies. Simulations were run with the experimental material used in a set of empirical studies on the processing of tone, chord, and key relationships. The activation levels of network units were interpreted as levels of stability. The more a unit (i.e., a chord unit, a tone unit) is activated, the more stable the musical event is in the corresponding context. The hypothesis was that the level of tonal stability affects performance in the tasks; for example, a higher activated, more stable, event would be more expected or be judged as being more similar to the previous event. The simulations showed that the trained self-organizing network behaved much as human participants do in experiments on tonal perception. This outcome suggests that activation spreading in tonal knowledge representation can be seen as a unifying mechanism underlying a range of cognitive tasks.

Perceived Relations between Chords. The network simulated results of similarity judgments on chord pairs [27], recognition memory [4] and harmonic expectation. The development of harmonic expectation has been studied in harmonic priming [6, 10, 41] and neurophysiological studies [34]. It has been reported that the processing of a target chord is facilitated if harmonically related to the preceding local and/or global context, and that the level of harmonic incongruity between target and context is reflected in the amount of positivity of event-related potentials (ERPs). When the experimental material was presented to the network, activation levels of chord units mirrored facilitation patterns in the priming task (e.g., with higher activation for chord units representing facilitated targets), and activation changes in chord units due to the target mirrored the amount of positivity in the ERP waveforms (e.g. with stronger activation changes for distant key targets).

Perceived Relations between Keys. Further simulations showed that the network accounts for listeners' implicit knowledge of key distances and reproduces listeners' changing sense of key [13, 26]. For example, [13] showed that listeners distinguish sequences that modulate only one step in distance on the circle of fifths from sequences modulating two steps, with a stronger changing sense of key for counterclockwise than for clockwise modulations. After the presentation of modulating sequences to the network, the activation pattern of key units changed more strongly for a two-step than for a one-step modulation (compared to the initial key of the musical sequence). The network also simulated the perceived asymmetry in modulation: for each distance, counterclockwise modulations caused a stronger change than clockwise modulations.

Perceived Relations between Tones. Even if the network was not trained with melodies, it simulated results on the perception of tones. For example, once a key context is instilled, listeners perceive a hierarchy of stability between tones [26]. Simulations realized with a one-chord context showed that the activation received by tone units after reverberation reflected the differences between in-key and out-of-key tones of the context. Due to reverberation, the model made this distinction also for tones that had not been present in the context. In addition to simulating similarity judgments on tone pairs [24], the network simulated melodic recognition memory [17]. The network did not learn the melody in itself, but it captured elements linked to tonal stability that influenced human participants in differentiating between targets and foils. The state of the network was, for example, more strongly correlated between items for which human participants showed lower performance.

6. Conclusion

The presented connectionist network simulates implicit learning of tonal regularities by self-organization and proposes activation spreading through tonal knowledge representation as mechanism underlying different cognitive tasks. Based on self-organization, the structure of the network adapted to tonal regularities through repeated exposure to tonal material. The trained network provides a low dimensional and parsimonious representation of tonal knowledge. For example, to account for the context dependency of the tonal system the network does not represent tones and chords several times, but contextual stability changes are reflected in activation levels. As a consequence, cognitive reference points (that also change with key contexts) are not stored separately, but emerge from activation. Finally, a further parsimonious feature is related to key induction. The network does not need to fall back on supplementary processes (e.g. template matching); rather, the underlying key emerges from activation in the key layer.

The network offers a framework for generating new testable predictions that relate to important issues in music perception. The experimental testing can be related to key identification (e.g., number of notes necessary to establish a key, disturbing effect of an unrelated event), to key modulation (e.g. how long the trace of a key remains) or to the eventual link between activation decay and musical short-term memory span. The self-organizing algorithm that allowed the network to learn underlying regularities conforms to principles of cortical information processing, such as the formation of spatial ordering in sensory processing areas. In auditory cortex, tonotopic organization has been shown among cells responding best to different frequencies [11, 45]. The outcome of the present simulations, together with aspects of cortical organization, leads to the question, whether there exist higher order maps such as a tonotopic organization of key centers.

Acknowledgments

This research was supported in part by a grant to B.T. from the Deutsche Akademische Austauschdienst DAAD, by grants to J.J.B. from the National Science Foundation (SBR-9601287) and NIH (2P50 NS17778-18), and by a grant to E.B. from the International Foundation for Music Research.

References

1. Altmann, G.T., Dienes, Z., & Goode, A. (1995). Modality independence of implicitly learned grammatical knowledge. *Journal of Experimental Psychology: LMC, 21*, 899-912.
2. Bharucha, J. J. (1984). Anchoring effects in music. *Cognitive Psychology, 16*, 485-518.
3. Bharucha, J. J. (1987). Music cognition and perceptual facilitation. *Music Perception, 5*, 1-30.
4. Bharucha, J.J., & Krumhansl, C.L. (1983). The representation of harmonic structure in music: Hierarchies of stability as a function of context. *Cognition, 13*, 63-102.
5. Bharucha, J.J., & Mencl, W. E. (1996). Two Issues in auditory cognition: Self-organization of octave categories and pitch-invariant pattern recognition. *Psychological Science, 7*, 142-149.
6. Bharucha, J.J., & Stoeckig, K. (1987). Priming of chords: Spreading activation or overlapping frequency spectra? *Perception & Psychophysics, 41*, 519-524.
7. Bharucha , J.J., & Todd, P. (1989). Modeling the perception of tonal structures with neural nets. *Computer Music Journal, 13,* 44-53.
8. Bigand, E., & Pineau, M. (1997). Global context effects on musical expectancy. *Perception & Psychophysics, 59*, 1098-1107.
9. Bigand, E., Perruchet, P., & Boyer, M. (1998). Implicit learning of an artificial grammar of musical timbres. *Current Psychology of Cognition, 17*, 577-600.
10. Bigand, E., Madurell, F., Tillmann, B., & Pineau, M. (1999). Effect of global structure and temporal organization on chord processing. *Journal of Exp. Psychology: HPP, 25*, 184-197.
11. Brugge, J. F., & Reale, R. A. (1985). Auditory cortex. In A. Peters & E. G. Jones, (Eds.), *Cerebral Cortex: Association and Auditory Cortices.* New York: Plenum Press.
12. Budge, H. (1943). *A study of chord frequencies.* Teacher College.
13. Cuddy, L. L., & Thompson, W. F. (1992a). Asymmetry of perceived key movement in chorale sequences: Converging evidence from a probe-tone analysis. *Psychological Research, 54,* 51-59.
14. Cuddy, L. L., & Thompson, W. F. (1992b). Perceived key movement in four-voice harmony and single voices. *Music Perception, 9*, 427-438.
15. Cuddy, L. L. & Badertscher, B. (1987). Recovery of tonal hierarchy: Some comparisons across age and levels of musical expertise, *Perception & Psychophysics, 41*, 609-620.
16. Dienes, Z., Broadbent, D., & Berry, D. (1991). Implicit and explicit knowledge bases in artificial grammar learning. *Journal of Experimental Psychology: LMC, 17*, 875-887.
17. Dowling, W.J. (1978). Scale and contour: Two components of a theory of memory for melodies. *Psychological Review, 85*, 341-354.
18. Elman, J.L., & McClelland, J. L. (1984). The interactive activation model of speech perception. In N. Lass (Ed.), *Language and speech.* (pp. 337-374), New York: Academic Press.
19. FrancËs, R. (1958).*La perception de la musique.* (2nd ed., 1984) Paris: Vrin.
20. Griffith, N. (1994). Development of tonal centers and abstract pitch as categorizations of pitch use. *Connection Science, 6*, 155-175.
21. Grossberg, S. (1970). Some networks that can learn, remember and reproduce any number of complicated space-time patterns. *Studies in Applied Mathematics, 49*, 135-166.

22. Hèbert, S., Peretz, I., & Gagnon, L. (1995). Perceiving the tonal ending of tune excerpts: The roles of pre-existing representation and musical expertise. *Canadian Journal of Exp. Psy.*, *49*, 193-209.
23. Kohonen, T. (1995). *Self-Organizing Maps*. Springer: Berlin.
24. Krumhansl, C. L. (1979). The psychological representation of musical pitch in a tonal context. *Cognitive Psychology*, *11*, 346-374.
25. Krumhansl, C. L. (1990). *Cognitive Foundations of Musical Pitch*. Oxford: University Press.
26. Krumhansl, C. L., & Kessler, E. (1982). Tracing the dynamic changes in perceived tonal organization in a spatial representation of musical keys. *Psychological Review*, *89*, 334-368.
27. Krumhansl, C. L., Bharucha, J. J., & Castellano, M. (1982). Key distance effects on perceived harmonic structure in music. *Perception & Psychophysics*, *32*, 96-108
28. Leman, M. (1995). *Music and Schema Theory*. Springer: Berlin.
29. Leman, M., & Carreras, F. (1997). Schema and Gestalt. (pp. 144-168), In: Leman, M. (Ed.) *Music, Gestalt, and Computing*. Springer: Berlin.
30. Lerdahl, F. (1988). Tonal Pitch Space. *Music Perception*, *5*, 315-345.
31. McClelland, J. L., & Rumelhart, D. E. (1981). An interactive activation model of context effects in letter perception: Part 1. An account of basic findings. *Psychological Review, 86*, 287-330.
32. Page, M. A. (1994). Modeling the perception of musical sequences with Self-Organizing neural networks. *Connection Science*, *6*, 223-246.
33. Parncutt, R. (1988). Revision of Terhardt's psychoacoustical model of the roots of a musical chord. *Music Perception*, *6*, 65-94.
34. Patel, A. D., Gibson, E., Ratner, J., Besson, M., & Holcomb, P. J. (1998). Processing syntactic relations in language and music. *Journal of Cognitive Neuroscience*, *10*, 717-733.
35. Reber, A. S. (1967). Implicit learning of artificial grammars. *Journal of Verbal Learning and Verbal Behavior*, *6*, 855-863.
36. Rumelhart, D. E., & Zipser, D. (1985). Feature discovery by competitive learning. *Cognitive Science*, *9*, 75-112.
37. Saffran, J. R., Newport, E. L., Aslin, R. N., Tunick, R. A., & Barrueco, S. (1997). Incidental language learning. *Psychological Science*, *8*, 101-105.
38. Sano, H. & Jenkins, B. K. (1991). A neural network model for pitch perception. In P. Todd & G. Loy (Eds.), *Music and Connectionism*. (pp. 42-49), Cambridge: MIT Press.
39. Seger, C. A. (1994). Implicit learning. *Psychological Bulletin*, *115*, 163-169.
40. Taylor, I., & Greenhough, M. (1994). Modeling pitch perception with adaptive resonance theory artificial neural networks. *Connection Science*, *6*, 135-154.
41. Tekman, H. G., & Bharucha, J. J. (1998). Implicit knowledge versus psychoacoustic similarity in priming of chords. *Journal of Experimental Psychology: HPP*, *24*, 252-260.
42. Thompson, W. F., & Cuddy, L. L. (1989). Sensitivity to key change in chorale sequences: A comparison of single voices and four-voice harmony. *Music Perception*, *7*, 151-168.
43. Tillmann, B., Bigand, E., & Pineau, M. (1998). Effects of local and global context on harmonic expectancy. *Music Perception*, *16*, 99-118.
44. Tillmann, B., Bharucha, J. J., & Bigand, E. (2000). Implicit Learning of Tonality: A Self-Organizing Approach. *Psychological Review. 107(4)*.
45. Wessinger, C. M., Buonocore, M. H., Kussmaul, C. L., & Mangun, G. R. (1997). Tonotopy in human auditory cortex examined with fMRI. *Human Brain Mapping*, *5*, 18-25.

Rules vs. Statistics in Implicit Learning of Biconditional Grammars

Bert Timmermans & Axel Cleeremans

Abstract

A significant part of everyday learning occurs incidentally — a process typically described as implicit learning. A central issue in this domain and others, such as language acquisition, is the extent to which performance depends on the acquisition and deployment of abstract rules. Shanks and colleagues [22], [11] have suggested (1) that discrimination between grammatical and ungrammatical instances of a biconditional grammar *requires* the acquisition and use of abstract rules, and (2) that training conditions — in particular whether instructions orient participants to identify the relevant rules or not — strongly influence the extent to which such rules will be learned. In this paper, we show (1) that a Simple Recurrent Network can in fact, under some conditions, learn a biconditional grammar, (2) that training conditions indeed influence learning in simple auto-associators networks and (3) that such networks can likewise learn about biconditional grammars, albeit to a lesser extent than human participants. These findings suggest that mastering biconditional grammars does not require the acquisition of abstract rules to the extent implied by Shanks and colleagues, and that performance on such material may in fact be based, at least in part, on simple associative learning mechanisms.

1. Introduction

Over development and learning, we acquire a considerable amount of information incidentally. Natural language offers perhaps the most striking example of such incidental learning: Infants do not need to be explained grammar rules in order to be able to communicate effectively and are presumably unaware of the fact that they are learning something at all. Adult speakers likewise "know" whether expressions of their native language are grammatically correct but can seldom explain why. Such dissociations between performance and ability to verbalize the relevant knowledge are often described as being subtented by processes of *implicit learning* (IL). Thus, the notion of *"implicit learning"* (IL) usually designates cases in which a person learns about the structure of a fairly complex stimulus environment, without necessarily intending to do so, and in such a way that the resulting knowledge is difficult to express [1]. IL is the ability to learn without awareness, as opposed to explicit learning, which is strategy- and/or hypothesis-driven, and of which one tends to be consciously aware. A considerable body of empirical evidence now suggests that people can indeed acquire information about the underlying structure of ensembles of stimuli in an incidental manner [5]. For instance, in a typical artificial grammar learning situation (e.g., [16]), Ss are asked to memorize a set of meaningless letter strings generated based on a simple finite-state grammar that specifies legal transitions between successive letters. Reber's main finding, now replicated hundreds of times, is that Ss are subsequently able to discriminate novel instances of grammatical strings from ungrammatical strings somewhat better than chance, despite remaining unable to verbalize the rules of the grammar. Based on these and other findings, Reber accordingly suggested that Ss must have unconsciously acquired abstract knowledge about the grammar. This early *abstractionist* account, however, has now become largely obsolete, based on (1) the fact that successful performance in this sort of task can be achieved without knowledge of the rule system (e.g., [2]), and on (2) the fact that when probed directly about the relevant knowledge, Ss often turn out to be able to express this knowledge [9], [23].

1.1 Implicit Learning and Abstraction

A central issue in this context is the question of whether the mechanisms through which implicit and explicit knowledge are acquired are best viewed as being subtended by separate processing systems or as being different manifestations of a single set of learning mechanisms. Early theories of IL (e.g. [16]) have tended to assume that it involves independent rule-based unconscious learning mechanisms. Today, based on issues raised by the complex measurement challenges associated with the assessment of awareness, as well as on the fact that many computational mechanisms can in fact perform in a *rule-like* manner without necessarily having acquired *rule-based* knowledge [17], many authors have proposed instead that performance in typical IL tasks is in fact best accounted for by assuming that Ss consciously learn either specific exemplars or fragments thereof during training. Performance at test can then be explained by simple mechanisms that compute the similarity between training and test exemplars, or that are sensitive to the overlap between fragments of the training items and the test items. From this perspective, the main distinction between implicit and explicit learning should thus not be one of awareness, but one of information-processing: Implicit learning would essentially involve incidental or episodic memory-based processes and result in conscious knowledge of exemplars and or fragments, whereas explicit learning would essentially involve active hypothesis testing and result in conscious knowledge of abstract rules. This position has been expressed most clearly by Shanks and colleagues [23], [22], [24]. However, while it is undeniable that humans are capable of abstract thought, the extent to which such processes are rooted in dedicated mechanisms remains unclear. Indeed, the debate about the nature of knowledge acquired in implicit learning situations finds an echo in recent research dedicated to various aspects of natural language learning, which we briefly discuss in the next section.

1.2 Implicit Learning and Natural Language

In a series of experiments modeled after the artificial grammar learning paradigm, Saffran et al. [18] exposed 6-7 years old children and adult Ss to a continuous speech flow such as *bupadapatubitutibudutabapidabu*. Ss were told that the experiment was about the influence of auditory stimuli on creativity. The only cues to word boundaries were the transitional probabilities between pairs of syllables (e.g., *bu-pa*), which were higher within words than between words. Afterwards, Ss heard two sets of sounds, each consisting of three syllable pairs, and were told to decide which one sounded more like the tape they had heard. Both adult and child Ss managed to perform well above chance, suggesting that learning about the deep structure of the material might proceed in the absence of intention to do so, and after only short exposure to the relevant material. Saffran et al. concluded that sensitivity to statistical structure is a fundamental process in language acquisition.

Marcus et al. [13], in stark contrast, claim that sensitivity to statistical structure is not sufficient to account for their data, and that 7-month-old infants can "represent, extract, and generalise abstract algebraic rules." The infants were exposed to artificial auditory "sentences" during a training phase, and were subsequently presented with test items instantiated with a novel set of sounds, half of which shared their abstract structure with the training items and half of which did not. For instance, infants habituated to *gatiti* or *linana* (both sharing an *ABB* structure) were subsequently presented with test sentences such as *wofefe* (familiar *ABB* structure) or *wofewo* (novel *ABA* structure). Despite the test material being instantiated over completely novel features, infants tended to listen more to the sentences instantiating a novel abstract structure. Marcus et al. concluded that infants had the capacity to represent "algebraic" rules, and that simple associative learning models such as connectionist networks would be unable to generalize as infants do. However, Marcus et al.'s claim that networks could not model the observed effect was disputed by several authors (e.g., [19], [15], [8]), essentially based on the fact that successful transfer need not necessarily be based on the overlap between features of the input patterns themselves. Instead "the relevant overlap of representations required for generalisation [...] can arise over internal representations that are subject to learning." ([15], p.2). Transfer and generalization therefore remain complex issues, in part because of the challenges associated with designing stimulus material that can *only* be learned through abstractive mechanisms. Shanks and colleagues [22], [11] have attempted to address precisely this issue in an interesting series of experiments described in the following section.

1.3 Biconditional AGL: Shanks et al. (1997)

As mentioned before, Shanks and St John [23] proposed to abandon the idea of a conscious/unconscious dichotomy in favour of a rule-based/instance-based dichotomy. The basic idea is that humans possess two learning systems capable of creating distinct forms of mental representation, one system consisting of symbolic rule-abstraction mechanisms and the other involving subsymbolic, memory-based, connectionist mechanisms (see [21] for a discussion). In this context, Shanks et al. considered transfer in AGL tasks to be at least to some extent mediated by abstract (rule-) knowledge and claimed that people systematically become aware of the relevant regularities in *those AGL tasks where only rule learning is possible.* To demonstrate, Shanks et al. exposed Ss to artificial grammar strings generated by a biconditional grammar (see also [14]). Biconditional grammars involve cross-dependency recursion (see [3]) such that letters that appear at each position before and after a central dot depend on each other. An example is given in Figure 1, where letter D is paired with F, G with L, and so on. Shanks et al. constructed biconditional grammar training strings as well as a set of grammatical and ungrammatical and test strings, in such a way that grammatical and ungrammatical test items could not be

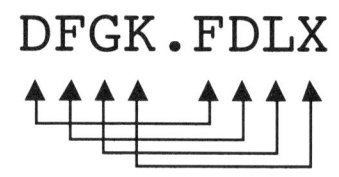

DFGK.FDLX

Figure 1: A biconditional grammar string as used by Shanks et al. (1997). Possible letters in each position before the dot are linked biconditionally with the letters that may appear after the dot.

distinguished — in contrast with the typical transitional grammars used in artificial grammar learning experiments — on the basis of their overlap with the training strings in terms of bigrams or trigrams (or any other n-gram). During training, two groups of Ss were shown strings one at a time on a computer screen and had to perform one of two tasks on each trial.

The *match* group Ss, who had been told that the task was about memory, were exclusively exposed to *grammatical* strings. On each trial, they first had to memorise a string displayed on the computer screen for a few seconds. Immediately thereafter, they had to identify this string among three possibilites (the string they had just memorized and two foils). The *edit* group Ss, in contrast, were exclusively exposed to *ungrammatical* strings. They were told that the strings had been constructed according to rules and that their task was to find them. On each trial, edit Ss were shown an *ungrammatical* string, and they had to indicate which letters they thought violated *(N)* or confirmed *(Y)* the rules. They were then given the correct string *and* the correct Y/N sequence as feedback. Shanks et al. showed a dissociation between the two groups: While the edit group performed well and most Ss extracted the rules, the match group performed at chance level, thus suggesting that "instance-memorisation and hypothesis-testing instructions recruit partially separate learning processes." ([22], p.243). Their basic claim is thus that discriminating between grammatical and ungrammatical biconditional strings *requires* abstract knowledge of the rule system, and that such knowledge *cannot* be learned by associative learning mechanisms such as instantiated in connectionist networks. In this paper, our goal is to explore the extent to which such networks can learn about biconditional grammars. To do so, we report on two simulation studies. Our first simulation study suggests that biconditional grammar learning can, under some conditions, be performed by networks developing representations based on frequency statistics. Our second simulation study was dedicated to exploring how differences between *match* and *edit* learning could be modelled without explicitly invoking a memory- versus rule-based distinction.

2. Simulation Study 1

In this first simulation study (see also [25]), we simply aimed to determine whether the Simple Recurrent Network (SRN; see Figure 2) was able to learn material from the Shanks et al. [22] experiments. The SRN, initially proposed by Elman (e.g. [10]; see also [6]) is one of the most influential connectionist models in the implicit learning and psycholinguistic literatures. SRNs are typically trained to predict the next element of sequences presented one element at a time to the network, and are therefore particularly appropriate to explore tasks involving sensitivity to sequential structure. To perform this prediction task, the network is presented, on each time

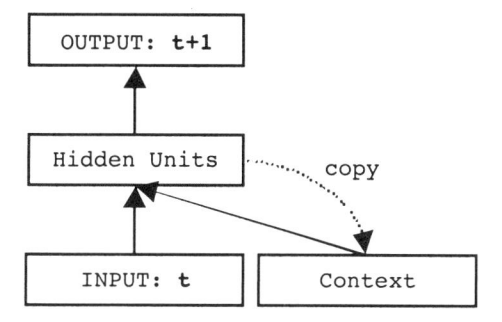

Figure 2: The Simple Recurrent Network as conceptualised by Elman (1990).

step, with element t of a sequence, and with a copy of its own internal state (i.e. the vector of hidden units activations) at time step t-1. Based on these inputs, the network has to predict element t+1 of the sequence. During training, the network's prediction responses are compared to the actual successor of the sequence, and the resulting error signal is then used to modify its connection weights using the back-propagation algorithm. As described in [20] and [4], the network progressively learns to base its predictions on the constraints set by an increasingly large and self-developed temporal window. This progressive incorporation of he statistical dependencies between successive elements of the sequence in the internal representations of the network eventually enables it to behave *as though* it had learned the relevant sequential rules. The SRN can thus, for instance, exhibit *perfect* generalization to an infinite number of novel sequences after (necessarily finite) training on a set of sequences generated from a finite-state automaton.

2.1 Network Architecture and Parameters

An SRN with 100 hidden units and local representations on its pools of input and output units was trained using backpropagation on the biconditional strings designed by Shanks et al. [22]. Strings were presented one element at a time to the network by activating the corresponding input unit (each of the 9 input units represented the letters D, F, G, L, K, X, the dot, the beginning, and the end of a string respectively) The learning rate was set to 0.15, and momentum was to 0.9. Context units were reset to zero after each complete string presentation.

2.2 Training Material

The training material consisted of the set of 18 strings designed by Shanks et al. [22] (List 1). The test material consisted of 18 novel grammatical and 18 ungrammatical strings respecting the following constraints: (1) Grammatical strings had to conform to the biconditional grammar: Letter position 1 is linked to 5, 2 to 6 and so on, with the linked letters being D–F, G–L, and K–X. (2) The use of the 6 letters was balanced, so that each letter appeared 3 times in each of the 8 letter locations. (3) Each training string differed from all other training strings by at least 4 letter locations. (4) Each training item had a grammatical similar item and an ungrammatical similar item that each differed from the training item by only 2 letter positions. Each training item was different from all other test items by at least 3 letter locations. The simulation was carried out on exactly these strings. A training epoch consisted of all 18 strings being presented once to the network, in a random fashion.

2.3 Procedure

Each of 9 networks initialized with different random weights was trained on the 18 training strings designed by Shanks et al. [22] for 3000 epochs. The networks were tested on seven different occasions during training. On each test, the networks were exposed to 18 novel grammatical strings and on 18 ungrammatical strings. Performance during test was assessed by recording the relative strength of the output unit corresponding to the actual successor of each element of each string. Different measurements of accuracy exist, of which we used the *Luce ratio* [12] — a simple measure of relative strength in which the activation of the target output unit is divided by the sum of the activations of all output units. These prediction responses were then averaged separately for each string so as to obtain a single measure of how well the networks were able to process each string. A high average luce ratio thus indicates that the network is successful in predicting each element of the corresponding string. Global measures of performance for each of the seven tests were obtained by averaging the mean luce ratios separately for grammatical and ungrammatical strings over the 9 networks.

2.4 Results

Figure 3 represents global prediction performance obtained during each of the 7 tests, and separately for training, novel grammatical, and ungrammatical strings. The figure clearly shows that the networks were able to discriminate between novel grammatical and ungrammatical strings. The training strings were learned almost perfectly from 100 epochs onwards. Further, the network clearly discriminates

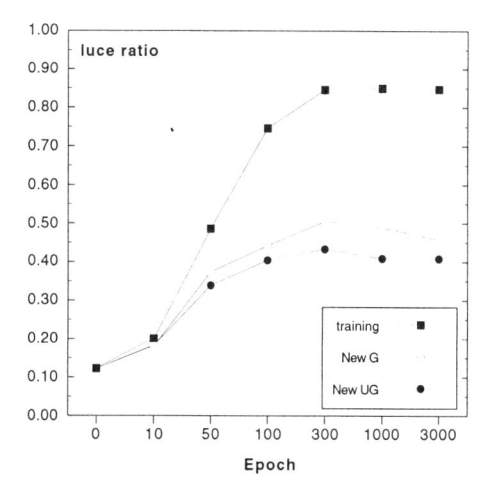

Figure 3: Average SRN prediction response strength, represented at various points during training, and plotted separately for training, novel grammatical and non-grammatical strings.

between novel grammatical and ungrammatical strings (i.e., better predictions for grammatical strings), even *before* it is completely successful in mastering the training strings. A MANOVA applied on these data confirmed that the networks successfully discriminated between novel grammatical and non-grammatical strings, $F(1, 8) = 97.08$, $p<.0001$. Further analyses aimed at ruling out that the networks had merely learned to predict the central dot or the end of the strings confirmed that letter-by-letter predictions were indeed better for grammatical than for ungrammatical strings, particularly for letters occurring after the central dot. Based on these findings, we can therefore conclude that contrary to what Shanks et al. claimed, the SRN can in fact distinguish between novel grammatical and ungrammatical strings generated by a biconditional grammar without making use of explicit rules.

It is important to note, however, that this result depends on the specific set of training strings used by Shanks et al. [22]. Indeed, the SRN exhibits well known specific difficulties in learning material that involves maintaining information across several times steps [20], as is the case here, and it would have failed had the stimulus material be perfectly balanced in terms of how frequently the different biconditional pairs occur in the stimulus set.

3. Simulation Study 2: The Match/Edit Distinction

In this second simulation study, our goal was to explore the effects of different training conditions on network performance. Recall that in [11], match Ss were given incidental learning instructions and were only exposed to grammatical strings. In contrast, edit Ss were informed that the strings instantiated a simple rule system and that their task consisted of uncovering this structure. Edit Ss were shown only ungrammatical strings, and had to indicate, on a letter-by-letter basis, which of the letters of each string they thought violated the rules. To do so, they typed a string of Y/Ns, endorsing or rejecting each letter of the string as grammatical. They were then shown the correct string, as well as the correct string of Y/N judgments.

3.1 Network Architecture and Parameters

To capture the match/edit distinction, we designed two simple feedforward networks. The *Match* networks were simple autoassociators that were trained exclusively on grammatical strings. The *Edit* networks in contrast, were exposed exclusively to ungrammatical strings during training, and were trained to produce both the correct string and the Y/N sequence as output, just as human participants.

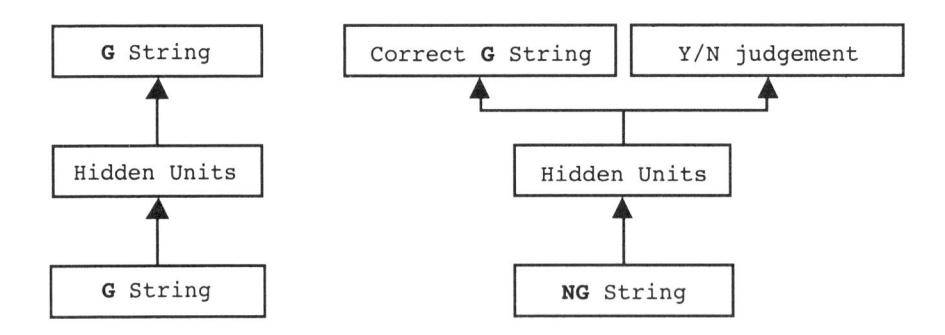

Figure 4: Match (left panel) and Edit (right panel) networks used in Simulation Study 2.

The networks are shown in Figure 4. Both networks used local representations on their pools of input and output units. Strings were presented by activating one of 6 units in each of 8 pools of units, each corresponding to the 6 letters that could occur in each of 8 positions within a string (the central dot was not represented). Edit networks were endowed with an additional pool of 8 units corresponding to the judgements about the grammaticality of each letter.

3.2 Procedure

A total of 9 networks in each condition were trained and tested in the same manner as described for Simulation Study 1. However, to assess performance in a way that more closely corresponds to human performance, we followed the procedure used by [7] so as to obtain percentages of correct classifications based on the networks' responses. To do so, we first computed average Luce Ratios for each test string, as described before (the activation of the nodes representing the Y/N input was not taken into account). Next, we computed the probability that each string would be classified as grammatical by entering its Luce Ratio in the following expression:

(1) $$p(\text{"grammatical"}) = 1/1 + e^{-k\,luce - T}$$

where k is a scaling parameter, *luce* is the average luce ratio for the string, and T is a threshold that was adjusted manually so as to yield equal numbers of "grammatical" and "ungrammatical" responses. The resulting individual probabilities were then averaged separately over grammatical and ungrammatical strings for each of the set of networks trained under match or edit conditions to yield global endorsement rates broken down by string type. Finally, based on these global endorsement rates, we computed the percentages of correct classifications expected for grammatical and ungrammatical strings in each condition.

3.3 Results

Results are shown in Figure 5. Following Shanks's analyses, Edit networks were classified as 'learners' and 'nonlearners' on the basis of the % correct responses at 1000, 2000 and 3000 epochs. The left panel shows the percentage of novel grammatical and ungrammatical strings that were endorsed by the networks as grammatical. The figure clearly shows that the Match networks fail to discriminate between G and NG strings, endorsing about 57% of each as grammatical. Edit Nonlearner networks perform better for most of the training period, but eventually likewise end up failing to discriminate between G and NG strings. In contrast, Edit Learner networks very quickly discriminate between G and NG strings, eventually endorsing about 57% of grammatical strings as grammatical, and correctly rejecting about 52% of the ungrammatical strings. The right panel of Figure 4 shows these data in a more compact form, representing the percentage of correct classifications produced by each type of network. Edit learner networks manage to achieve 57% of correct classifications overall. This result is well in line with standard results in the artificial grammar learning literature, but falls far short of the 95% correct classifications reported by [11]. Further manipulations of the simulation parameters and architecture will explore the extent to which this significant discrepancy can be reduced, but at this point, one can nevertheless conclude the following: First, the simulations were successful in showing that training the networks under "match" conditions indeed results in their failing to learn the biconditional grammar. Mere exposure to grammatical instances of biconditional strings does not seem to be sufficient for the auto-associator networks to become sensitive to the structure of the grammar. Second, we observed, like [11], that some Edit networks fail to learn while others succeed. Third, the simulations again suggest, consistently with Simulation Study 1, that biconditional grammars can be learned to some extent

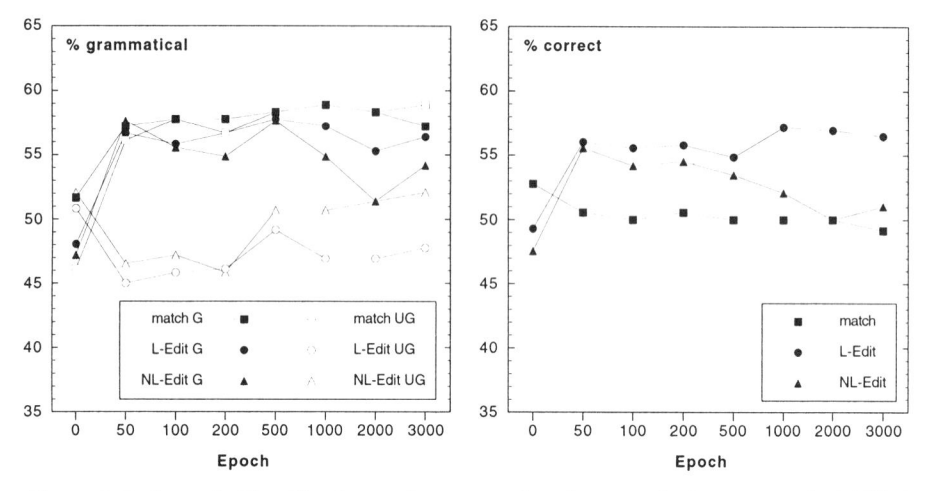

Figure 5: Left panel: Classification performance, plotted separately for grammatical (filled symbols) and ungrammatical (open symbols) strings, and for match (diamonds), non-learning edit (triangles) and learning edit (circles) networks. Right panel: Percentage of strings classified correctly, plotted separately for match (diamonds), non-learning edit (triangles) and learning edit (circles) networks.

through purely associative learning mechanisms. We discuss the implications of these findings in the general discussion that follows.

4. General Discussion

The goal of this paper was to explore the extent to which simple networks can learn about biconditional grammars. These grammars, in contrast to typical finite-state grammars, cannot be learned based on surface similarity, to the extent that neither memorized instances or fragments of the training strings contain cues about the grammatical status of a test item. Our main finding is that simple networks such as the SRN or some of the auto-associators networks used in Simulation Study 2 can actually learn to discriminate between novel grammatical and ungrammatical instances of biconditonal grammar strings. This outcome does not entail that rule-based learning never occurs (as it obviously does for some Ss in Shanks et al.'s experiments), but simply (1) that biconditional grammars might not address all the issues involved in efforts to dissociate rule-based vs. memory-based learning processes in the implicit learning literature, and (2) that abstraction might, at least on the larger portion of a representational continuum extending from pure instance-based representations to fully abstract, propositional representations, be a *graded* dimension. In this respect, connectionist models are particularly striking examples of the graded character of abstraction, to the extent that their internal representations can span most of the underlying continuum depending on task demands. Hence, while genuine abstraction may ultimately involve dedicated mechanisms closely tied to awareness and language, we believe that simple learning mechanisms based on functional similarity are often surprisingly powerful in the critical steps of developing ensembles of relevant sub-symbolic representations upon which further processes can then operate.

Acknowledgments

Axel Cleeremans is a Research Associate of the National Fund for Scientific Research (Belgium). This work was supported by a grant from the Université Libre de Bruxelles in support of IUAP program #P/4-19 and by European Commission Grant HPRN-CT-1999-00065. Bert Timmermans is supported by a grant from the Vrije Universiteit Brussel.

References

1. Berry, D.C. and Dienes, Z., (1993). *Implicit learning: Theoretical and empirical issues*. Hillsdale: Lawrence Erlbaum Associates.
2. Brooks, L.R. (1978). *Non-analytic concept formation and memory for instances* In *Cognition and Concepts*, (E. Rosch and B. Lloyd, eds). Hillsdale: Lawrence Erlbaum Associates, pp. 16-211.
3. Christiansen, M. and Chater, N. (1999). Toward a connectionist model of recursion in human linguistic performance. *Cognitive Science*, 23, 157-205.
4. Cleeremans, A., (1993). *Mechanisms of implicit learning: Connectionist models of sequence processing*. Cambridge: MIT Press.
5. Cleeremans, A., Destrebecqz, A., and Boyer, M. (1998). Implicit learning: News from the front. *Trends in Cognitive Sciences*, 2, 406-416.
6. Cleeremans, A., Servan-Schreiber, D., and McClelland, J.L. (1989). Finite state automata and simple recurrent networks. *Neural Computation*, 1, 372-381.
7. Dienes, Z. (1992). Connectionist and memory-array models of artificial grammar learning. *Cognitive Science*, 16, 41-79.
8. Dienes, Z., Altmann, G., and Gao, S.-J. (1999). Mapping across domains without feedback: A neural network model of transfer of implicit knowledge. *Cognitive Science*, 23, 53-82.
9. Dulany, D.E., Carlson, R.A., and Dewey, G.I. (1984). A case of syntactical learning and judgment: How conscious and how abstract? *Journal of Experimental Psychology: General*, 113, 541-555.
10. Elman, J.L. (1990). Finding structure in time. *Cognitive Science*, 14, 179-212.
11. Johnstone, T. and Shanks, D.R. (submitted). Abstractionist and processing accounts of implicit learning. .
12. Luce, R.D. (1963). Detection and Recognition. In *Handbook of Mathematical Psychology*, (R.D. Luce, R.R. Bush and E. Galanter, eds). Wiley: New York.

13. Marcus, G.F., Vijayan, S., Bandi Rao, S., and Vishton, P.M. (1999). Rule learning by seven-month-old infants. *Science*, *283*, 77-80.
14. Mathews, R.C., Buss, R.R., Stanley, W.B., Blanchard-Fields, F., Cho, J.R., and Druhan, B. (1989). Role of implicit and explicit process in learning from examples: A synergistic effet. *Journal of Experimental Psychology: Learning, Memory and Cognition*, *15*, 1083-1100.
15. McClelland, J.L. and Plaut, D. (1999). Does generalization in infant learning implicit abstract algrebra-like rules? *Trends in Cognitive Sciences*, *3*, 166-168.
16. Reber, A.S. (1967). Implicit learning of artificial grammars. *Journal of Verbal Learning and Verbal Behavior*, *6*, 855-863.
17. Redington, M. and Chater, N. (1996). Transfer in artificial grammar learning: A reevaluation. *Journal of Experimental Psychology: General*, *125*, 123-138.
18. Saffran, J.R., Newport, E.L., Aslin, R.N., and Barrueco, S. (1997). Incidental language learning: Listening (and learning) out of the corner of your ear. *Psychological Science*, *8*, 101-105.
19. Seidenberg, M.S. and Elman, J.L. (1999). Do infants learn grammar with algebra or statistics? *Science*, *14*, 433.
20. Servan-Schreiber, D., Cleeremans, A., and McClelland, J.L. (1991). Graded State Machines: The representation of temporal contingencies in simple recurrent networks. *Machine Learning*, *7*, 161-193.
21. Shanks, D.R. (1998). Distributed representations and implicit knowledge. In *Knowledge, Concepts and Categories*, (K. Lamberts and D.R. Shanks, Eds). Hove: Psychology Press, pp. 197-214.
22. Shanks, D.R., Johnstone, T., and Staggs, L. (1997). Abstraction processes in artificial grammar learning. *The Quarterly Journal of Experimental Psychology*, *50*, 216-252.
23. Shanks, D.R. and St. John, M.F. (1994). Characteristics of dissociable human learning systems. *Behavioral and Brain Sciences*, *17*, 367-447.
24. St John, M.F. and Shanks, D.R. (1997). *Implicit learning from an information-processing standpoint* In *How implicit is implicit learning?*, (D.C. Berry, Ed.). Oxford: Oxford University Press, pp. 124-161.
25. Timmermans, B. and Cleeremans, A. (2000). Rules versus statistics in biconditional grammar learning: A simulation based on Shanks el al. (1997) In *Proceedings of the Twenty-Second Annual Conference of Cognitive Science Society*. Hillsdale: Lawrence Erlbaum Associates, pp. 947-952.

Hidden Markov Model Interpretations of Neural Networks

Ingmar Visser, Maartje E.J. Raijmakers & Peter C.M. Molenaar

Abstract

Simple recurrent networks (SRN) can learn languages generated by finite state automata (FSA) [5]. The reverse process, i.e., extracting rules from neural networks in order to get FSAs, has also been explored. Rules from neural networks are generally extracted by partitioning the hidden state space of the network. Hidden Markov models (HMM) can also be used to extract FSAs from neural networks. The difference with other approaches is that it is not necessary to use the hidden state space activities of the network to extract the FSA: only the input-output relations of the network are required in fitting a HMM. Nonetheless, equivalent automata can be extracted. HMMs can thus be used to provide interpretations for the representations of neural networks.

1. Introduction

Modeling cognitive tasks with (back-propagation) neural networks has become increasingly popular over the last decade. Neural networks have been applied successfully in various areas of research such as memory, vision and learning. This success is, at least partially, due to the ease of use of neural networks as a means of modelling. It is fairly straightforward to apply neural networks to a given cognitive task: simply feed the network with the input a subject would get in an experiment, and provide it with feedback about its errors.

Another advantage of neural networks over classical models is that learning is a natural part of the system. Not only can a neural network represent certain cognitive abilities, it also reveals how such representations are built from scratch during the learning process.

One of the problems with neural networks is that, although they are successful at replicating certain data from human subjects, they do not really reveal how this is accomplished. One could defend the view that in applying a neural network to a given cognitive task one just creates another black box, just like the human mind one wants to model in the first place. This extreme view is not held by many cognitive scientists but the core of the argument is not to be taken lightly. More specifically, the problem is that the weights of a neural network and the internal states it has during responding to a certain input are, in general, hard to interpret.

Several researchers [9, 5, 13, 16] have tried to overcome this problem by interpreting the dynamics of neural networks in terms of FSAs. In the present paper we also extract FSAs from neural networks. The main difference between our approach and previous approaches is that we only use the overt behavior of the network, that is input-output relations, to build a finite state representation. We do not discretize the hidden state space of the neural network. This paper is organized as follows: first we outline the general approach used by other authors to extract finite state machines from networks. Then we contrast this with our approach which is based on HMMs. In the next section, we present simulation results obtained by extracting FSAs from recurrent neural networks using HMMs. Finally we discuss our results and consider how they can shed light on the aforementioned interpretational problems that exist with neural networks.

2. Extracting Finite State Automata from Networks

The main reason for extracting FSAs from neural networks is for purposes of interpretation. An FSA formalizes the activities of the hidden state space of a neural network into a set of clearly expressible rules. These rules can, in general, be stated in the form of (stochastic) dependencies between successive responses.

The common starting point in extracting FSAs from neural networks is to record the activities of the hidden state space during a training or testing session of the network. If the network has N hidden units, the result of this recording is a (time-ordered) trajectory through an N-dimensional state space. From this point, there are several ways to continue.

Giles et al. [9] partition this state space into hypercubes. These hypercubes are numbered and they represent the states of the FSA[1]. Transitions between states in the FSA are added whenever there is a corresponding transition in the recorded state space trajectory. The transition is labeled with the input symbol which is fed to the network at that point in the trajectory. The derived FSA is then minimized using a standard algorithm [10].

The second approach to constructing FSAs from networks involves clustering. Again, the starting point is a recorded trajectory through the state space of the hidden nodes of a neural network that has learned a finite state grammar. Cleeremans & McClelland [5] use clustering to identify states of an FSA that represents a neural network. They do not, however, actually construct FSAs from their trained neural networks.

Tino & Köteles [16] also use a form of clustering on the state space trajectory. In particular, they use a procedure similar to K-means clustering on the hidden state space activations that are recorded during a test run of the network. In contrast with Giles et al. [9], the FSAs that Tino & Köteles [16] extract from their network are actually stochastic FSAs: the transitions between states have probabilities associated with them.

In the present study we use HMMs to extract finite state representations from (recurrent) neural networks. This approach differs from the aforementioned approaches in several respects. First, we do not use recorded hidden state space trajectories but only input-output pairs from the trained networks. Therefore we do not have to make

---
[1] In general, only a fraction of the hypercubes actually become states in the FSA because the recorded trajectory does not pass through all the possible hypercubes.

any assumptions about this state space like, for example, in K-means clustering where the number of clusters has to be decided beforehand. The optimal number of states of the HMM can be determined by several goodness of fit statistics that will be discussed later. An advantage over the approach by Giles et al. [9] is that no minimization is required after extraction of the (stochastic) FSA. Further discussion of the similarities and differences between clustering, partitioning and HMMs as a means of extracting finite state machines is given below.

3. Recurrent Neural Networks

In this paper we chose to study SRNs for several reasons. First, SRNs have been shown to be able to learn languages generated by finite state automata [5]. Second, SRNs have been successful in modeling cognitive tasks in the domain of language learning [8] and in implicit learning [4]. As Elman [8] points out, the SRN is successful in these domains, because the recurrent connections of the network allow it to have a *memory* for past events, usually stimuli that have been presented to the network at earlier trials.

The architecture of the SRN that we used is given in [8]. The network, depicted in Fig. 1, consists of three layers of input, hidden and output nodes. In addition, the network has a context or recurrent layer that keeps a copy of the hidden unit activity at time $t - 1$. The SRNs behavior is generated by feedforward dynamics and learning is achieved with back-propagation [8].

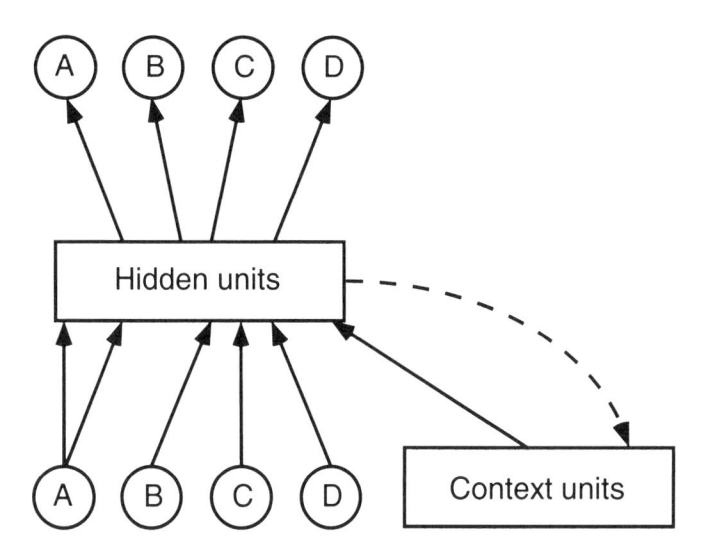

Figure 1: Simple recurrent network from Elman [8]. The network has four input nodes that each correspond to a different input symbol. The network is trained with back-propagation. The dashed arrow indicates that the activation of hidden units is copied to the context units. In contrast to the other connections, these are not trainable weights.

Because of its recurrent connections the SRN keeps a copy of its internal representation at $t-1$, that is, its combined representation of the stimuli that were presented to it at previous trials. This memory makes the SRN especially suitable to learn to represent sequentially structured data.

4. Hidden Markov Models

HMMs are mainly used in the area of speech recognition, c.f. [15, 3, 14], but have also found their way into psychological modeling, for example, of human action learning [18] and of coding of recall data [6]. Other applications are in physiology [2] and in computational genetics [11].

HMMs are stochastic finite state machines[2]. Formally an HMM consists of the following elements (notation from [14]):

1. a set of states S_i, $i = 1, \ldots, N$
2. a set V of observation symbols V_k, $k = 1, \ldots, M$
3. a matrix A of transition probabilities a_{ij} for moving from state S_i to state S_j
4. a matrix B of observation probabilities $b_j(k)$ of observing symbol V_k while being in state S_j
5. a vector π of initial state probabilities π_i corresponding to the probability of starting in state S_i at $t = 1$

The equations describing the dynamics of the model are as follows:

$$S_{t+1} = A\,S_t + \zeta_{t+1} \qquad (1)$$
$$O_{t+1} = B\,S_t + \xi_{t+1}. \qquad (2)$$

Equation (1) describes the hidden process S_t and (2) the observed process O_t; ζ_{t+1} and ξ_{t+1} are zero mean martingale increment processes (see [7, p. 20] for further details). In Fig. 2 an example of an HMM is depicted and compared with a normal Markov model.

The set of parameters $\lambda = (A, B, \pi)$ can be used to express certain notions about HMMs[3]. For example, the HMM can be used to generate a series of observations. Given specific values of the above parameters, it is easy to compute the likelihood $L(O|\lambda)$ of such a series by taking the product of the individual probabilities of moving from one state to the next and producing the observations O_t, $t = 1, \ldots, T$, $O_t \in V$, in those states. More interestingly, given a series of observations, it is possible to compute the parameter values of an HMM that optimize this likelihood. Rabiner [14] describes a version of the EM algorithm to maximize the likelihood $L(O|\lambda)$.

---

[2] HMMs come in many different guises and have many different designations: latent Markov models, (analytically tractable) Boltzmann machines, (analytically tractable) belief networks and (analytically tractable) state space models.

[3] It should be noted that the number N of hidden states is also a (hyper-)parameter of the model.

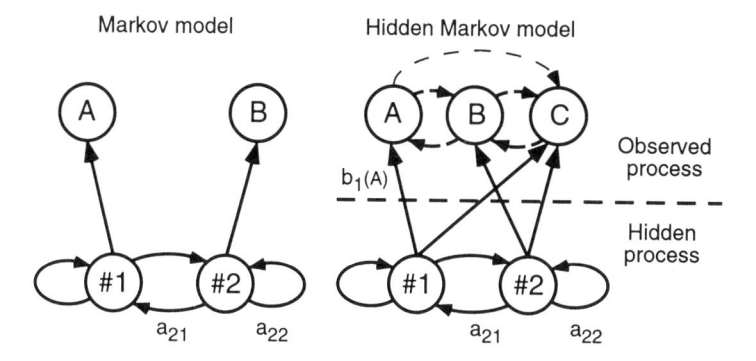

Figure 2: Comparison of a Markov and HMM. In the normal Markov model each state is linked with one and only one indicator. In the HMM each state can have multiple indicators. Therefore, from an observed symbol we can not immediately infer the state in which the process is at a given time. Alternatively, different states can share the same indicators, which leads to the same ambiguity. Observation probabilities are denoted with $b_1(A)$, transition probabilities with a_{ij}. The dashed arrows in the HMM denote observed transitions between symbols.

5. Simulations

5.1 Method

With the FSA depicted in Fig. 3 we produced a string of 420.000 symbols used for training the network. The training of the network is a replication of [5], using the same parameter settings and the same length of the training sequence. During training the network is presented with consecutive symbols of this string. Its task is to predict the next symbol in the string, hence the error between this target and the actual prediction of the network is backpropagated. The SRN had 3 hidden nodes. The weights of the network had random starting values between -0.5 and +0.5; the learning rate was set to 0.01 and the momentum term to 0.5. The criterion for learning was that the (normalized) activation of the node corresponding with the next symbol was over 0.3.

At regular intervals during the training the SRN was made to predict a string of symbols by itself. This was done with the following steps: (1) the network was presented with the begin symbol B, (2) the output activations of the network were normalized, (3) the normalized activations were treated as a probability distribution from which the next symbol was drawn (4) this next symbol was then given to the network as input. These steps were repeated until a sequence of 1000 symbols was generated. In this generation or prediction phase no feedback was given to the network and hence no learning took place.

The generated sequences were used to extract FSAs. HMMs with increasing numbers of states were fitted on these sequences until an optimal model was found according to a number of statistical criteria. First we considered the loglikelihood: if the loglikelihood does not increase anymore following an increase in the number of states, this means that adding more states to the HMM does not make for a better model. The loglikelihood however does not differentiate very well between models that are very

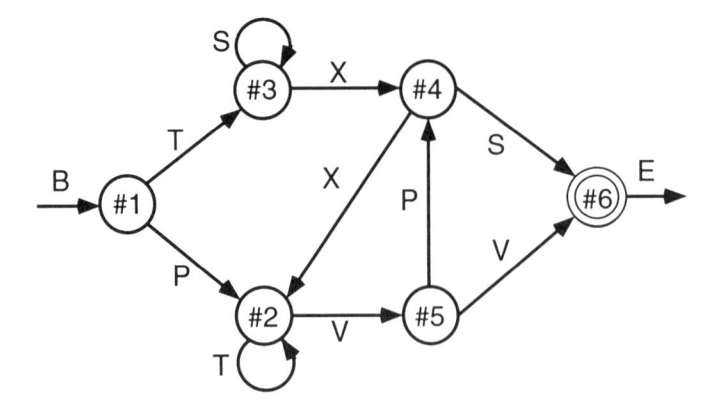

Figure 3: Representation of the finite state grammar that Reber we trained the SRN with. Sentences are generated by this grammar by starting at the symbol B (begin). With equal probability one of the arcs leaving state # 1 is chosen and the corresponding symbol is the next symbol in the string. This process of selecting arcs and noting the corresponding symbol continues until the end state # 6 is reached whence the process starts in state # 1.

similar. The loglikelihood keeps increasing, albeit just a little, when more states are added to the model and it is difficult to see a point where it levels off. As a second goodness of fit measure for the models, a kind of prediction error was used. The prediction error E_p is defined as:

$$E_p(n) = \sum_{\omega \in S_n} \frac{\left(F(\omega) - \hat{F}(\omega)\right)^2}{\hat{F}(\omega)}, \qquad (3)$$

where ω is a subsequence (word) of length n; S_n is the set of all words of length n; $F(\omega)$ is the observed frequency of ω, that is, the number of times this word occurs in the sequence under consideration; $\hat{F}(\omega)$ is the expected frequency of ω given the parameter values of the fitted HMM. For each n, $E_p(n)$ is χ^2-distributed with the degrees of freedom df equal to the number of terms in the above sum. A third goodness of fit measure, that is strongly related to the prediction error, is entropy.

Entropy is a measure of information density that is designed for languages or sets of sentences. The definition is [16]:

$$h = \lim_{n \to \infty} h_n, \qquad (4)$$

where h_n is the entropy rate which is defined as:

$$h_n = \sum_{\omega \in S_n} -p(\omega) \log(p(\omega)), \qquad (5)$$

where ω is a word (string of symbols) and S_n is the collection of grammatical sentences of length n. Entropy is an interesting invariant of symbolic sequences, because it measures a global property of (a set of) such sequences. Since we are not interested in exact reproduction of the behavior to be represented, but in the global statistical properties of that behavior, entropy is an excellent measure to use.

To be able to compute entropy we would need the complete language that an HMM (or SRN) can produce. However, we do not have a complete language at our disposal so we need an approximation of the entropy based on a finite sample of the language. The approximation we use is the Lempel-Ziv entropy. Suppose we have a string S=ABDACDABDADACABDC. We parse this string into substrings by taking the smallest substring that has not yet occurred. Since we start with an empty set of substrings A is the first substring that is parsed. Continuing this procedure, until the end of the string is reached, results in the following set of substrings: A, B, D, AC, DA, BD, AD, ACA, BDC. The Lempel-Ziv estimate of the entropy of the string is:

$$h_{LZ} = \frac{c(S) \log c(S)}{N}, \tag{6}$$

where $c(S)$ is the number of substrings resulting from the parsing procedure above and N is the total length of the string S [16].

Our use of this measure of fit involves comparing the Lempel-Ziv entropy of the sequence to which an HMM is fitted with the Lempel-Ziv entropy of the HMM itself. By doing this we know how closely the HMM matches the frequency distribution of the sequence it is supposed to model.

5.2 Results

We trained 20 SRNs with random starting values, two of which came to represent the grammar perfectly within two epochs of training. A third SRN came to learn the grammar perfectly after seven epochs. At twelve points before, during and after training we recorded a sequence of predictions by one of the successful SRNs. On each of these sequences we fitted HMMs with an increasing number of hidden states until an optimal model was found. We used both prediction errors and entropy for model selection. Specifically, we selected models that had the smallest number of states, but still performed well on the prediction error measures. After this selection, we verified whether the entropy estimates of the model were close to the estimate of the sequence. By checking the prediction error measures we avoid fitting an HMM which has the same entropy as the sequence, but nonetheless has a very different model structure. Figure 4 shows the entropies of the consecutive models.

There are a number of noteworthy findings in these results. First, the first HMM to be fitted has three states. That is, before the SRN has had any training it already has a three state HMM that best describes its production of sequences. This is surprising since a truly random model would correspond to a one state HMM. Hence an SRN with *random* starting values does *not* produce completely random behavior[4]. This is

---

[4] It is possible that the non-random behavior is an artifact of the generation procedure, namely that the same short (random) sequence is repeated again and again. This is not the case, however. This possibility is excluded by treating the output activations of the network as a probability distribution. Since every symbol has a nonzero activation it can always be selected as a successor in the generated sequence and hence repeating sequences are prevented.

Lempel-Ziv entropy

[Graph showing Lempel-Ziv entropy values decreasing from about 2.45 down to approximately 1.71 across consecutive models with state counts: 3, 5, 7, 8, 8, 8, 8, 11, 10, 11, 11, 12, 12. Horizontal lines indicate "Maximal entropy" near the top and "True entropy" near 1.70.]

 # states of consecutive models

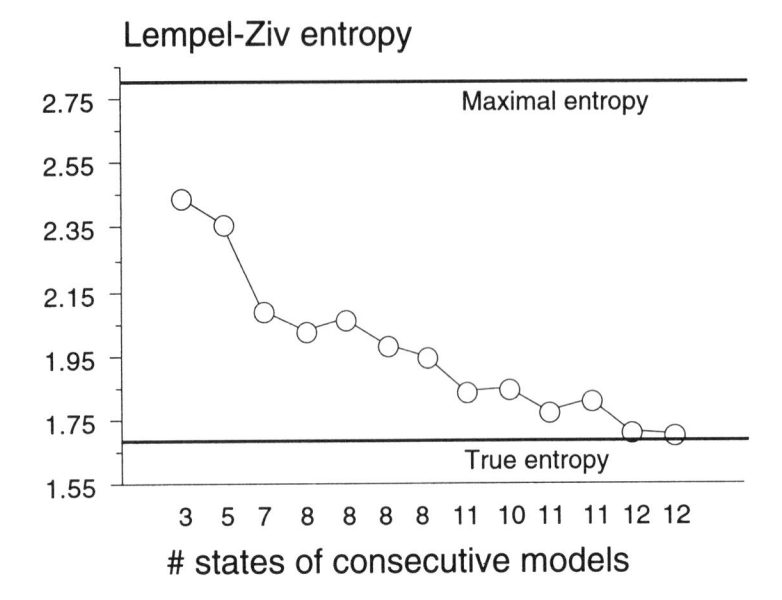

Figure 4: Below the ×-axis the number of states of the consecutive models is indicated. The first model, that is the model before the start of training, has three states, the last model has 12 states, the same as the true model, that is the model that was used to generate the data to train the SRN. Lempel-Ziv entropy is detailed in the text.

confirmed by the Lempel-Ziv entropy estimate of the sequence which has a value of 2.45 which is well below the theoretical estimate of 2.804 for a completely random sequence with 7 symbols. Second, the final HMM, i.e. the HMM that is fitted to the final sequence generated by the SRN has a Lempel-Ziv entropy of 1.709 which is very close to the Lempel-Ziv entropy of the true model which is 1.704. Finally, the consecutive models have nicely decreasing values for their Lempel-Ziv entropies which in the end converge to the entropy of the true model.

6. Discussion

HMMs can be applied successfully to extract automata from networks. This can be done without having to record the hidden state space activities of the neural network. This is an advantage over the method proposed by [9] because it means that the same procedure can be used for extracting (stochastic) FSAs from data generated by human subjects. For example, this can be done on generation data that subjects produce in implicit learning experiments (see e.g. [12]).

Once we have extracted HMMs from recurrent neural networks, we have interpretable models of the networks behavior. The description of a network's behavior in terms of an HMM provides explicit rules that describe its behavior. Moreover, since fitting of HMMs is done with maximum likelihood, all advantages of that approach are available: exact inference about parameter values, standard errors for the parameter estimates [17], the possibility of fitting HMMs with (equality) constraints on the parameters et cetera. Another advantage resulting from the maximum likelihood framework is the availability of goodness of fit measures.

In comparison with the partitioning method from [9] using HMMs to extract (stochastic) FSAs has the advantage that construction of an FSA and minimization is done together in one optimization procedure. When discretizing the state space, as [9] do, there are no obvious criteria available for deciding whether the discretization is fine-grained enough or whether further discretization is necessary. Some excellent work has been done by [1] to develop such criteria.

The use of HMMs for extracting FSAs is closely related to the approach developed by [16]. Tino & Köteles [16] used a version of K-means clustering to extract FSAs from neural networks. They introduced Lempel-Ziv entropy to test goodness of fit of their extracted FSAs. Using HMMs expands their method: maximum likelihood estimation of parameters provides exact inference about parameter estimates. Moreover, FSA extraction with HMMs is done in one optimization procedure instead of their two-step procedure of clustering and adding connections between clusters.

Acknowledgements

We wish to thank two anonymous reviewers for useful comments on an earlier draft which have helped clarifying some points, notably about entropy and random behavior of SRNs. We also thank Bram Bakker for fruitful discussions about many issues in this paper and Conor Dolan for proofreading this paper.

References

1. Bakker, B. & Jong, M. de. (In press). The epsilon state count. In *From animals to animats 6: Proceedings of the sixth international conference on the simulation of adaptive behavior, SAB 2000.*
2. Becker, J. D., Honerkamp, J., Hirsch, J., Fröbe, U., Schlatter, E. & Greger, R. (1994). Analysing ion channels with hidden Markov models. *European Journal of Physiology*, 426, 328–332.
3. Chien, J. T. & Wang, H. C. (1997). Telephone speech recognition based on bayesian adaptation of hidden Markov models. *Speech Communication, 22(4)*, 369–384.
4. Cleeremans, A. & McClelland, J. L. (1991). Learning the structure of event sequences. *JEP: General, 120*, 235–253.
5. Cleeremans, A., Servan-Schreiber, D. & McClelland, J. L. (1989). Finite state automata and simple recurrent networks. *Neural Computation, 1*, 372–381.
6. Durbin, M. A., Earwood, J. & Golden, R. M. (2000). Hidden Markov models for coding story recall data. In L. R. Gleitman & A. K. Joshi (editors), *Proceedings of the twenty-second annual conference of the cognitive science society*, 113–117. Lawrence Erlbaum Associates.
7. Elliott, R. J., Aggoun, L. & Moore, J. B. (1995). *Hidden Markov models: Estimation and control.* New York: Springer Verlag.
8. Elman, J. L. (1990). Finding structure in time. *Cognitive Science, 14*, 179–211.

9. Giles, C. L., Miller, C. B., Chen, D., Chen, H. H., Sun, G. Z. & Lee, Y. C. (1992). Learning and extracting finite state automata with second-order recurrent neural networks. *Neural Computation*, *4*, 393–405.
10. Hopcroft, J. & Ullman, J. (1979). *Introduction to automata theory, languages and computation.* Redwood City (CA): Addison-Wesley.
11. Krogh, A. (1998). An introduction to hidden Markov models for biological sequences. In S. L. Salzberg, D. B. Searls & S. Kasif (editors), *Computational methods in molecular biology*, 45–63. Elsevier.
12. Nissen, M. J. & Bullemer, P. (1987). Attentional requirements of learning: Evidence from performance measures. *Cognitive Psychology*, *19*, 1–32.
13. Omlin, C. & Giles, C. (1996). Extraction of rules from discrete-time recurrent neural networks. *Neural Networks*, *9(1)*, 41–51.
14. Rabiner, L. R. (1989). A tutorial on hidden Markov models and selected applications in speech recognition. *Proceedings of IEEE*, *77(2)*, 267–295.
15. Schmidbauer, O., Casacuberta, F., Castro, M. J. & Hegerl, G. (1993). Articulatory representation and speech technology. *Language and Speech*, *36(2)*, 331–351.
16. Tino, P. & Köteles, M. (2000). Extracting finite state representations from recurrent neural networks trained on chaotic symbolic sequences. *IEEE Transactions on Neural Networks*. (In press)
17. Visser, I., Raijmakers, M. E. J. & Molenaar, P. C. M. (In press). Confidence intervals for hidden Markov model parameters. *British journal of mathematical and statistical psychology*. (Preprint available from: ingmar@dds.nl)
18. Yang, J., Xu, Y. & Chen, C. S. (1997). Human action learning via hidden Markov model. *IEEE Transactions on Systems, Man and Cybernetics*, *27(1)*, 34–44.

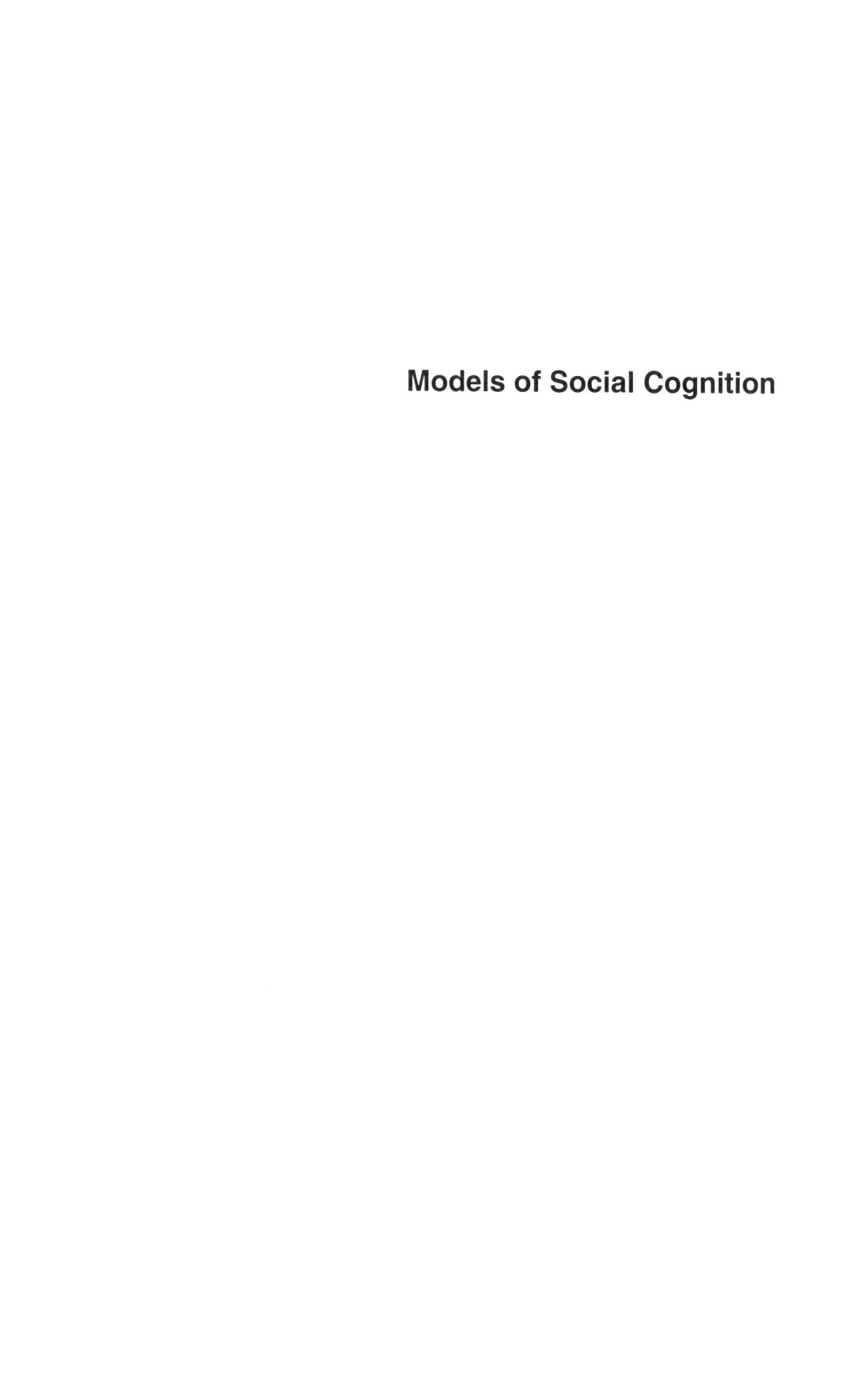

Models of Social Cognition

A Connectionist Model of Person Perception and Stereotype Formation

Christophe L. Labiouse & Robert M. French

Abstract

Connectionist modeling has begun to have an impact on research in social cognition. PDP models have been used to model a broad range of social psychological topics such as person perception, illusory correlations, cognitive dissonance, social categorization and stereotypes. Smith and DeCoster [28] recently proposed a recurrent connectionist model of person perception and stereotyping that accounts for a number of phenomena usually seen as contradictory or difficult to integrate into a single coherent conceptual framework. While their model is based on clearly defined and potentially far-reaching theoretical principles, it nonetheless suffers from certain shortcomings, among them, the use of misleading dependent measures and the incapacity of the network to develop its own internal representations. We propose an alternative connectionist model - an autoencoder - to overcome these limitations. In particular, the development of stereotypes within the context of this model will be discussed.

1. Introduction

Until recently, connectionist models have had only a marginal impact on research in social psychology. However, in the last five years, researchers have tried to account for certain well established phenomena in the social cognitive literature using connectionist models. Among these phenomena are illusory correlation [7], cognitive dissonance [26, 30], person perception [28], impression formation [16, 17], and causal attribution [22, 29]. Read & Miller [23] brought together these disparate models in a review book dedicated to connectionist models in social psychology. Like Smith [27], we believe that connectionist modeling in social psychology may lead to a major theoretical integration of our understanding of social behavior and cognition. After a long period of conflict between the 'social cognitive' and the 'social identity' approaches, a connectionist approach to certain areas of social psychology could shed light on our understanding of stereotyping, prejudice, discrimination and other intergroup processes.

Smith & DeCoster [28] recently proposed what is, to our knowledge, the only connectionist model of social perception and stereotyping. They use a recurrent network based on the McClelland & Rumelhart's model of learning and memory [20]. They use a nonlinear activation update, bounded real-valued activations, and the delta learning rule. While their model is based on clearly defined and potentially far-reaching theoretical principles, it nonetheless suffers from a number of shortcomings and has, we believe, several methodological problems. The most important limitation of their model is that it suffers from a linearity constraint in pattern learning: their model can learn a set of patterns only if the external input to each unit can be predicted perfectly by a linear combination of the activations of all other units, across the entire set of patterns. Smith & DeCoster suggest adding a hidden layer, something that is done in the model we propose here. There are a number of secondary problems with their model.

First, to account for the fact that people are able to generalize from multiple presentations of the same pattern, they train the network on a single pattern which is repeated 200 times (supposedly a frequently encountered individual), plus 1000 patterns randomly picked from a normal distribution (these are supposed to be the general background knowledge encountered by the social perceiver). To test a potential generalization, they probe the network with a corrupted version of the repeated exemplar. This task is unrealistic because people are exposed to many frequently encountered individuals. The problem with learning a single, often repeated exemplar is that this produces a very large basin of attraction for this pattern. Any starting pattern would likely reach the only available attractor state.

Second, we have had difficulty in reproducing their results. Using the equations they used, it would appear that the network has trouble finding a stable state in the activation update process. Formal analyses have shown that recurrent networks, in spite of their interesting dynamical properties, can also exhibit chaotic or oscillatory trajectories and cycle attractors [15]. This raises the question of when to modify the weights if a stable activation state has not been reached.

In this paper, we show that a simple multi-layer connectionist model - an autoencoder - can account for many robust phenomena in the social psychological literature. The rationale of our simulations was to develop a single model capable of qualitatively accounting for a wide range of well-known phenomena rather than fine-tuning one model to precisely reproduce the results of a single experiment.

2. Target Phenomena for these Simulations

We will test our model on a number of uncontroversial data patterns that can be found in the major handbooks and reviews of social cognition [9, 11], namely:

- *Exemplar-based inference*: Numerous studies [1, 18] have shown that people can learn particular properties of specific individuals (friends, family members, etc). With this knowledge, people can make inferences, often unconsciously, about unobserved traits or characteristics of a newly encountered individual.
- *Group-based stereotyping*: People can acquire stereotypes through social learning (gossip, media) and direct exposure to group members. They extract regularities in the traits of the people encountered and can apply this knowledge to draw inferences about either unobserved or perceived features (perceptual change in order to confirm the stereotype) of a new individual [14].
- *Concurrent exemplar-based inference and group-based stereotyping*: Our intent here is to demonstrate that a simple autoassociative memory can account for these two ways of processing that are generally difficult to integrate into a single framework. Traditional models [3, 8] have trouble integrating these two processes without resorting to a number of ad hoc hypotheses.

- *Development and formation of stereotypes*: We show that our model can account for many aspects of the development and formation of stereotypical knowledge without recourse to other factors, such as motivation, attention, cognitive load or norms. The use of stereotypical knowledge is seen as an increasing function of experience. Therefore, in our model, stereotyping can be conceived as a functional property in order to reduce the complexity of the social environment.

Only exemplar-based inference and group-based stereotyping were target phenomena in Smith & DeCoster's simulations. Most importantly, this model provides a unified theoretical framework for a fairly large number of phenomena related to stereotyping and social perception and can make novel predictions that can be tested in a traditional experimental setting.

3. Specific Aspects of the Model

Because of humans' necessary interactions with their social environment, the human brain has evolved in a such a way that social perceivers are able to cope with the intrinsic complexity of the social world. From this interaction and evolution have emerged, among other cognitive abilities, efficient face recognition, cheater detection, and one's own ingroup recognition. These abilities certainly provided an adaptive advantage during our evolutionary past. As a result, human brains are now able to recall a large number of individuals and events that allow them to deal with complex social situations. One of the most effective means of achieving this is to segregate the world into categories. Even if social and natural categories do not share the same properties, it is likely that the cognitive mechanisms by which we acquire them are similar. In both cases, categories could be extracted by a perceiver through statistical learning of the regularities in the world. Clearly, this is not an adequate explanation of all human learning but a large part of what a human being learns is probably implicit and requires no explicit rules or teachers.

In order to model certain cognitive social phenomena, particularly those involved in preconscious perceptual stages of conceptual interpretation, we chose to use an autoencoder whose task is to autoassociate a pattern via a hidden layer. This layer acts as a bottleneck and yields compressed representations of patterns. This network can learn without rules by observing exemplars, can automatically generalize, and can store precise information with a high degree of accuracy. Furthermore, autoencoders, unlike recurrent networks, do not suffer from the intrinsic problem of non-convergence to an attractor state. They also have a hidden layer which allows the network to overcome the problem of linear separability of the patterns to be learned. This hidden layer allows the network to develop its own internal representations, which it is certainly an essential feature of the human memory.

The results below are based on the performance of a 10-8-10 feedforward network. Activation values were either -1, 0, or $+1$. The rationale behind the coding is as follows: $+1$ could be conceived as the presence of an attribute or a trait, -1 as the absence of a trait, 0 was used to mean "impossible to determine whether the trait is present or not" (i.e. 0 is a "don't know" state. This coding fits the logic of social interactions because it is frequently impossible to say if a person has a trait or not. Each pattern presented to the network represents an individual that a social perceiver could encounter. We never explicitly present a group in itself. The basic rationale of the simulations depicted here was inspired by a study by Mareschal & French [19] on early infant categorization. We use the standard backpropagation learning rule with momentum. The learning rate was set to 0.0001 and the momentum to 0.9. A Fahlman correction of 0.1 was applied. Networks were trained for a maximum of 100 epochs or until a error criterion of 0.2 for all outputs for all patterns was reached. The particular details of each simulation are given below.

4. Simulations

4.1 Simulation 1: Exemplar-Based Influence

The goal of this simulation is to see if the network is able to store 4 different frequently encountered patterns. Moreover, we want to assess the network's ability to generalize to novel exemplars. Although this property is a relatively well known feature of this class of networks, we performed this simulation in order to reproduce the Smith & DeCoster's simulations scheme as closely as possible.

Method & Results

The network was always given the same four bit-strings of length 10. However, to introduce variability, any bit could be randomly set to zero (i.e. "don't know" state) with a probability of 0.1. Each "participant" saw an equal number of each bit-string. We simulated 10 participants. The order of presentation of these 4 exemplars was randomized for each run.

We tested the network's memory for these exemplars by presenting two kinds of patterns. First, we probed the network with each "pure" exemplar (i.e. no 0's) to see if it learned to autoassociate the exemplar. Second, we probed the network with degraded versions of each exemplar to see how it fills in the blanks. For each exemplar and for each run, we present 10 "2-bit", 10 "3-bit", and 10 "4-bit" corrupted versions (i.e., 2, 3, 4 bits were randomly set to 0). For each of these two procedures, we computed an error measure consisting of the discrepancy between the actual output and the pure exemplar. When we probe with the pure exemplar, we expect to see a decrease of the error compared to the error level before training. Moreover, when we probe with new patterns, close to the original patterns, we expect a slight increase in error but significantly below the initial error for unlearned patterns. In other words, well learned exemplar representations influence the way new patterns, close to the originally learned patterns, are perceived.

The network performs as expected. Figure 1 shows the mean initial error score, the mean error score (after training) for the pure exemplars, and the mean error score (after training) for the approximate versions. The error is averaged over the 4 exemplars.

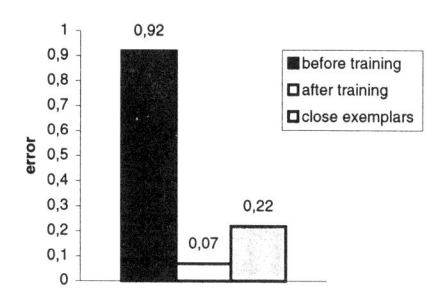

Figure 1: Mean error scores (Exemplar-based inference)

After learning, error is significantly lower, suggesting that the network has developed a reliable internal representation of the 4 exemplars. The generalization error rises slightly but stays well below the initial error, suggesting that the perception of similar exemplars is deeply influenced by the frequently encountered exemplars.

4.2 Simulation 2: Group-Based Influence

A stereotype is defined as a cognitive structure containing the social perceiver's knowledge, beliefs, and expectations about a group, learned through direct experience with individual group members and through social learning. Therefore, stereotypes affect inferences about newly encountered individuals [14] and these effects are often unintended and unconscious [5, 12]. Associations between a social category and a trait are most likely formed when they co-occur frequently and without too much variability [6].

Method & Results

We intended to show that the network is able to extract regularities in the presented patterns and can develop a prototypical representation of a group. Moreover, the network should be able to use this emergent knowledge to make inferences about new group members (i.e. patterns sharing some features with the prototypical group membership).

Instead of presenting the same bit-strings to the network, we presented variations of a prototype, ensuring that the network never encountered the prototypical group member. We continue to use the "don't know" state with a probability of 0.1. We simulate 2 groups, each consisting of 50 patterns. We first defined two bit-strings that will serve as stereotypes for the test phase. These two stereotypes were designed as follows: Five of the ten units were chosen as the "defining" features of the group because they are assumed to co-occur frequently. To introduce more variability, the probability that one of these units had the stereotypical feature was arbitrarily set to 0.8. The 5 other units were picked at random from randomly assigned values of 1 and –1. We simulated 10 participants. The order of presentation of the 100 patterns was randomized for each run.

We tested the network with the never-encountered stereotype to see if the network had extracted it from the repeated presentation of members. Second, we probed the network with incomplete versions of the stereotype to see how the network would fill in the blanks. For instance, does providing two stereotypical features allow the network to infer other stereotypical attributes ? For each stereotype and for each run, we presented 10 "1-bit", 10 "2-bit", and 10 "3-bit" corrupted versions (i.e., 1, 2, 3 bits set to 0). For each of these two procedures, we computed an error measure consisting of the discrepancy between the actual outputs and the desired outputs for the 5 "stereotypical" units.

Figure 2 shows the mean initial error score, the mean error score (after training) for the stereotype, and the mean error score (after training) for the previously unseen group members. Errors were averaged over both stereotypes.

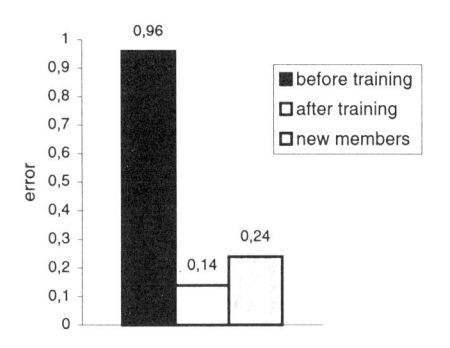

Figure 2: Mean error scores (group-based stereotyping)

After learning, the error is lower suggesting that the network has extracted regularities from the environment and has developed stereotypical knowledge. The generalization error is slightly higher but remains well below the initial error, suggesting that the network uses its newly acquired knowledge to infer stereotypical characteristics for new group members.

4.3 Simulation 3: Concurrent Exemplar- and Group-Based Influences

To investigate if the results of the two previous simulations were not artifacts, we test the network to see if it can exhibit both exemplar- and group-based processes simultaneously. A real social perceiver can simultaneously have a stereotype of a specific group and an accurate representation of exemplars which could be subtyped. The perception of newly presented individuals could be influenced either by the exemplar's representation or by the stereotypical one. Does the present network have these properties ?

Method & Results

We build a set of patterns consisting of a single exemplar presented 50 times and 50 group members presented each once. The exact procedure was the same as in the previous simulations. We simulate 10 participants. After training, we probe the network both with patterns close to the exemplar or with "new group member" patterns. We expect that, in both cases, errors will decrease compared to their initial state.

After training, the network exhibits both exemplar and group-based learning. Depending on the "person" (i.e., new pattern) encountered, the network is influenced both by frequently encountered exemplars and by emergent stereotypical patterns. Figure 3 shows the decrease in error, both for the "close-exemplar" and for the "stereotyped members" patterns.

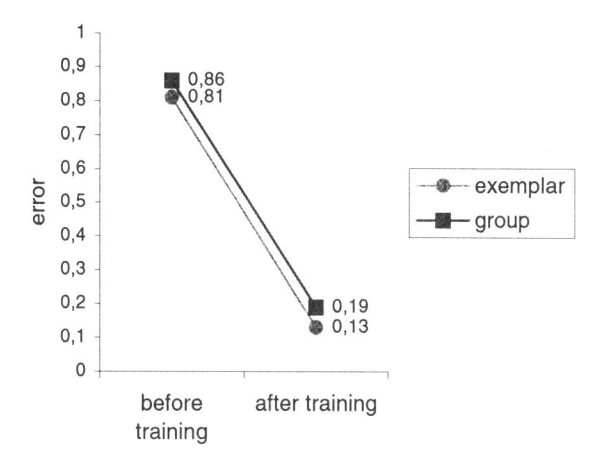

Figure 3: Mean error scores (concurrent exemplar-based and group-based stereotyping)

4.4 Simulation 4: Development of Stereotypical Knowledge

One important aspect of this model is that it can account for the development of mental representations of stereotypes. Sherman [24] noted that one of the most important factors influencing perceivers' reliance on exemplars or abstracted prototypes is the amount of experience perceivers have with the target to be judged. It is assumed that early encounters with a target will be of disproportionate importance because only a limited number of exemplars have been encountered from which to extract useful abstract knowledge. But as the number of encountered exemplars increases, a stereotypical representation can emerge, that then serves as the basis for subsequent inferences [25]. In this model, this property arises as a natural consequence of interacting with the complexity of the environment.

Method & Results

We took one of the stereotypes used in the second simulation and created a single exemplar from it. In the first phase, we used the same design as in the first simulation but with only this single exemplar instead of four. We then tested the network by computing two error measures: one with respect to the exemplar and the second with respect to the stereotype. The exemplar error was low and the stereotype error high. (See Figure 4). In the second phase, we presented the network with a second set of patterns, which consisted of group members. This composition reflects the particular experimental design used by Sherman [24]. After this second phase, the same error measures were computed as before. We observe an increase in the exemplar error and a simultaneous decrease in the group-based error. These results are consistent with Sherman [24]. All results were averaged over ten runs and are shown in Figure 4.

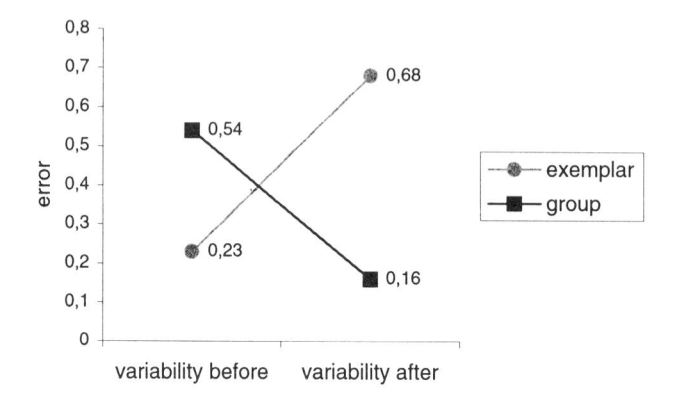

Figure 4: Mean error scores (Development of stereotypes)

5. Discussion

Compared to Smith & DeCoster's model, the autoencoder model is more powerful due to the presence of a hidden layer. This main improvement could be further tested in future studies (e.g. analysis of internal representations). In addition, autoencoders extract eigenvectors and are very close to PCA extractors. But they are more than PCA tools, insofar as they allow for learning, development and dynamical extrapolation of representations. However, although this model exhibits some interesting properties and can reproduce a number of important effects, it suffers from certain limitations. One of the most appealing properties of connectionist networks is their ability to do pattern completion, a property that is important in the simulations reported here. But the down side of this property is that the network will *always* fill in missing information whereas humans sometimes do not. This is certainly a drawback for current connectionist models of person perception. Sparse coding might be able to overcome this problem by producing a special "no recognition" state [4]. A second limitation of the model described here is catastrophic interference. Although learning is conceived somewhat differently in our simulations than in traditional cognitive psychology paradigms, the fact that catastrophic forgetting [21] can occur in these models is a real shortcoming for any model of human memory and cognition. This model, as any standard feedforward or Hopfield network models suffers from the "stability-plasticity" dilemma [13]. However, humans certainly do not suffer from catastrophic interference, especially with stereotypes. In fact, a stereotype is often particularly resistant to change. Modular computational architectures have been developed based on the brain's hippocampal-neocortical division of labor to overcome this problem [2, 10]. Finally, by using an autoencoder, we lose the dynamic properties of an attractor network and, in particular, we lose the ability to study the evolution of attitudes or impression formation over time. Nevertheless we could undoubtedly overcome this limitation by adding recurrent links to the present model.

6. Conclusion

We have presented a simple model that captures certain properties of the early unconscious stages of social perception and stereotyping. The autoencoder, like humans, develops a relatively accurate representation based on single exemplars that can be automatically used to make inferences on newly encountered exemplars similar to those already encountered. Moreover, the network is able to reproduce these effects even in presence of variable inputs. This means that a stereotypical representation of a group can be extracted from repeated presentations of different members of the group. We do not claim that this is the only mechanism for stereotype formation and stereotyping but this statistical interpretation can arguably account for the early stages of these complex processes. We also show that exemplar-based inference and group-based stereotyping can be exhibited by a single autoencoder simulating the way humans store this type of knowledge. This network also offers a potential model of the development of stereotypical representations.

Acknowledgments

Christophe Labiouse is a Research Fellow of the National Fund of Scientific Research (Belgium). This work was supported in part by a Camille Hela grant awarded to C. Labiouse by the University of Liege, and by a research grant from the European Commission (HPRN-CT-1999-00065).

References

1. Andersen, S., & Cole, S. (1990). "Do I know you ?": The role of significant others in general perception. *JPSP, 59*, 384-399.
2. Ans, B., & Rousset, S. (1997). Avoiding catastrophic forgetting by coupling two reverberating neural networks. *Académie des Sciences de la vie, 320*, 989-997.
3. Brewer, M. (1988). A dual process model of impression formation. In T. Srull & R. Wyer (Eds.), *Advances in social cognition, Vol. 1* (pp. 1-36). Hillsdale, NJ: LEA.
4. Buhmann, J., Divko, R., & Schulten, K. (1989). Associative memory with high information content. *Physical Review A, 39*, 2689-2692.
5. Devine, P. (1989). Stereotypes and prejudice: Their automatic and controlled components. *JPSP, 56*, 5-18.
6. Dijksterhuis, A., & van Knippenberg, A. (1999). On the parameters of associative strength: Central tendency and variability as determinants of stereotype accessibility. *Personality and Social Psychology Bulletin, 25*, 527-536.
7. Fielder, K. (2000). Illusory correlations: A simple associative algorithm provides a convergent account of seemingly divergent phenomena. *Review of General Psychology, 4*, 25-58.

8. Fiske, S., & Neuberg, S. (1990). A continuum of impression formation, from category-based to individuating processes. *Advances in Exp. Social Psychology, 23*, 1-74.
9. Fiske, S., & Taylor, S. (1991). *Social cognition (2nd edition)*. New York: McGraw Hill.
10. French, R. (1997). Pseudo-recurrent connectionist networks: An approach to the "sensitivity–stability" dilemma. *Connection Science, 9*, 353-379.
11. Gilbert, D., Fiske, S., & Lindzey, G. (Eds.) (1998). *Handbook of social psychology (4th edition)*. Boston, MA: McGraw-Hill.
12. Greenwald, A. & Banaji, M. (1995). Implicit social cognition: Attitudes, self-esteem, and stereotypes. *Psychological Review, 102*, 4-27.
13. Grossberg, S. (1982) *Studies of Mind and Brain: Neural Principles of Learning, Perception, Development, Cognition, and Motor Control*. Boston: Reidel
14. Hamilton, D., & Sherman, J. (1994). Stereotypes. In R. Wyer, & T. Srull (Eds.), *Handbook of social cognition* (2nd edition, Vol. 2, pp. 1-68). Hillsdale, NJ: LEA.
15. Hertz, J. (1995). Computing with attractors. In M. Arbib (Ed.), *Handbook of brain theory and neural networks* (pp. 230-234). Cambridge, MA: MIT Press.
16. Kashima, Y., Woolcock, J., & Kashima, E. (2000). Group impressions as dynamic configurations: The tensor product model of group impression formation and change. *Psychological Review, 107*, 914-942.
17. Kunda, Z., & Thagard, P. (1996). Forming impressions from stereotypes, traits, and behaviors: A parallel-constraint-satisfaction theory. *Psychological Review, 103*, 284-308.
18. Lewicki, P. (1985). Nonconscious biasing effects of single instances on subsequent judgments. *JPSP, 48*, 563-574.
19. Mareschal, D., & French, R. (1997). A connectionist account of interference effects in early infant memory and categorization. In *Proceedings of the 19th Annual Cognitive Science Society Conference* (pp. 484-489), Hillsdale, NJ: LEA.
20. McClelland, J., & Rumelhart, D. (1986). A distributed model of human learning and memory. In J. McClelland, & D. Rumelhart (Eds.). *Parallel Distributed Processing* (Vol. 2, pp. 170-215). Cambridge, MA: MIT Press.
21. Ratcliff, R. (1990). Connectionist models of recognition memory: Constraints imposed by learning and forgetting functions. *Psychological Review, 97*, 285-308.
22. Read, S., & Montoya, J. (1999). An autoassociative model of causal reasoning and causal learning: reply to Van Overwalle's (1998) critique of Read and Marcus-Newhall (1993). *JPSP, 76*, 728-742.
23. Read, S., & Miller, L. (Eds.) (1998). *Connectionist models of social reasoning and social behavior*. Hillsdale, NJ: LEA.
24. Sherman, J. (1996). Development and mental representations of stereotypes. *JPSP, 70*, 1126-1141.
25. Sherman, J., & Klein, S. (1994). The development and representations of personality impressions. *JPSP, 67*, 972-983.
26. Shultz, T., & Lepper, M. (1996). Cognitive dissonance reduction as constraint satisfaction. *Psychological Review, 103*, 219-240.
27. Smith, E. (1996). What do connectionism and social psychology offer each other ? *JPSP, 70*, 893-912.
28. Smith, E., & DeCoster, J. (1998). Knowledge acquisition, accessibility, and use in person perception and stereotyping: Simulation with a recurrent connectionist network. *JPSP, 74*, 21-35.
29. Van Overwalle, F. (1998). Causal explanation as constraint satisfaction: A critique and a feedforward connectionist alternative. *JPSP, 74*, 312-328.
30. Van Overwalle, F. (submitted). A feedforward connectionist model of cognitive dissonance: An alternative to Shultz and Lepper (1996).

Learning about an Absent Cause: Discounting and Augmentation of Positively and Independently Related Causes

Frank Van Overwalle & Bert Timmermans

Abstract

Standard connectionist models of pattern completion like an auto-associator, typically fill in the activation of a missing feature with internal input from nodes that are connected to it. However, associative studies on competition between alternative causes, demonstrate that people do not always complete the activation of a missing feature, but rather actively encode it as missing whenever its presence was highly expected. Dickinson and Burke's revaluation hypothesis [4] predicts that there is always forward competition of a novel cause, but that backward competition of a known cause depends on a consistent (positive) relation with the alternative cause. This hypothesis was confirmed in several experiments. These effects cannot be explained by standard auto-associative networks, but can be accounted for by a modified auto-associative network that is able to recognize absent information as missing and provides it with negative, rather than positive activation from related nodes.

1. Introduction

In connectionist models that produce pattern completion like an auto-associator, whenever a feature is missing, its activation is filled up with internal input from nodes that are connected to it [9]. However, in human induction, we so often do not complete the activation of a missing feature, but rather actively encode it as missing whenever its presence was highly expected on the bases of related features at input. This phenomenon of missing features has recently been studied in associative learning research, in the context of backward competition between alternative causes [4, 7, 15, 16]. Because standard connectionist networks [9] cannot explain backward competition, our aim is to provide an alternative connectionist account of this effect by modifying the standard auto-associative network.

1.1 Forward and Backward Competition

Competition refers to the tendency to alter the perceived strength of a cause or cue in the face of alternative or competing causal explanations. Perhaps the most well known illustration of competition is discounting (or blocking), which refers to the tendency to disregard or underestimate potential causes when the facilitatory influence of an alternative cause is already established. For instance, a person's success is attributed less to internal abilities given evidence of additional external aid. The opposite effect, augmentation (or superconditioning) refers to the tendency to overestimate the strength of a focal cause when it overcomes the inhibitory influence of an alternative cause. For instance, success is more strongly attributed to internal capacities when the task was hard.

Competition can have an effect on a cause that is either known or novel. When the alternative cause is already known and exerts its influence on a novel focal cause, this is called forward competition. For instance, when we know that someone is a hell of a good tennis player (A+, see Table 1) and when that player wins a doubles game with a novel partner (AT+), we tend to discount the contribution of the novel partner in the win. Conversely, when we know that someone is a poor player (A-), we tend to augment the contribution of the novel partner in the win (AT+).

Conversely, when a novel alternative cause exerts its influence afterwards, that is, on a known focal cause, this is called backward competition [4]. For instance, when we only recently learn that one of two partners of a well-known successful doubles tennis team (AT+, see Table 1) is now winning all his or her single games (A+), we tend to discount our initial high estimation of the other partner. Conversely, when one partner of a successful doubles team (AT+) is losing all his or her single games (A-), we are now likely to augment our initially evaluation of the other partner.

	Phase 1	Phase 2
Forward		
Discounting of T	A+	AT+
Augmentation of T	A-	AT+
Backward		
Discounting of T	AT+	A+
Augmentation of T	AT+	A-

Table 1: Schematic illustration of forward and backward discounting and augmentation. Note. T = target cause that is discounted or augmented, A = alternative cause that produces discounting or augmentation, + = focal outcome (e.g., winning a game), - = opposite outcome (e.g., losing a game).

1.2 Backward Competition and Missing Features

Previous associative theories of causal induction such as the popular model of Rescorla and Wagner [12] — which is identical to the delta learning algorithm embedded in a feedforward network [9] — are unable to account for backward competition. The reason is that in the original Rescorla-Wagner model, the absent cause on which competition (i.e., discounting or augmentation) should be exerted is assumed to have zero activation. Hence, its causal weight cannot be adjusted.

Van Hamme and Wasserman [13] therefore proposed to give absent causes negative activation that would result in backward competition. Dickinson and Burke [4] further hypothesized that such a negative activation of an absent cause would occur only if this absence was unexpected, for instance, after the cause had formed a consistent relation with the alternative cause. Thus, it was predicted that backward discounting and augmentation are more likely when "target and competing cues were consistently paired during compound training" [4, p. 73]. Note that this differs from forward competition, which occurs regardless of the relation between causes in line with the original Rescorla-Wagner model [12].

To elucidate people's reaction to missing features, in research on causal competition, the relation between causes was manipulated [4, 7, 15, 16]. This relation was either consistent (always paired together) or varied (paired with many other causes), resulting in a positive or independent statistical relation respectively between the causes. Once this relationship was established, researchers investigated the effect of deleting one of the causes from the input (i.e., backward competition). A schematic example of such a procedure used by Van Overwalle and Timmermans [15] is given in Table 2.

The results revealed either discounting when the effect occurred even in the absence of a focal cause, or augmentation when the effect reversed in the absence of a focal cause. More importantly, this only occurred when the relation between the causes was positive (consistent pairing) as predicted by Dickinson and Burke's revaluation hypothesis. These findings have been replicated in several experiments [4, 7, 15, 16].

2. A Modified Auto-Associative Network

In this section, we develop a connectionist approach to the backward revaluation hypothesis of Dickinson and Burke [4] by introducing a modification to the standard auto-associative network [9]. The key idea is that this modification recognizes absent but expected input as missing and provides it with negative, rather than positive activation from related nodes.

To account for backward revaluation, Van Hamme and Wasserman [13] proposed that absent causes could take on negative activation when their absence is unexpected. Dickinson and Burke [4] further refined this proposal and asserted that the expectation that a cause should be present depends on the relationship with other causes: "Only the omission of an expected cue should generate a ... negative activation ... it is the formation of within-compound associations during the first stage of training that provides the basis for this expectation" [4, p. 63]. To allow such within-compound connections (i.e., between all nodes including nodes representing a causal input), we used a recurrent network and more in particular, an auto-associator [9].

Block & Relation	Trials		
Block 1: Target & Alternative Causes ("compound" trials)			
Consistent			
	A & B	*A* & B	*A* & B
	R & S	*R* & S	*R* & S
	X & Y	*X* & Y	*X* & Y
Varied			
	A & B	*A* & S	*A* & Y
	R & B	*R* & S	*R* & Y
	X & B	*X* & S	*X* & Y
Block 2: Alternative Cause Only			
	B	B	B
	S	S	S
	Y	Y	Y

Table 2: Illustration of a Backward Competition Design. Note. Each row represents three trials. Target causes A, R, X are in italic. In forward competition, the order of Blocks 1 and 2 was reversed. In the control conditions, Block 2 was omitted. To illustrate, a trial from Block 1 reads as follows: "[Target and alternative names] won their doubles tennis game", and from Block 2 reads as follows: "[Alternative name] won her singles tennis game".

In an auto-associative network, causes and outcomes are represented in nodes that are all interconnected. The perceived influence of a cause on an outcome is reflected in the weight connecting the cause with the outcome. Processing information in this model takes place in two phases. In the first phase, the activation of the nodes is computed, and in the second phase, the weights of the connections are updated. We describe the auto-associative model in some more detail below together with a discussion of our modifications.

2.1 Activating Present and Absent Nodes

During the first phase of information processing, each node in the network receives activation from external sources. Because the nodes are all interconnected, this activation is then spread throughout the network where it influences all other nodes.

The activation coming from the other nodes is called the internal input. Together with the external input, this internal input determines the final pattern of activation of the nodes, which reflects the short-term memory of the network.

In mathematical terms, every node i in the network receives external input, termed ext_i. In the auto-associative model, every node i also receives internal input int_i which is the sum of the activation from the other nodes j in proportion to the weight of their connection, or

$$int_i = \Sigma (activation_j * weight_{ij}), \qquad (1)$$

for all $j \neq i$. Typically, activations and weights range between -1 to $+1$. The external input and internal input are then summed to the net input, or

$$net_i = E\, ext_i + I\, int_i, \qquad (2)$$

where E and I reflect the degree to which the net input is determined by the external and internal input respectively. Typically, in a recurrent network, the activation of each node i is updated during a number of cycles until it eventually converges to a stable pattern that reflects the network's short-term memory. According to the linear activation algorithm, the updating of activation is governed by the following equation:

$$\Delta activation_i = net_i - D * activation_i, \qquad (3)$$

where D reflects a memory decay term. To simplify the recurrent model and to demonstrate its relationship with the original Rescorla-Wagner model of associative learning, in the present simulations, we used only one internal updating cycle and the parameter values $D = I = E = 1$. Given these simplifying assumptions, the final activation of node i reduces simply to the sum of the external and internal input, or:

$$activation_i = net_i = ext_i + int_i \qquad (3')$$

A typical emergent feature of an auto-associative model is that an absent input stimulus m will be assimilated. That is, missing input will generally be 'filled up' by similar activation from the internal input. This is so, because although an absent cue does not receive any external input $(ext_m = 0)$, it still receives internal input from all other nodes with which it is connected.

However, to simulate Dickinson and Burke's backward revaluation hypothesis [4], some modifications were introduced to this typical assimilation effect of a standard auto-associative network. Essentially, this modification prescribes that during learning by the network, an absent stimulus m that is clearly anticipated through a connection with another stimulus j, will not be 'filled up' by the activation coming from node j, but rather will receive an opposite activation resulting in a contrast or correction. Specifically, when the activation from one of the other nodes j exceeds some 'missing threshold' (denoted by μ), then the missing stimulus m accrues *missing input* rather than internal input.

In mathematical terms, for any node m where $ext_m = 0$ and $|activation_j * weight_{mj}| > \mu$, Equation 1 is replaced by

$$miss_m = \Sigma (activation_j * weight_{mj}). \qquad (4)$$

Note that the missing input is summed only for those nodes j where the missing threshold μ is exceeded. For the final activation pattern, the missing input is then subtracted from the external input (instead of being added as in Equation 2), effectively resulting in a contrast or correction effect. Hence, once the missing threshold of an absent input m is exceeded, Equation 2 is replaced by

$$net_m = E\ ext_m - I\ miss_m. \quad (5)$$

After making the simplifying assumptions of one internal updating cycle and parameter values $D = I = E = 1$, this becomes

$$activation_m = ext_m - miss_m = -miss_m. \quad (5')$$

2.2 Weight Updating

After this first phase, the auto-associative model then enters in its second learning phase, where the short-term activation is consolidated in long term weight changes to better represent and anticipate future external input. Here we follow the standard assumption of the auto-associator [9, p. 166]. Basically, weight changes are driven by the discrepancy between the internal input from the last but one updating cycle of the network and the external input received from outside sources, formally expressed in the delta algorithm:

$$\Delta weight_{ij} = \varepsilon * (ext_i - int_i) * activation_j, \quad (6)$$

where ε is a learning rate that determines how fast the network learns.

In summary, the proposed recurrent network can reproduce not only the typical assimilation effect of standard auto-associative networks but also contrastive revaluation as proposed by Dickinson and Burke [4]. When the presence of a missing stimulus is not greatly anticipated (because of weak interconnections with other nodes), a typical assimilation effect will ensue. In contrast, when the absence of a missing stimulus is very much unexpected (because of strong connections with other nodes), a contrast effect will result.

3. Simulations

The predictions of the modified recurrent network were tested with the data from Van Overwalle and Timmermans [15]. The results of these experiments, and of the simulation (to be explained shortly) are shown in Figure 1.

As predicted by Dickinson and Burke's revaluation hypothesis [4], there was significant discounting in comparison with the control conditions (see top panel), except with the backward varied condition. Likewise, there was also significant augmentation in comparison with the control conditions (see bottom panel) but less so with the backward varied condition. Although the predicted attenuation of augmentation was less clear-cut (but see [4, 7] for stronger effects), with a backward order, augmentation was still significantly stronger in a consistent than varied relation, while this difference failed to reach significance with a forward order as predicted.

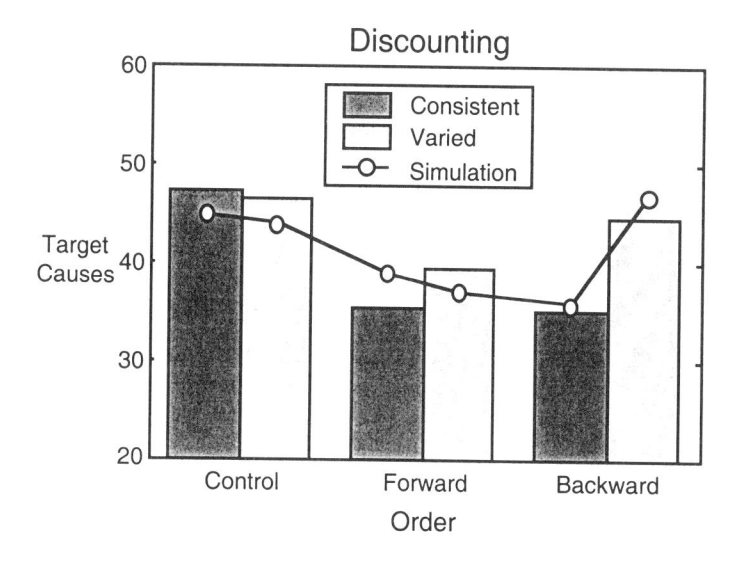

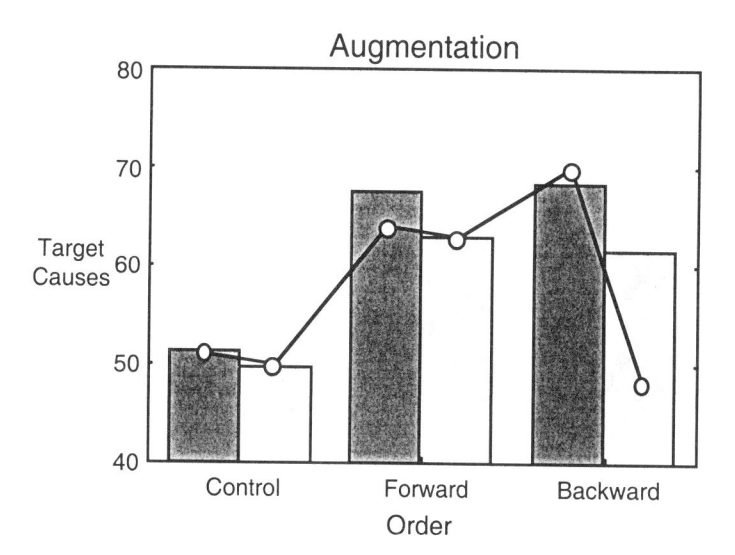

Figure 1: Discounting and Augmentation: Causal ratings in function of Order and Relation (based on Van Overwalle & Timmermans [15]).

In our simulations, we used an auto-associative network architecture with a single node for each cause or outcome, and the modified activation updating algorithm as explained above. The model was run using exactly the same order of trials and blocks as in the experiments. Because trial order was randomized within blocks, we ran 50 simulations with a random trial order for each block, and averaged the results.

The presence of a cause at a trial was encoded by setting the external input to +1; otherwise the external activation remained at resting level 0. Likewise, a target outcome was encoded by an external input of +1, and the opposite outcome by an external input of –1. The weights of the connections were updated after each trial. In the simulations that we report, we explored different values for the missing threshold parameter μ until one was found that resulted in the best visual fit with the observed data. The other model parameters were kept constant $(\varepsilon = .10, E = I = D = 1, cycles = 2)$. These parameters were chosen so as to increase the similarity with the original Rescorla-Wagner associative model on which Dickinson and Burke's [4] backward revaluation hypothesis rests.

At the end of each simulated trial history, the causal influence of the target and alternative stimuli was tested by turning the external input of the corresponding nodes to +1 and reading off the resulting activation of the output node. During this testing phase, the revaluation process was not active (i.e., standard Equation 2 was executed instead of the 'missing' correction of Equation 5). The obtained simulation values were projected onto the observed ratings using linear regression (with slope > 0), to demonstrate visually the fit of the simulations.

The full lines in Figure 1 show the results. As can be seen, consistent with the data and Dickinson and Burke's backward revaluation hypothesis, the simulations predict an attenuation of discounting and augmentation for the target ratings in the backward varied conditions only. The simulations matched the discounting data (top panel) very well, as can be verified also from the correlation between simulated and observed means, $r = .981$ $(\mu = .15)$. The observed mean ratings included not only the target causes as shown in Figure 1, but also the alternative causes not shown. However, the fit was less adequate for the augmentation data (bottom panel, as noted earlier, these data showed less of the predicted pattern), although the correlation was still high, $r = .938$ $(\mu = .20)$. These simulation results were not obtained with a standard recurrent model without our revaluation modification.

4. Discussion

To account for backward competition, learning theories must take into account the order of presentation at encoding and the interrelations between causes or cues. Because probabilistic and other rule-based models of causality are not sensitive to order or inter-cause relationships [2, 3, 6, 11], they cannot account for the present data. It is not easy to see how these two fundamental flaws can be amended in a reformulation of these theories. Alternative theories that posit that all competition takes place at the time of testing [1, 10] also fail to account for backward competition.

The backward revaluation data also pose difficulties for standard feedforward connectionist models because they fail to incorporate interrelations between causes [12, 14]. Although recurrent auto-associative models include interrelations between causes [9], we demonstrated that they need an additional corrective mechanism to reflect the revaluation hypothesis, that is, to create a special status for positively related causes that are expected but "missing". Currently, these recurrent models only assume that that absent concepts are "filled up" with activation from related concepts in memory, which leads to less rather than more discounting and augmentation, contrary to the revaluation hypothesis. Graham [5] developed a similar corrective revaluation mechanism into a recurrent network that is, however, not so successful in replicating our data.

Our network model of backward revaluation can also account for related phenomena that are also potentially driven by a revaluation mechanism. An obvious case is conditioned inhibition, which is exactly the opposite of augmentation. Conditioned inhibition refers to the tendency to increase the inhibitory strength of a focal cause when it attenuates the facilitatory influence of an alternative cause. For instance, failure is more strongly attributed to lack of internal capacities when the task was easy. The mechanism that explains backward conditioned inhibition is identical to that for augmentation, the only difference being the negative direction of the outcome. Simulations of our proposed modified auto-associator with the backward conditioned inhibition data of Larkin et al. [7, exp. 4] showed a high fit.

Another illustrative case is backward competition between causes that occur sequentially (and thus are, in statistical terms, negatively related). For instance, as suggested by the proverb that new brooms sweep clean, novel leaders who replace older management often get the sole credit for success, although their success often depends in part on prior managerial decisions. This indicates that initial causes can be discounted by subsequently presented alternative causes. Matute and Piñeño [8] recently documented such backward discounting. According to our proposed corrective mechanism for missing nodes, because the initial causes are expected given their prior pairings with the outcome, their omission is very salient and results in downward revaluation. Simulations of our modified auto-associator with the data of Mutate and Piñeño [8, exp. 1] again showed a substantial fit.

References

1. Bouton, M. E. (1993). Context, time, and memory retrieval in the inference paradigms of Pavlovian learning. *Psychological Bulletin, 114*, 80—99.
2. Busemeyer, J. R. (1991) Intuitive statistical estimation. In N. Anderson (Ed.) *Contributions to Information integration theory, vol. 1: Cognition*. Hillsdale, NJ: Erlbaum.
3. Cheng, P. W., & Novick, L. R. (1990). A probabilistic contrast model of causal induction. *Journal of Personality and Social Psychology, 58,* 545—567.
4. Dickinson, A. & Burke, J. (1996). Within-compound associations mediate the retrospective revaluation of causality judgments. *Quarterly Journal of Experimental Psychology, 49B*, 60-80.
5. Graham, S. (1999). Retrospective revaluation and inhibitory associations: Does perceptual learning modulate our perceptions of the contingencies between events? *Quarterly Journal of Experimental Psychology, 52B*, 159-185.

6. Hogarth, R. M. & Einhorn, H. J. (1992) Order effects in belief updating: The belief-adjustment model. *Cognitive Psychology, 24*, 1—55.
7. Larkin, M. J. W., Aitken, M. R. F., Dickinson, A. (1998). Retrospective revaluation of causal judgments under positive and negative contingencies. *Journal of Experimental Psychology, Learning, Memory, and Cognition, 24*, 1331—1352.
8. Matute, H. & Pineño, O. (1998). Stimulus competition in the absence of compound conditioning. *Animal Learning and Behavior, 26*, 3-14.
9. McClelland, J. M. & Rumelhart, D. E. (1988). *Explorations in parallel distributed processing: A handbook of models, programs, and exercises.* Cambridge, MA: Bradford.
10. Miller, R. R. & Matzel, L. D. (1988). The comparator hypothesis: A response rule for the expression of associations. In G. H. Bower (Ed.), *The psychology of learning and motivation* (Vol. 22, pp. 51—92). San Diego, CA: Academic Press.
11. Morris, M. W. & Larrick, R. P. (1995). When one cause casts doubt on another: A normative analysis of Discounting in causal attribution. *Psychological Review, 102*, 331—355.
12. Rescorla, R. A. & Wagner, A. R. (1972). A theory of Pavlovian conditioning: Variations in the effectiveness of reinforcement and nonreinforcement. In A. H. Black & W. F. Prokasy (Eds.) *Classical conditioning II: Current research and theory* (pp. 64–98). New York: Appleton-Century-Crofts.
13. Van Hamme, L. J. & Wasserman, E. A. (1994). Cue competition in causality judgments: The role of nonpresentation of compound stimulus elements. *Learning and Motivation, 25*, 127—151.
14. Van Overwalle, F. (1998) Causal Explanation as Constraint Satisfaction: A Critique and a Feedforward Connectionist Alternative. *Journal of Personality and Social Psychology, 74*, 312-328.
15. Van Overwalle, F. & Timmermans, B. (2000). *Discounting and Augmentation in Attribution: The Role of the Relationship between Causes.* Manuscript submitted for publication.
16. Wasserman, E. A. & Berglan, L. R. (1998). Backward blocking and recovery from overshadowing in human causal judgment: the role of within-compound associations. *Quarterly Journal of Experimental Psychology, 51B*, 121—138.

Evolution

# Exploring the Baldwin Effect in Evolving Adaptable Control Systems

John A. Bullinaria

Abstract

A neural network model is presented which is an abstraction of many real world adaptable control systems that seems to be sufficiently complex to provide interesting results, yet simple enough that computer simulations of its evolution can be carried out in weeks rather than years. Some preliminary explorations of the interaction between learning and evolution in this system are described, together with some suggestions for future research in this area.

1. Introduction

The idea of evolution by survival of the fittest is now widely accepted. However, whilst it is clear that many human abilities have become innate as a result of evolution, others still need to be learned or modified during an individual's lifetime, so the nature-nurture debate rages on in many areas. It has been known for some time that there will be an interaction between learning and evolution (commonly called the Baldwin Effect [1, 2, 4]), but realistic explicit simulation is difficult due to the enormous computational resources required. Training a non-trivial fully dynamical neural network takes time, and consequently, training a changing population of such neural networks over many generations with reasonable procedures for procreation, mutation and survival is still barely feasible. In this paper I shall present a neural network model that is an abstraction of many real control systems (e.g. reaching, pointing, oculomotor control) which seems to be sufficiently complex to provide interesting results, yet simple enough that computer simulations of its evolution can be carried out in weeks rather than years. A range of preliminary explorations of the Baldwin effect in this system will be described, together with some suggestions for future research in this area.

2. The Simplified Control Model

The simplified generic control system that will form the basis of the study presented in this paper is shown in Figure 1. Its inputs are fairly accurate near cues and less accurate far cues. The network's response to these cues is generated by an initial approximate open loop signal based on the far cues, followed by a more accurate closed loop signal based on the near cues. These signals feed into integral and proportional controllers, the outputs of which are added to bias and tonic signals, and

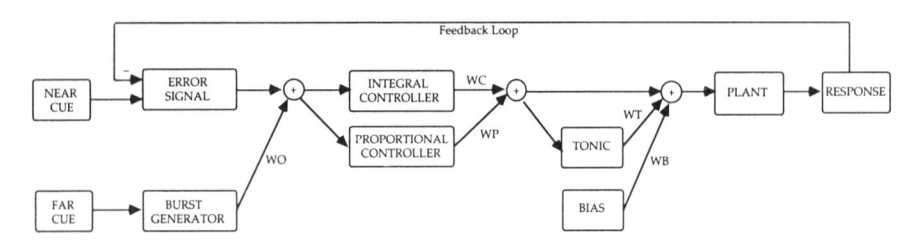

Figure 1: Simplified control model with five learnable parameters.

fed into the plant to produce the required response. The bias provides an appropriate resting state, and the tonic allows short time-scale adaptation of the resting state during periods of constant demand. The whole system can be regarded as a fully dynamical network of leaky integrator neurons. In the human accommodation (eye focussing) system, for example, we have blur and proximal cues being processed to generate appropriate signals for the ciliary muscles in the eye [3]. The model has five adjustable parameters (weights WO, WC, WP, WT, and bias WB) which are learnt by a simple gradient descent algorithm that minimizes a cost function consisting of response error and regularization (smoothing) components which will be readily available to the system. Corresponding to these learnable weights, each instantiation of the model has five fixed initial weights (iWO, iWC, iWP, iWT, iWB) and five fixed learning rates (eWO, eWC, eWP, eWT, eWB). The model also has various other parameters (neuron time constants, plant characteristics, feedback time delay, and so on) which we take to be the same for all instantiations. Such a system that has learnt/evolved a good set of parameters will produce appropriate damped responses to arbitrary discontinuous output requirements (steps) and smooth pursuit of arbitrary continuous output changes (ramps) as illustrated in Figure 2.

3. Evolving the Model

To simulate an evolutionary process for our model we take a whole population of individual instantiations of the model and allow them to learn, procreate and die in a manner approximating these processes in real (living) systems. The genotype of each individual will depend on the genotypes of its two parents, and contain the initial weights and learning rates. Then during its life the individual will learn from its environment how best to adjust its weights to perform most effectively. Each individual will eventually die, perhaps after producing a number of children.

In realistic situations, the ability of an individual to survive or reproduce will rely on a number of factors which can depend in a complicated manner on that individual's performance on a range of related tasks (food gathering, fighting, running, and so on). For the purposes of our simplified model, we shall consider it to be a sufficiently good approximation to assume a simple linear relation between our single task fitness function and the survival or procreation fitness. In fact, any monotonic relation will result in similar evolutionary trends.

We shall follow a more natural approach to procreation, mutation and survival than many evolutionary simulations [2]. Rather than training each member of the

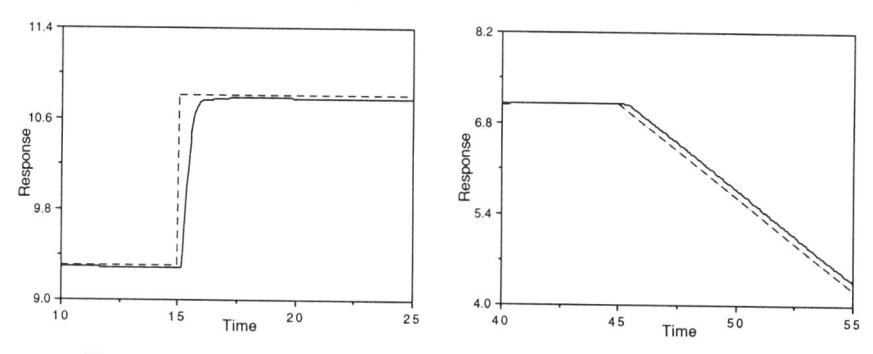

Figure 2: Typical system responses to arbitrary step and ramp stimuli.

whole population for a fixed time and picking the fittest to breed and form the next generation, our populations contain competing learning individuals of all ages, each with the potential for dying or procreation at each stage. During each simulated year, each individual learns from their own experience with a new randomly generated common environment (i.e. set of training/testing data) and has their fitness measured. A biased random subset of the least fit individuals, together with a flat random subset of the oldest individuals, then die. These are replaced by children, each having one parent chosen randomly from the fittest half of the population who randomly chooses their mate from the rest of whole population. Each child inherits characteristics from both parents such that each innate free parameter is chosen at random somewhere between the values of its parents, with sufficient noise (or mutation) that there is a reasonable possibility of the parameter falling outside the range spanned by the parents. Ultimately, our simulations might benefit from more realistic encodings of the parameters, concepts such as recessive and dominant genes, learning and procreation costs, different inheritance and mutation details, different survival and procreation criteria, more restrictive mate selection regimes, offspring protection, different learning algorithms and fitness functions, and so on, but for the purposes of this paper, our simplified approach seems adequate.

4. The Baldwin Effect

Since the Lamarckian idea of inheriting acquired characteristics is now known not to happen in real biological systems, it is commonly assumed that lifetime learning and evolution are independent processes. However, they are actually tied together by the so-called Baldwin effect [1]. This synergy comes about in two stages:

1. If a mutation (e.g. a change in learning rate or initial weight) can be used by the learning process to allow the system to acquire better properties, then it will tend to proliferate in the population.
2. If the learning has an associated cost (e.g. requires time or energy), then its results will tend to be incorporated into the genotype and the learned behaviours will become innate.

Put another way, evolution first creates a population that can learn good properties, and then removes the need for learning whenever it is possible to do so. In this way we have genetic assimilation of the learnt behaviour. Although this "new factor in evolution" dates back to the nineteenth century, it was not until the work of Hinton and Nowlan [3] that it became widely known to the connectionist community [2]. They demonstrated explicitly, for a particular simplified system, that the ability to learn a behaviour was able to improve the rate of evolution of that behaviour, without the learnt characteristics being passed between generations in the genotype. Their task, however, involved a very localized fitness function that was particularly difficult for evolution to search on its own. They noted that "for biologists who believe that evolutionary search space contains nice hills (even without the restructuring caused by adaptive processes) the Baldwin effect is of little interest". This, of course, depends on what interests you. It might be of little interest if you just want to demonstrate that learning can speed evolution, but there remains much of interest if you wish to explore genetic assimilation, the nature-nurture debate, individual differences, and such like.

In this paper I wish to begin to explore the Baldwin effect in its broadest sense. As with any modelling endeavor, there are numerous system design choices that can potentially have crucial consequences for the results, and in any complete study these will have to be explored systematically. Many of these details will, as in real systems, ultimately be determined as a result of the evolutionary process itself. For the purposes of this preliminary study, however, we shall concentrate on exploring the interaction of learning and evolution for the particular hand-crafted system described above. It is clearly important for any realistic developmental model to know how much of the system's behaviour is innate and how much has to be learnt by interaction with its environment. Assuming that learning does have a significant cost, a fixed system in a stable environment can be expected to evolve so that its optimal behaviour is completely innate. However, if the system really does need to retain the ability to learn, for example to adapt to unknown or changing conditions, or to compensate for aspects of its own natural maturation, then we may only get partial assimilation. We can still expect evolution to result in an efficient learning system that has minimal associated cost, but the appropriate innate properties may no longer correspond to a final learned behaviour. Moreover, if learning allows individuals with different genotypes to perform equally well, this will reduce the ability of natural selection to discriminate between them, and we will be left with a considerable range of individual differences. The remainder of this paper explores these effects by explicit simulations of the model presented above.

5. Simulation Results

Even after we have made the basic system design decisions discussed above, there are still a number of parameter choices to make. Clearly much further work will be required to justify particular choices for realistic models of human evolution. For concreteness here, all the fixed network parameters (neuron time constants, plant characteristics, environment characteristics, and so on) were chosen to match those of the human oculomotor control system [3]. The main evolutionary parameter values in this study were largely forced on us by limited computational resources. Choosing a fixed population size of 200 was a trade-off between maintaining genetic

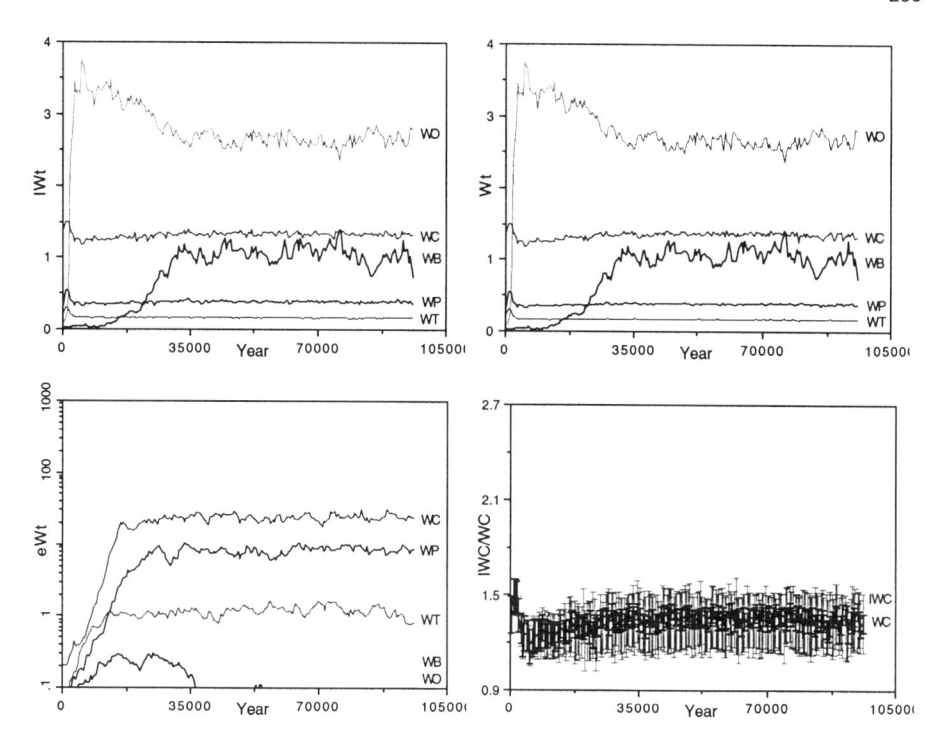

Figure 3: Evolution of weights and learning rates from zero.

diversity and running the simulations reasonably quickly. The death rates were set to produce reasonable age distributions. This meant about 5 deaths per year due to competition, and another 5 individuals over the age of 30 dying each year due to old age. The fixed population size meant each individual then produced on average two children. The mutation parameters were chosen to speed the evolution as much as possible without introducing too much noise into the process. The amount of training data presented to each individual per year also had to be restricted to minimise the computation times. All these compromises led to coarser simulations than one would like, but otherwise the simulations would still be running.

The natural starting point is to begin with a population consisting of individuals that have all their weights and learning rates zero, and see if they can evolve into a population that can perform well at their given task. Figure 3 shows how such a population's mean initial weights (iWt) and learning rates (eWt) evolve in this case. It also shows the mean actual weights (Wt) across all the individuals, which appear little different to the mean initial weights. The difference is clear, however, in the fourth graph which shows the means and standard deviations of the iWC and WC weights for mature individuals (i.e. those at least 10 years old). We see that the learning process has the effect of tightening the weight distribution from that generated by the inaccurate procreation process towards the optimal values.

As noted above, there is a crucial distinction between systems that really do need to learn, for example to adapt to changing conditions during their lifetime, and those which could perform perfectly well without learning if given appropriate innate

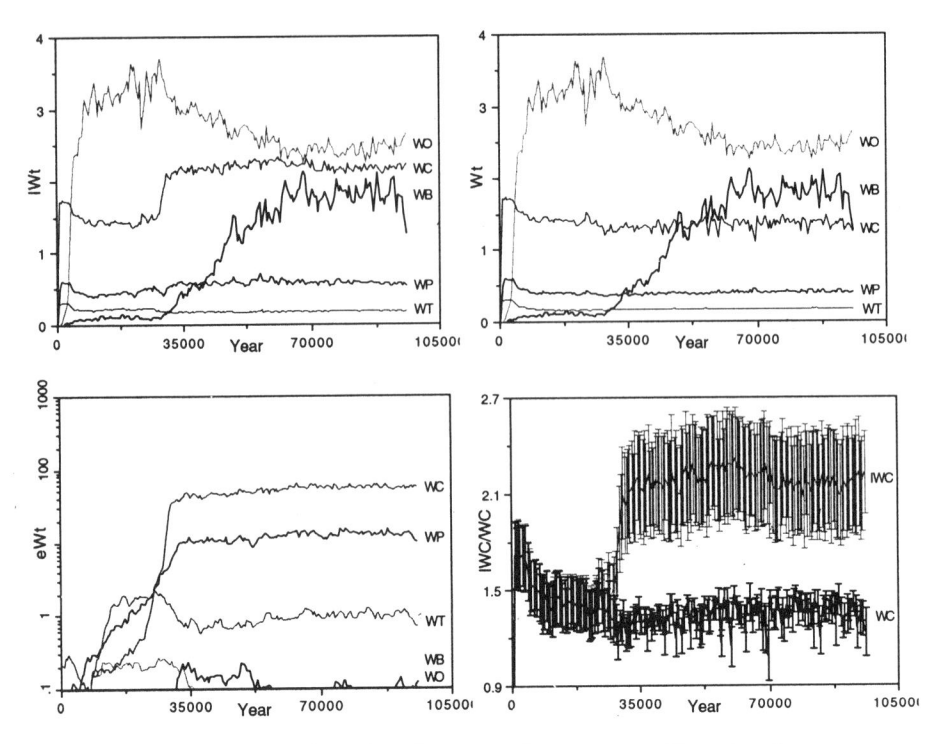

Figure 4: Evolution when the system has to cope with maturation.

values for all the necessary parameters. We have seen that there is a certain necessity for our models to adapt due to the variability that is built into the procreation process, but we also need to investigate the effect of needing to adapt to changes that take place during an individuals lifetime, in which case good adult weights are not necessarily appropriate initial weights. For example, in oculomotor control, the relationship between the eye rotations and the object's distance will need to vary as the inter pupil distance grows during childhood [3]. To incorporate such an effect into the simulations, a simple maturational scale factor was applied to the output, varying linearly from 0.5 to 1.0 between the ages of 0 and 10 years. The evolution of this new system is shown in Figure 4, and it does indeed differ significantly from that seen in Figure 3. In particular, we see larger evolved learning rates and, once a significant amount of learning has evolved, there is a clear difference between the mean initial and mature weights WC and WP. Comparing the distributions of the initial and mature WC weights we see that, once learning has evolved, there is a clear separation between the initial weight distribution and the tighter mature weight distribution (cf. the corresponding graph of Figure 3).

To understand this system better, Figure 5 presents two further graphs which show important general characteristics of our models. On the left we see how the mean, minimum and maximum individual fitnesses evolve, with clear improvements corresponding to the onset of learning seen in Figure 4. Notice that the maximum fitness often decreases. This is because even the fittest individuals eventually die, and they will not necessarily produce children as good as themselves, and even when

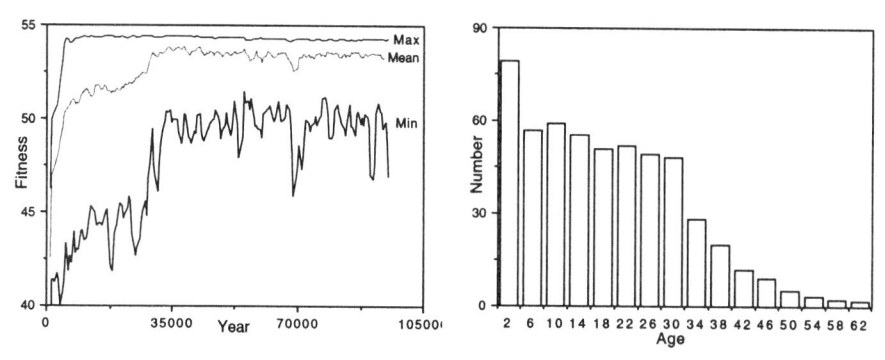

Figure 5: Evolution of fitness and evolved age distribution.

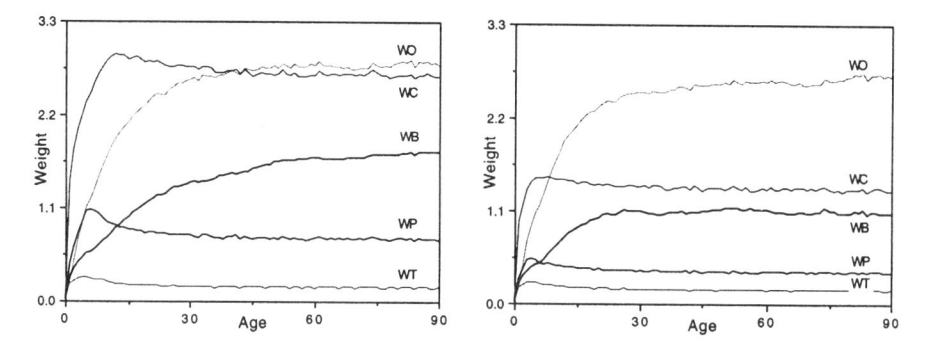

Figure 6: Learning weights appropriate for a newborn and an adult.

they do, those children are not necessarily going to experience learning environments as good as their parents'. As a result, the evolution is slower, but it is the price we pay for realism. This is one area where we might want to improve upon the human system when building artificial systems. On the right of Figure 5 we have the age distribution of our evolved population. We see an initial sharp drop corresponding to a high infant mortality rate, followed by a slower fall until the age of 30 after which individuals start to die of old age.

To assess the initial weights and learning rates that our systems evolve, we really need to know what the optimal newborn and mature weights would be to produce a good average performance across the whole (infinite) distribution of training data. Fortunately, these can easily be determined sufficiently accurately using the model's own learning algorithm whilst keeping the maturational scale factor fixed at either 0.5 or 1.0. Figure 6 shows how such learning alone leads to appropriate weights for a newborn and an adult. We end up with the same final weights using a range of initial weights and learning rates, indicating that these represent the true global minima which we might expect our evolving populations to reach. Note that, because of the linearity of our model and the way we have chosen to parameterize it, the maturational output scale factor can be compensated by an inverse scaling of WB, WC and WP, whilst leaving WO and WT alone. This is seen clearly in the optimal weights from Figure 6, which are in broad agreement with the mean evolved initial weights in Figures 3 and 4. The expected equality of the evolved mature

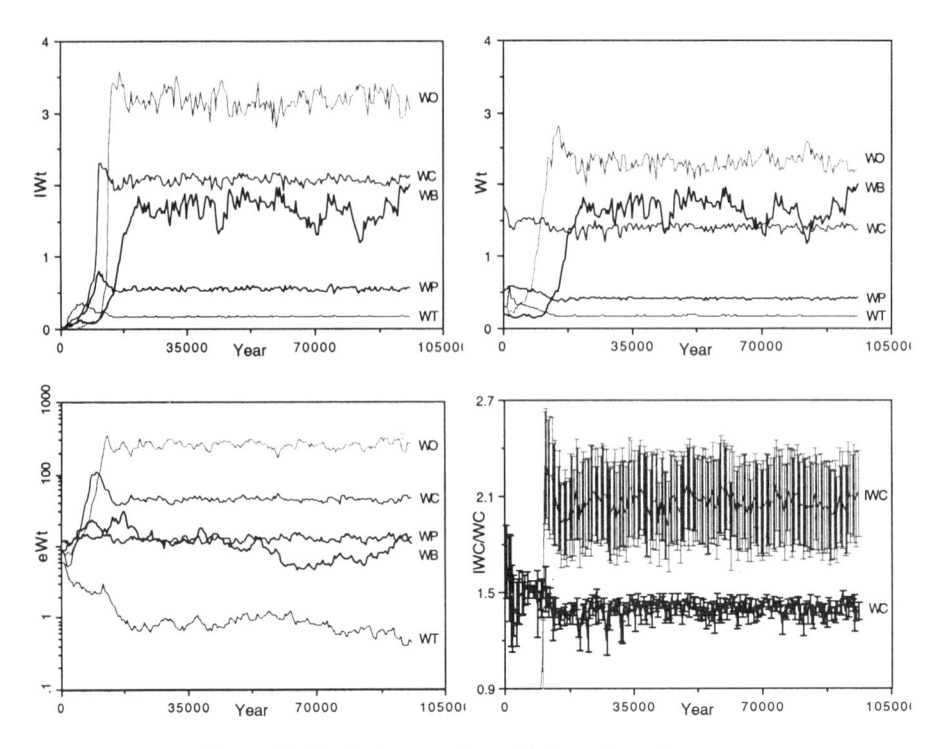

Figure 7: Evolution starting with large learning rates.

weights between Figures 3 and 4 is also found, apart from the bias *WB*, which we would expect to have evolved a significant learning rate so that it could compensate appropriately for the maturation. The reason for this discrepancy is something that will be investigated further later.

The next variation to consider, now that the basic model has been simulated and examined, is the specification of the initial population. Starting all the initial weights and learning rates from zero makes sense, but so does starting them off with a large range of random values. Figure 7 shows the evolution that results when all the learning rates start off randomly distributed in the range 0.0 to 20.0. There are several interesting differences from the corresponding graphs in Figure 4. First, having large learning rates right from the start *delays* the evolution of the innate weights, because individuals are able to *learn* the weights necessary to perform well. In due course however, genetic assimilation does occur, and the *WC* learning rate can be seen to fall dramatically as *iWC* increases. Eventually, the system settles down with mature weights near the optimal values observed before. For some reason, however, the *WO* innate weights and learning rates become unexpectedly large, and the *WB* weight again fails to adjust in line with the maturation.

This leads us to consider the case in which both the initial weights and the learning rates are randomly distributed in the range 0.0 to 20.0. Figure 8 shows how this affects the evolution. Naturally the initial stages are different from those seen in Figures 4 and 7, but the system soon settles down into the same pattern observed for the case of zero starting weights in Figure 7.

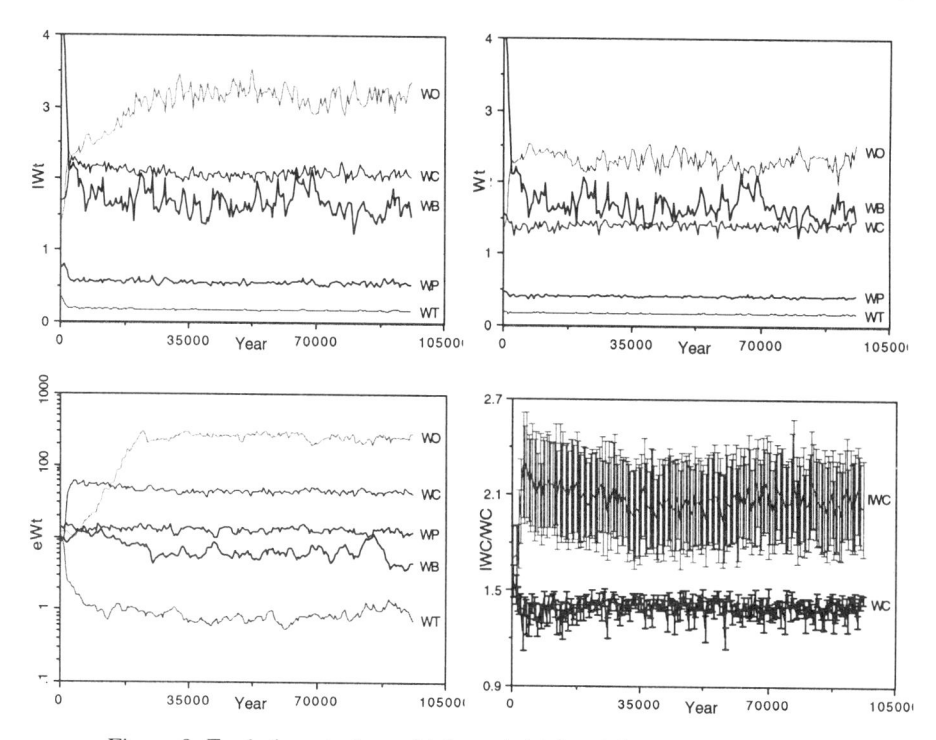

Figure 8: Evolution starting with large initial weights and learning rates.

Finally, to ensure a fair assessment of how learning affects evolution, Figure 9 shows how the system evolves when there is no ability to learn. The unused learning rates are plotted to show the kind of random walks that result when there is no evolutionary pressure. The initial and mature weights in this case are identical and constitute a compromise that allows the individual to survive the immature stage whilst allowing reasonable performance when mature.

6. Discussion

To a large extent our simulations have behaved how we would hope and expect them to, with appropriate evolved values for the innate weights and learning rates which lead to appropriate mature learned behaviour. However, we have noticed that the bias *WB* consistently fails to take on optimal values, and that when we start with large learning rates, the innate values of the weight *WO* appear larger than optimal. Clearly, some further investigation is required here.

One thing that is particularly evident from the simulations is that the weights are not all behaving in the same way – some are evolving faster than others, some are evolving larger learning rates, and so on. A convenient way to explore this in more detail is to test the sensitivity of the cost function to changes in each weight while all the other weights are kept fixed. This will give an indication of the forces acting on each weight during learning and evolution. Figure 10 presents the relevant results corresponding to the simulation shown in Figure 7. On the left we have the mean

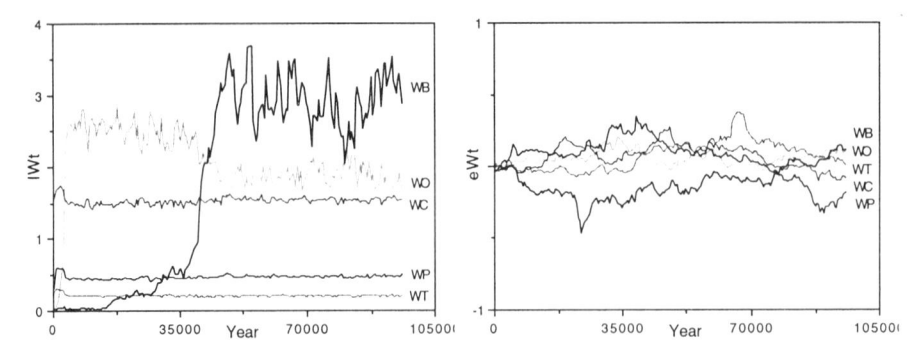

Figure 9: Evolution when there is no ability to learn.

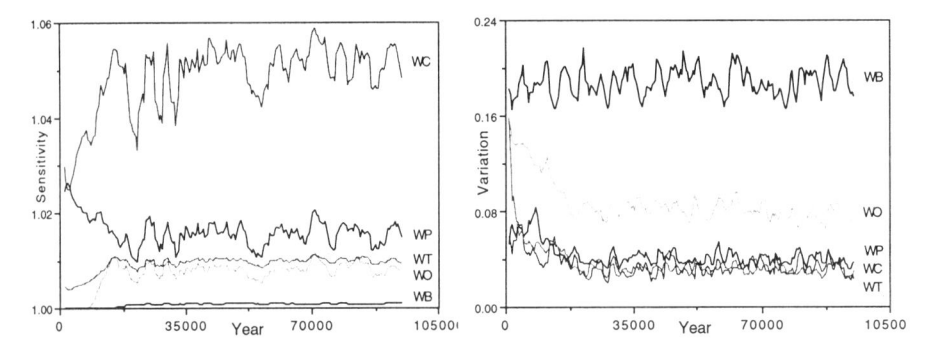

Figure 10: Weight sensitivities and individual variations.

proportional changes in the cost that is caused by increasing or decreasing each weight by 20% at each stage of the system's evolution. From this it is now clear why the bias WB is so slow to evolve and take on its optimal values during maturation. The cost function is so insensitive to it that the drive for it to change is being lost in the noise. It is quite possible that simulating a longer evolutionary period, or performing a less noisy simulation (e.g. by using a larger population size or more training data per simulated year), will allow the drive to manifest itself, but for the present we must be satisfied with understanding what is happening.

On the right of Figure 10 is shown the variability (i.e. standard deviation over mean) of each weight across the population during the system's evolution. We see that the bias WB stands out as having the particularly large range of individual differences we would expect given its small influence on the fitness. Here it is worth remembering that an important relevant feature of the models is that the weights within each individual are not independent. The optimal value for each weight will depend on the values of the other weights. There will exist an overall optimal set of weights, but if for some reason one weight does not take on its optimal value, then the others can partially compensate. For example, in each of Figures 4, 7, 8 and 9, we can see that before the innate weight WO evolves, the values of WC, WP, WT are larger. Similarly, low values of WB seem to coincide with higher values of WO. This anti-correlation of WO with the poor bias WB signals may well explain the unexpected behaviour of WO in some situations.

A realistic feature of our models is that each individual not only needs to perform well on average, but must also be able to cope if an unusual/extreme environmental condition arises (such as requiring a particularly large response change, or a particularly rapid sequence of response changes). To do this may require a set-up that is less than optimal for normal conditions. So, whilst it is reassuring to see that our evolved individuals are broadly in line with the optimal performance parameters, this is not something we can rely on happening in general. The extent to which this is responsible for the sub-optimal parameters that have evolved in our models is clearly worthy of further investigation in the future.

Having understood the factors underlying what has been happening in our simulations, there are several further issues worth commenting on. First, for the models, survival is determined by competition among the individuals, rather than by competition of the individuals against a hostile environment. Moreover, children are always produced to replace the dead, however young and unfit the parents might be. Consequently, whilst a non-adaptive population appears to evolve faster than their adaptive counterparts (e.g. Figure 4 compared with Figure 7), the poor fitness of individuals during the early stages of evolution (seen in Figure 5) may, under more realistic circumstances, result in the population not surviving long enough for the evolution to happen at all. It is in this sense that learning will assist evolution. The traditional Baldwin Effect [1, 4] corresponds to the comparison of the evolution in Figure 7 not with Figure 4, but with that of a system which has to wait an enormous time for a rare mutation to create a non-adaptable individual that is able to survive long enough in its environment to procreate.

Finally, note that the cost of learning is not encoded explicitly into any of our fitness functions, but it is implicit in the sense that if any unnecessary learning is required, then individuals are at a disadvantage until the learning is completed. The population can compensate for this to a certain degree by having individuals learn quickly, but if the learning rates are constant throughout each individual's lifetime, this may lead to instability. A sensible strategy, that occurs in real systems and will be explored further for our models elsewhere, might be to evolve initially high plasticities that decrease with age. Alternatively, a parent or population may evolve a propensity to protect its offspring until they have acquired the ability to fend for themselves, as many real species do. This is also worthy of a study in its own right, but for the present simulations, we simply protect the offspring until they have completed their first year of learning, and expect them to compete with adults after that. The net effect will be that, where possible, evolution will result in the need to learn being replaced by innate behaviour. For now, we leave open the question of whether a more explicit learning cost might be more beneficial or realistic.

7. Conclusions

We have seen how it is possible to simulate the interaction of learning and evolution in a class of simplified neural network control models. The processes whereby appropriate innate connection weights and learning rates evolve can be understood, and are broadly in line with what we might expect. However, it is also clear how the need to cope in unusual/extreme environmental conditions may result in individuals evolving in such a way that their performance is sub-optimal under normal conditions. We have also seen how some weak effects can easily get lost in the noise and result in the evolution of less than optimal solutions, which in turn may result in various compensatory effects generating unexpected properties.

We already knew that in order to understand real developmental processes, it is important to understand how those developmental systems have been constrained by their evolution [3]. The simulations presented in this paper now indicate that to understand the evolutionary process we also need to take into account the history of the evolutionary environment, as there are dependencies on the evolutionary initial conditions and we cannot rely on an optimal developmental system from evolving within a given time-scale. Moreover, fitness insensitivity to particular parameters can lead to considerable ranges of individual differences that can be difficult to predict without explicit simulation. In the case of oculomotor control, for example, individuals exhibit a wide range of cross-link strengths between accommodation and vergence with little effect on their normal performance [3].

Having seen how fundamental questions about the nature-nurture debate and individual differences may be answered, the next stage of this work will be to check the extent to which our results change as we vary the details of the models and the simplifications we have made to the real evolutionary process. We also need to test the approach against some real systems. This will be complicated by the fact that, in practice, real control systems will evolve alongside their plant, rather than independently with a fully operational plant. It is also quite likely that the initial population will arise as mutations of some other existing system (possibly adaptable, possibly not) and this will surely affect what evolves. There is clearly some way to go to achieve reliable simulations of real human or animal evolution. Alternatively, we could consider the approach to be an appropriate technique for developing efficient artificial control systems for real world problems. In which case, we have some way to go in another direction.

Acknowledgements

This work was begun many years ago while at Edinburgh University supported by the MRC. It is currently supported by the EPSRC as part of a project with Patricia Riddell on models of the development of accommodation and vergence.

References

1. Baldwin, J.M. (1896). A New Factor in Evolution. *The American Naturalist, 30*, 441-451.
2. Belew, R.K. & Mitchell, M. (Eds) (1996). *Adaptive Individuals in Evolving Populations*. Reading, MA: Addison-Wesley.
3. Bullinaria, J.A. & Riddell, P.M. (2000). Learning and Evolution of Control Systems. *Neural Network World, 10*, 535-544.
4. Hinton, G.E. & Nowlan, S.J. (1987). How Learning Can Guide Evolution. *Complex Systems, 1*, 495-502.

Borrowing Dynamics from Evolution: Association using Catalytic Network Models

Harald Hüning

Abstract

Catalytic networks are abstracted from chemistry, and have recently been used to study cooperation in molecular evolution. Here catalytic networks are regarded as a connectionist model with sigma-pi units in a recurrent dynamics. This paper partly presents previous work on the learning and generalisation in an association task. A particular type of architecture for catalytic networks is put forward here. The architecture of doubly-linked chains has been found to provide useful dynamics for association, while it allows for a range of different chain lengths. The chains can also branch, giving for example a tree structure. The nodes at the branching points perform logical AND or OR, depending on their connection parameters. An example is demonstrated where words are associated with letters, and each letter can have alternative (localist) representations. The main features of this model are the high sensitivity to single inputs and the two-way association. Possible extensions to using stochastic dynamics and distributed representations are discussed.

1. Introduction

A new kind of connectionist model, a catalytic network, is applied here to the disambiguation of letters using the top-down activation from words. This has previously been modelled by McClelland & Rumelhart [11]. These authors had to make an 'extended search' [11, pp. 386] for suitable parameters of their model, so automatic methods would be favourable. Other problems in current neural models are for example that most often the backpropagation algorithm is the only method of finding internal representations, and most models are lacking self-structuring mechanisms or other ways to improve their scalability.

So there is a need for novel models with organisational principles that are capable of building more complex networks, and of finding parameters automatically. This paper puts forward an application of Eigen & Schuster's [4, 5] theory of self-organisation in neural models. Eigen & Schuster consider populations of molecules in a competitive environment (selection pressure). Self-organisation refers to the population of selected types of molecules becoming high, while the population of other molecules goes towards zero.

The mechanism that leads to self-organisation in population systems is either the self-replication of molecules or the replication of several types of molecules together by cooperation. The mechanism for cooperation in molecular evolution is catalysis, which means that a substrate enhances a reaction without being consumed itself. Sometimes the cooperation feeds back to the same set of molecules that act as catalysts, for example in a cycle. Such a cycle or other closed catalytic networks can provide the positive feedback that lets selected molecules grow. For example, when molecule A takes part in a reaction that produces molecule B, and B in turn supports the production of A, a cycle is formed.

The effect of this cooperation is important for producing copies of identical molecules, and thus increase their population. The molecules (or species, in general) that exhibit a high population number due to the catalytic support are called an autocatalytic set. The appearance of autocatalytic sets has also been investigated in artificially defined chemistries [6, 1, 3, 10]. It is typical that the part of the reaction network where reactions keep taking place is much smaller than the whole reaction network, i.e. all possible reactions in the chemistry. This poses a problem to the problem-dependent design of autocatalytic sets, as it is intended in this study.

How can the self-organisation of autocatalytic sets be used for association networks or other neural networks? In this approach (Hüning, 1998b) the dynamics are used as a way of decision making. In order to know what is being decided, one needs a full understanding of the dynamics, or use restricted architectures where the behaviour is known. The current paper briefly reviews this work and points out how catalytic networks can be used for association and generalisation.

2. Deterministic Catalytic Network Model

Eigen & Schuster have used a deterministic model of catalytic networks to explain cooperation and competition in the area of molecular evolution [5]. Their model is modified here mainly by using the maximum instead of a sum. The system of differential equations for the population variables x_i is

$$\dot{x}_i = g_i(\mathbf{x}) - \frac{x_i}{0.5} \max(g_1(\mathbf{x}), g_2(\mathbf{x}), \ldots g_N(\mathbf{x})) \tag{1}$$

with the growth terms

$$g_i(\mathbf{x}) = \max(b_{i1}\, x_1^{w_{i11}} x_2^{w_{i21}} \cdots x_s^{w_{is1}},\, b_{i2}\, x_1^{w_{i12}} x_2^{w_{i22}} \cdots x_s^{w_{is2}},\, \ldots \\ \ldots, b_{im_i}\, x_1^{w_{i1m_i}} x_2^{w_{i2m_i}} \cdots x_s^{w_{ism_i}}) \tag{2}$$

for $i = 1$ to N, the total number of equations. The competition term to the right of the minus sign in (1) only serves for limiting the values of the state variables. The original model [5] uses a sum, which keeps the sum of the state variables constant. The maximum used here keeps the maximum state at a defined level.

All couplings of the equations are contained in the growth terms g_i. Weighting exponents w_{ijk} are introduced here as a new kind of connection parameter to be used in product terms [8, 9]. The products come from the model of catalysis [5]. The responses of these product terms are more sensitive to single inputs in contrast to the weighted sums most often used in neural models. Several product terms are combined by taking their maximum. This then gives the growth term g_i for one equation. The g_i from all equations appear in the competition term of the system equation (1).

Altogether the growth terms are parameterised by the exponents w_{ijk}, the biases b_{ik}, and the number of product terms m_i that limits the range of $k = 1 \ldots m_i$. The index $j = 1 \ldots s$ refers to the visible states x_j, of which there are N internal state variables. In addition there are external input states, so each product term has $s \geq N$ factors and exponents. The inputs are not applied as initial states as in some recurrent neural networks, but they stay available as external states while the catalytic network settles.

For the simulations, a normalised version of the weighting exponents has been used: $\dfrac{x^w}{x^x}$ bounded by the maximum of 1. All states x_j have a lower bound of 0.001.

3. Architecture for Association

The dynamics of catalytic networks selects groups of variables called autocatalytic sets. Each equation of the model can be considered as a *node*, which is connected to other nodes and some inputs wherever there is a non-zero weighting exponent. The dynamics of autocatalytic sets can be considered as association, i.e. recall of certain sets of active nodes. When activity propagates along the autocatalytic set, the decision of what variables become high or low can be used as output activity.

In what general ways can these autocatalytic sets be composed? The problem that needs to be avoided here is that autocatalytic sets are often smaller than the connected reaction network. So one cannot just link nodes together in an arbitrary way, but a suitable architecture needs to be found. A comparison of several possible architectural components of catalytic networks has been made. The stability of autocatalytic sets has been analysed by fixed-point analysis, see Table 1. It has been found that the architecture of doubly linked chains (bidirectional links along a chain) is flexible in the number of variables that can be linked. When for example three or four variables are in a doubly linked chain, the dynamics has a stable point, and even more than four variables can be linked by using exponents smaller than one. So this type of architecture can be used for association, and activity can propagate in both ways along the chains.

System	Connectivity	fixed point $x_i > 0$
$\dot{x}_1 = x_2 - x_1 \sum \ldots$ $\dot{x}_2 = x_1 x_3 - x_2 \sum \ldots$ $\dot{x}_3 = x_2 - x_3 \sum \ldots$	$x_1 \quad x_2 \quad x_3$	$\lambda_1 = -1.38$ $\lambda_{2,3} = -0.618$ stable

Table 1: Fixed-point analysis of the autocatalytic set in a doubly linked chain of three variables. The negative eigenvalues indicate a stable point.

Furthermore, the doubly linked chains can be combined by branches, and relatively arbitrary tree structures can be implemented. Branches can have simultaneous activity like a logical AND, but normally not too many variables can be multiplied. Exponents smaller than one solve this problem, so one can link many branches in an AND-like fashion. Similarly, branches that are alternatively active like logical OR can be realised. In this case, there are several product terms in an equation, of which the maximum is taken.

The tree structures are specified by means of the parameters in the equations of the previous section.

4. Learning in Catalytic Networks

In order to create autocatalytic sets according to some training patterns, the fine tuning of suitable exponents is solved by a learning rule here. The standard competitive learning rule is used [7]. While in neural networks with competitive learning only one winner neuron is allowed to learn, here several nodes learn (a node corresponds to an equation of the catalytic network). After the catalytic network dynamics has settled, a threshold is applied to the activations, and all nodes with an activation above the threshold are subject to the following learning rule, where n is a counter of the learning events [7].

$$\Delta w_{ijk} = \frac{1}{n}(x_j - w_{ijk})$$

Also, a rule is applied to adjust the biases with a learning rate α

$$\Delta b_{ik} = \alpha(\frac{1}{g_i} - 1)$$

More details can be found in [9].

5. Association

A network architecture is set up manually to demonstrate the performance on association. Another paper [8] deals with how to build this type of architecture automatically from the data. The task is to associate words with their constituent letters, and for each letter there are alternative representations possible. Figure 1 shows the initial architecture of the catalytic network.

There are subdivisions such that the subnetworks dealing with individual letters only communicate via the subnetwork for the word level. The subnetwork for the level of the words has input 1. For example, the node 'W1' responds to the first bit in input 1, see Table 2. The three subnetworks for the letter positions are called slots, which have one letter each of input 2. Every link in Figure 1 corresponds to one exponent being above zero, which is set up corresponding to the words and letters from the training items in Table 2.

Figure 2 shows the time course of the activities during the training. The responses for the four training items are approaching a uniform level. An example of the resulting exponents is shown in Table 3. One can see from the Table that the exponent

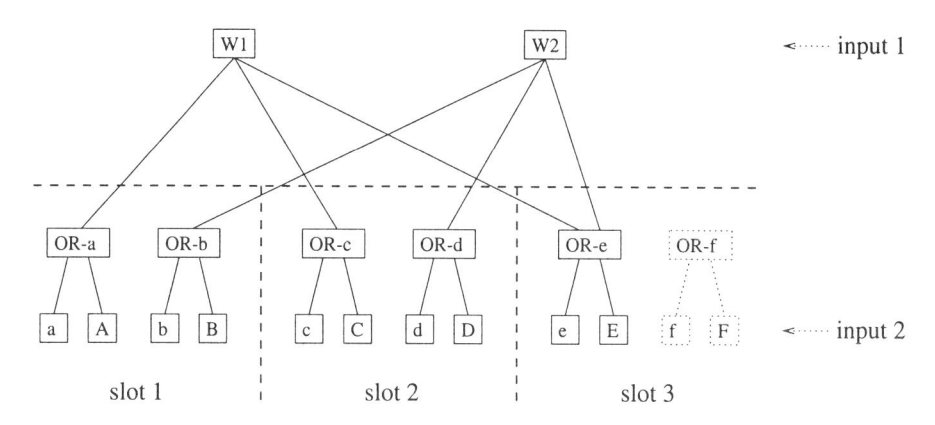

Figure 1: Architecture for learning words with two alternative representations of the letters. Each box denotes a node, and the label indicates when this node will be active. There are three letter positions, and the subnetworks for the letters ('slots') are only coupled through the word level (W1 and W2). Input states are not shown explicitly, but during training those nodes in the rows marked as 'input 1' and 'input 2' are always simultaneously active with the corresponding input symbol, a letter or a word. Initially, the architecture only requires one-directional connections, and the bi-directional connections are completed by learning.

Data item	Input 1	Input 2
1	W1 = 10	ace = 100000 100000 100000
2	W2 = 01	bde = 010000 010000 100000
3	W1 = 10	ACE = 000100 000100 000100
4	W2 = 01	BDE = 000010 000010 000100

Table 2: Training data for learning words. Combinations of letter symbols are presented to the network as binary external states using a one-of-N code. A symbol in input 1 stands for a word (W1, W2) and input 2 presents three slots with one letter each. There are six bits reserved for the letters. The letters are coded differently in each slot.

		\multicolumn{6}{c}{connected node}
	input	a	A	b	B	OR-a	OR-b
W1	1	0	0.21	0.21	0	0	0.42	0

Table 3: Resulting exponents in node W1. Only the exponents for the input and one of the three slots are shown. The exponents for node 'a' and 'A' have half of the value of the exponent for 'OR-a'.

for the input from the node 'OR-a' is twice as high as the exponents for 'a' and 'A'. The exponents for the single letters are by-passing the doubly linked chains. These exponents have a lower value, because the single letters are active only half of the time that the word node measures their activity. The many low exponents are a disadvantage of applying the learning rule uniformly to all nodes, but the learning of the bias compensates for this. The learning also completes the bi-directional connections.

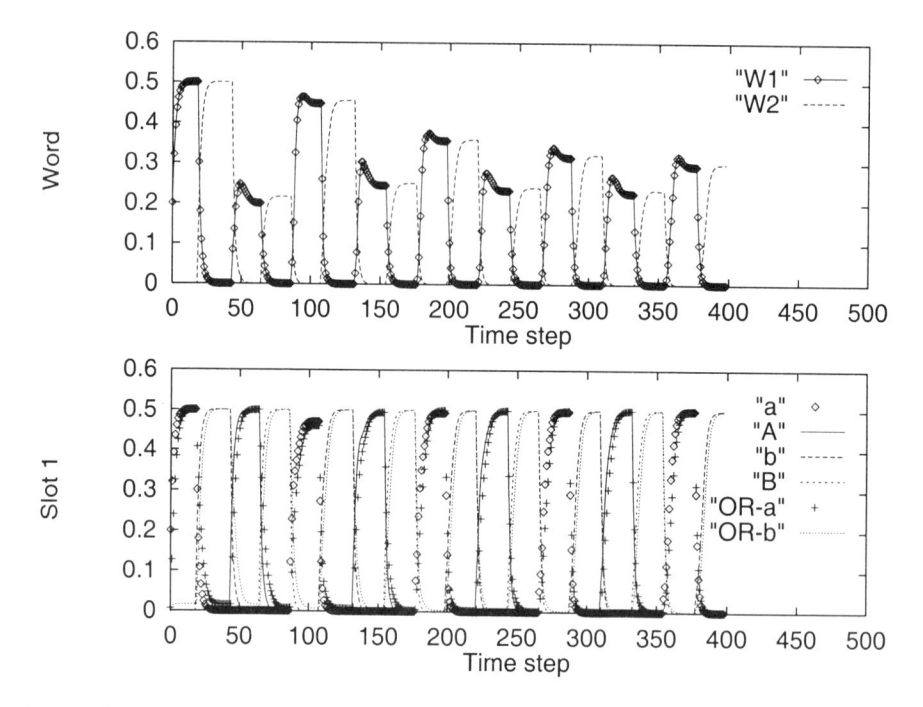

Figure 2: Simulation of the learning rule for the association of words and letters. Only the subnetworks for the word and slot 1 are shown. The learning rules are applied at stable points (horizontal time-course). The labels are as in Fig. 1. The time steps are chosen such that the discretisation of the continuous dynamics is sufficiently accurate.

6. Generalisation

Having set up the network and completed the tuning of the exponents by learning, one can test the recognition of the learned patterns. The three items in Table 4 are applied to the network, and the responses are shown in Figure 3. The first test item has uncertainty at the first letter, the second test item is a word, and the third item is a non-word. An intermediate state level of 0.5 is chosen to represent an unknown symbol.

Data item	Input 1	Input 2
1	W2 = 01	xde = 0.5 0.5 0.5 0 0 0 010000 100000
2	WX = 0.5 0.5	ace = 100000 100000 100000
3	WX = 0.5 0.5	ade = 100000 010000 100000

Table 4: Test data for an incomplete word, a word that has been trained and a word that has not been trained.

At the presentation of the first test item there is a lower response than usual for the letter 'b'. The word that is responding here is 'W2' (trained as 'bde'). Top-down activation from the word makes slot 1 respond with the correct letter, while its input

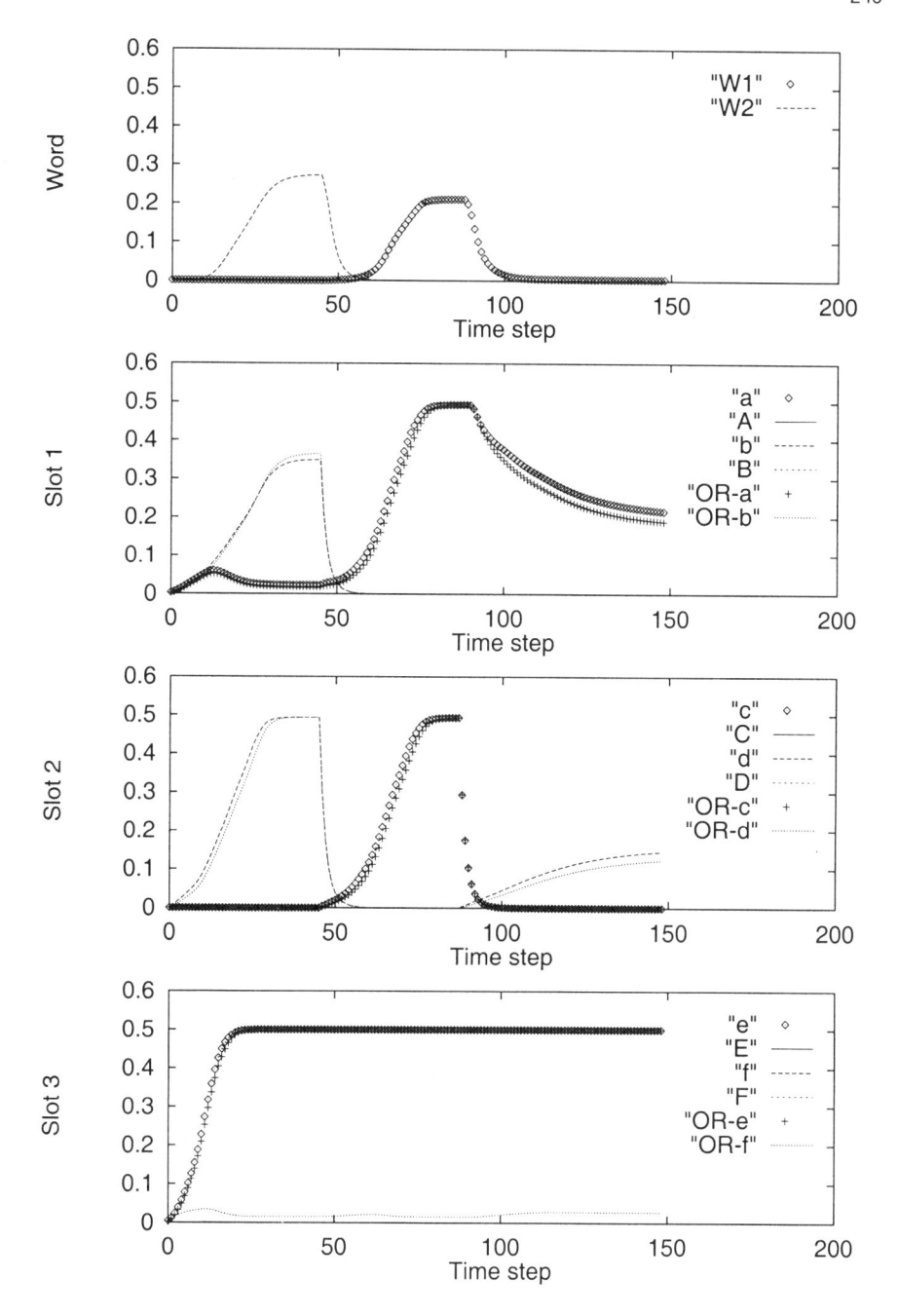

Figure 3: Testing of unknown patterns. Responses are shown for all nodes in all four subnetworks. The first item shows the disambiguation of a letter, the second item (around time step 80) shows the recognition of a word, and the last item shows rejection (no word responds).

is uncertainty (0.5 equally for all possible letters). This may be interpreted as context dependent perception of letters [11].

At the second test item all letters have a high response, but the word (W1) has a slightly lower response than usual. The input to the word subnetwork is uncertainty (0.5 everywhere), so the response is only driven by the letters. Thus the word W1 is recognised. No word responds for the last test item. This could be expected, because the letter combination 'ade' had not been trained. So the network rejects this pattern from recognition.

7. Using Distributed Representations

A different way of implementing catalytic networks employs stochastic dynamics, which is closer to simulating individual chemical reactions. Reactions are considered to take place after stochastic collisions of the molecules involved. In simulations, the collisions are spread out in time such that only one is considered at a time step [3]. This method of simulation allows taking into account more detail of individual molecules than the deterministic model. The population now consists of individual representations (of molecules) and no longer just the number of copies. It is convenient to consider binary strings to represent the individuals. A potential benefit from using the detailed representations is to make use of the similarity of these representations.

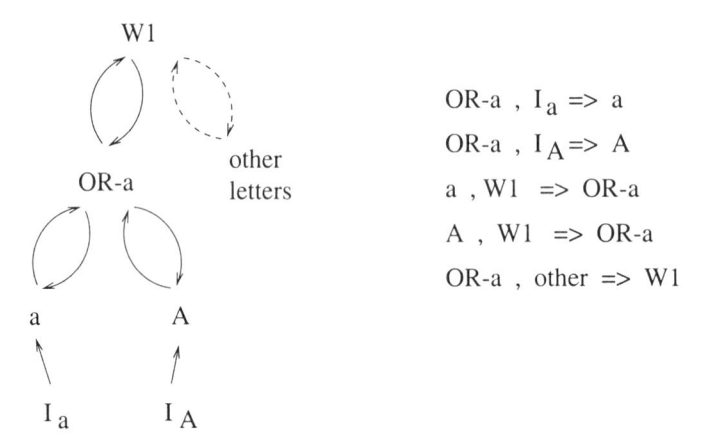

$$\text{OR-a}, I_a \Rightarrow a$$
$$\text{OR-a}, I_A \Rightarrow A$$
$$a, W1 \Rightarrow \text{OR-a}$$
$$A, W1 \Rightarrow \text{OR-a}$$
$$\text{OR-a}, \text{other} \Rightarrow W1$$

Figure 4: Individual rules to implement the architecture for association in a stochastic way. The architecure (**left**) and the rules (**right**) are exemplified for one slot. I_a and I_A denote an input bit, e.g. as from Table 2 input 2. The other symbols can be substituted by unique strings. The population now consists of individuals (not shown). As an example, when a collision of 'OR-a' and 'I_a' is detected, the first rule will put an 'a' into the population.

Detailed representations can be introduced into the network for association by using a population of strings. The couplings from the association network can be transformed into the rules shown in Figure 4. There is one rule for each product term from the earlier model ('OR' nodes have two product terms). All symbols used so far can now be substituted for unique corresponding strings. For the moment, these strings are hand-coded.

If a rule matches the strings in a collision, its right side replaces a randomly chosen string in the population. The stochastic dynamics corresponds on average to the deterministic model, so it can be used for association. There may be additional randomness in tight decision cases.

Once that strings and rules are used to implement the association network, some method of finding the strings can be investigated. One possibility is to use another population dynamic way, for example by coupling several systems for self-organising binary strings [10].

8. Conclusion

The dynamics of catalytic networks have been used for association, as an alternative to for example McClelland & Rumelhart's interactive activation model [11]. The architecture of doubly linked chains has been found suitable for composing association networks with localist representations. Association works in both directions along each chain, so top-down and bottom-up activations can be combined. The doubly linked chains can branch, so tree structures can be formed with a choice of logical AND and OR at the branching points. This kind of architecture is also suitable for automatic structuring methods [8].

A learning rule has been used for the fine tuning of the weighting exponents, which first completes the doubly linked chains automatically, and second maximises the responses in cases where more exponents are high than inputs are active. Simulations have shown the performance of word recognition, rejection and disambiguation of letters. The rejection shows that in contrast to weighted neural models, the product terms used for recognition have a higher sensitivity to single inputs.

Furthermore, the association model can be implemented in a stochastic way. This is a first step towards using distributed representations for each letter or word. The stochastic implementation reduces the product terms to simple coincidence detection. This may be an alternative to Calvin's [2] view how some mechanisms underlying evolution, here the selection dynamics, may take place in the living brain.

References

1. Bagley, R. J. & Farmer, J. D. (1992). Spontaneous Emergence of a Metabolism. In C. G. Langton, C. Taylor, J. D. Farmer & S. Rasmussen (Eds.) *Artificial Life II*. Santa Fe Institute Proceedings Vol. X, (pp. 93-140). Redwood City, CA: Addison-Wesley.
2. Calvin, W. H. (1996). The cerebral code. Cambridge, MA: MIT Press.
3. Dittrich, P. & Banzhaf, W. (1998). Self-Evolution in a Constructive Binary String System. *Artificial Life* **4**(2), 203-220.
4. Eigen, M. (1971). Selforganization of Matter and the Evolution of Biological Macromolecules. *Die Naturwissenschaften* **58**, 465-523.
5. Eigen, M. & Schuster, P. (1978). The Hypercycle. A Principle of Natural Self-Organization. Part B: The Abstract Hypercycle. *Die Naturwissenschaften* **65**, 7-41.

6. Farmer, J. D., Kauffman, S. A. & Packard, N. H. (1986). Autocatalytic replication of polymers. *Physica D* **22**, 50-67.
7. Hertz, J., Krogh, A. and Palmer, R. G. (1991). *Introduction to the theory of neural computation.* Redwood City, Addison-Wesley.
8. Hüning, H. (1998a). Learning Decompositional Structures in a Network of Max-Π Units with Exponents as Connection Strengths. In L. Niklasson, M. Boden & T. Ziemke (Eds.) *ICANN'98*, Proceedings of the 8th International Conference on Artificial Neural Networks, Skövde, Sweden, 2-4 September 1998. London: Springer Verlag 1998.
9. Hüning, H. (1998b). An Approach to Learning in Autocatalytic Sets in Analogy to Neural Networks. Published on CD-ROM: I. Aedo-Cuevas, C. Bousono-Calzon, F. Diaz-de-Maria, A. R. Figueiras-Vidal, C. Martin-Pascual & A. Rodriguez-de-las-Heras (Eds.) Workshop *Learning 98*, Universidad Carlos III de Madrid, Getafe, Madrid, Spain, 16-18 September 1998.
10. Hüning, H. (2000). A search for multiple autocatalytic sets in artificial chemistries based on boolean networks. In: M. A. Bedau, J. S. McCaskill, N. H. Packard & S. Rasmussen (Eds.) *Artificial Life VII: Proceedings of the seventh international conference on artificial life,* Portland, OR, USA. Cambridge, MA: MIT Press.
11. McClelland, J. L. & Rumelhart, D. E. (1981). An interactive activation model of context effects in letter perception: Part 1. An account of basic findings. *Psychological Review* **88**(5), 375-407.

Evolving Modular Architectures for Neural Networks

Andrea Di Ferdinando, Raffaele Calabretta & Domenico Parisi

Abstract

Neural networks that learn the What and Where task perform better if they possess a modular architecture for separately processing the identity and spatial location of objects. In previous simulations the modular architecture either was hardwired or it developed during an individual's life based on a preference for short connections given a set of hardwired unit locations. We present two sets of simulations in which the network architecture is genetically inherited and it evolves in a population of neural networks in two different conditions: (1) both the architecture and the connection weights evolve; (2) the network architecture is inherited and it evolves but the connection weights are learned during life. The best results are obtained in condition (2). Condition (1) gives unsatisfactory results because (a) adapted sets of weights can suddenly become maladaptive if the architecture changes, (b) evolution fails to properly assign computational resources (hidden units) to the two tasks, (c) genetic linkage between sets of weights for different modules can result in a favourable mutation in one set of weights being accompanied by an unfavourable mutation in another set of weights.

1. Modularity as a Solution to the Problem of Neural Interference

Neural networks that learn only one task can have simple architectures and may not need modularity. However, real organisms generally have not one task but many different tasks to accomplish in order to survive and reproduce. Hence, their nervous systems tend to be organized with anatomically and functionally distinct modules. Using neural networks that have to learn different tasks can help understand why organisms develop modular nervous systems.

Why are neural modules useful? One possible answer is that modules allow a neural network to solve the problem of neural interference. Learning consists in progressively modifying an initial set of weights in such a way that at the end of learning the network produces the desired output in response to each input. Consider a single one of these weights. During learning the weight's initial value is gradually changed in such a way that at the end of learning it will be the correct value, i.e., the weight value that together with the values of the other connection weights produces the correct output. If the network has only one task to learn, the problem can be solved reasonably easily. However, if the network has to learn two different tasks and the particular connection weight we are considering has a role in both tasks, i.e., the output the network must generate in response to the input depends on the value of this weight both when the network is engaged in one task and when it is engaged in the other task, then the situation may become more complicated. The correct accomplishment of the first task may require that the initial weight value of the connection be increased during learning while the correct accomplishment of the second task may require that the initial value be decreased. This will lead to some sort of interference or conflict between the two tasks. (The problem encountered by nonmodular architectures learning multiple tasks is called "cross-talk" by Plaut and Hinton [11] and Jacobs *et al.* [7]. Cross-talk refers to contradictory messages arriving to a neural network's hidden unit, interference to contradictory messages arriving to a network's connection weights. But the two are more or less equivalent.)

Modularity solves the interference problem. If the network architecture is such that no single connection weight has a role in determining the network's output for both tasks, there will be no interference between the two tasks. The weight value of each particular connection will be changed during learning to satisfy the requirements of the single particular task in which the connection plays a role and it will never happen that the same connection will have to respond to contradictory pressures to change its weight value. All the connections that play a role in one particular task and in no other task constitute a module. The connections of a module are "proprietary", i.e., they are dedicated exclusively to the accomplishment of a single task. Their weight value can be adjusted during learning without interfering with, and being interfered by, other tasks.

An example of the problem of multiple tasks is represented by organisms that must recognize both the identity (What) and the spatial location (Where) of visually perceived objects. Nervous systems that must learn this What and Where task have two separate neural pathways, a ventral (temporal) pathway for recognizing the identity of the object and a dorsal (parietal) pathway for identifying its location [14]. (The dorsal pathway can be concerned with "How" to accomplish a physical movement with respect to the object rather than with "Where" the object is, but the two interpretations can be considered as equivalent for our purposes.) Rueckl *et al.* [13] have taught the What and Where task to both modular and nonmodular networks using the backpropagation procedure and have found that modular networks learn much better the task than nonmodular ones. In Rueckl *et al.*'s simulations the network architectures are hardwired by the researchers and what is investigated is how different network architectures give different results. In biological reality it is not the researcher but nature that creates network architectures. Hence, it might be interesting to study how modular network architectures may spontaneously arise as part of a process of development in individual networks or evolution in a population of networks.

Jacobs and Jordan [8] have described simulations in which modular architectures for the What and Where task emerge as part of an individual's development. Their model is based on a preference for establishing short rather than long connections between pairs of neurons during brain development. During development individual neurons reach particular positions in the physical space of the brain. When the neurons grow their axons and establish connections with other neurons, it is more probable that a connection will be established between two spatially close neurons than between two more distant neurons. Using this simple developmental rule they were able to show that modular architectures rather than nonmodular ones tend to emerge as a result of the development of the brain. However, Jacobs and Jordan [8] seem to be able to obtain this result only because they hardwired the spatial location of units in such a way that separate modules for the What and the Where task tend to emerge developmentally. In other words, in their model nature has replaced the researcher only partially. Modular architectures emerge because of decisions taken by the researcher, not truly spontaneously. One could simulate the entire process of the emergence of modular architecture during brain development by using a genetic algorithm to find out evolutionarily the appropriate locations of units in the physical space of the nervous system and then have the preference for short connections generate the appropriate network architecture during development. This would be more appropriately called development since it would consist in changes during life in which inherited genetic information has a critical role. (For a simulation of brain development in which both the physical location of individual neurons and the establishment of connections, especially short connections, between neurons emerge spontaneously, see [4]).

Another possibility is to imagine that biological evolution takes care of the problem of finding the appropriate modular architecture. Networks that must learn two distinct tasks are born with a genetically inherited modular architecture which has been shaped during the course of evolution. Network architecture emerges not during an individual's life but during a succession of generations in a population of individuals. (Murre [10] also has suggested to use the genetic algorithm to design modular network architectures.)

We have conducted two sets of simulations using the genetic algorithm as a model of evolution to develop neural networks that are able to accomplish the What and Where task. In a first set of simulations we used the genetic algorithm to evolve both the network architecture and the connection weights but we were unable to solve the task using this approach. In a second set of simulations the genetic algorithm was used to evolve the network architecture but the connection weights were learned by the individual networks during their 'life' using the backpropagation procedure. This second approach gave the desired solution.

2. The What and Where Task

The What and Where task requires a neural network to recognize both the identity and the spatial location of perceived objects. In Rueckl *et al.* [13] the neural network is presented in each cycle with one of 9 different objects that can appear in one of 9 different positions on a retina for a total of 9x9=81 possible inputs. The network has two distinct sets of 9 output units each for indicating the identity and the location of the presented object, respectively, and a single layer of 18 hidden units. In the nonmodular architecture all the hidden units are connected with both the What output units and the Where output units. Various modular architectures are tried out. The modular architecture that performs much better than the nonmodular architecture has 14 hidden units connected only with the What output units and the remaining 4 hidden units connected only with the Where output units (Figure 1). The reason for the success of this particular architecture is that the What subtask is more difficult than the Where subtask. The networks learn using the backpropagation procedure.

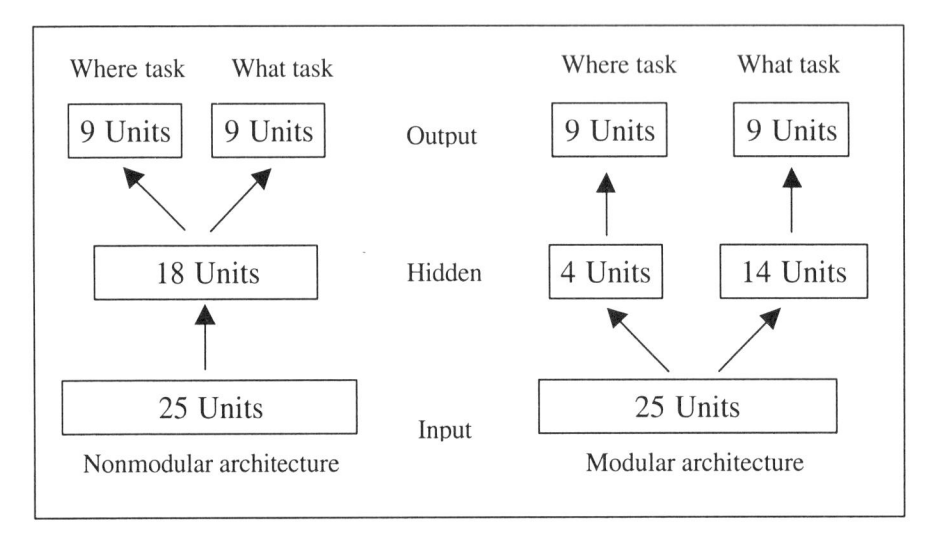

Figure 1: Nonmodular (left) and modular (right) network architectures for the What and Where task.

In Rueckl *et al.*'s simulations the network architecture is imposed by the researcher. The networks' task is to find the appropriate connection weights given a certain architecture. However, the What and Where task could also be solved at the population level. Imagine an entire population of networks, each different from all the others. Individual networks are born with genetically inherited information that specifies the network architecture and, possibly, also the connection weights. This genetically inherited information is the result of a process of biological evolution which takes place in successive generations of individuals and is based on the selective reproduction of the most successful individuals and the constant addition of new variants to the population's genetic pool.

We describe two sets of simulations. In the first set both the network architectures and the connection weights evolve and are genetically inherited. In the second set the network architectures evolve and are genetically inherited but the appropriate connection weights are learned during life by each individual.

3. Simulations

3.1 Using the Genetic Algorithm to Evolve Both the Network Architecture and the Connection Weights

Imagine a population of organisms living in an environment in which the reproductive chances of each individual depend on the individual's performance in the What and Where task. The individuals that have a smaller error on the What and Where task are more likely to reproduce than the individuals with a larger error.

An individual is born with an inherited genotype which is divided up into two parts. One part specifies the architecture of the individual's neural network and the other part the network's connection weights. Some general features of the architecture are fixed and identical in all individuals (and therefore are not encoded in the genotype and do not evolve). All architectures have three layers of units with 25 input units (encoding a 5x5 retina), 18 hidden units, and 18 output units (9 for indicating the identity of the perceived object and 9 for indicating the object's spatial location in the retina). In all architectures each input unit is connected with all the hidden units. What can vary from an architecture to another are the connections between the hidden units and the output units. The portion of the genotype which encodes the network architecture contains 18 genes, one for each hidden unit. Each of these architectural genes has three possible values that specify if the corresponding hidden unit is connected (a) to all the What output units, (b) to all the Where output units, or (c) to both the What and the Where output units. The third possibility, (c), is included to allow for the evolution of nonmodular architectures. The other part of the genotype encodes the connection weights and it includes one gene for each possible connection weight (weight genes). The weight genes are 774 because there is a maximum of 774 connection weights in a nonmodular network. Modular architectures have less than 774 genes and in this case some of the weights may remain unexpressed. The weight genes are encoded as real numbers.

At the beginning of the simulation a population of 100 individuals is created and each individual possesses a genotype with random values for both the architectural and the weight genes. The values of the weight genes are randomly chosen in the interval between -0.3 and +0.3. Each individual is presented with the 81 input patterns of the What and Where task and an individual's fitness is greater the lower its summed squared error on these patterns. The 20 best individuals are selected for reproduction. Each of these individuals generates 5 offspring which inherit the genotype of their single parent with the addition of some random mutations. The architectural genes are mutated by replacing the value of a gene with a new randomly chosen value with a probability of 5%. The weight genes are mutated by adding a quantity randomly chosen in the interval between -1 and +1 to 10% of the genes. The simulation is terminated after 10,000 generations. Ten replications of the simulation were run with randomly chosen initial conditions.

The results of the experiment show that the genetic algorithm is unable to solve the What and Where task if both the architecture and the connection weights are subject to evolution and are genetically inherited. The total error is about 40 after 10,000 generations. In Rueckl *et al.*'s simulations using the backpropagation procedure the terminal error is practically zero for the best architecture. While the performance in the Where task is good enough but not as good as in Rueckl *et al.* (error = 7), the performance in the What task is very poor (error = 33).

These negative results appear to be caused by the difficulty on the part of the genetic algorithm to evolve an appropriate set of weights if the network architecture is evolving at the same time. Each architecture has its own appropriate set of weights and, therefore, changing an architecture can be destructive from the point of view of the weights. A given set of weights which is appropriate for a given architecture may be completely inappropriate if the architecture changes. In our simulations the genotype specifies the weights of all possible connections even if some of these connections are not expressed in the phenotype. Therefore, when an unexpressed connection get expressed as a result of a mutation, its value is not zero. But adding even a single connection with its value already specified can destroy the equilibrium of the connectivity pattern. The same applies if the weight value of the new connection is randomly generated or if a previously expressed connection is canceled by a mutation together with its connection weight.

This interpretation is supported by the results obtained by manipulating the mutation rate. If the mutation rate of the network architecture is increased (10%), the final error increases (60). If it is reduced (1% and even 0.1%) the final error decreases although it never approaches zero (25 for 1% mutation rate; 22 for 0.1% mutation rate).

However, the bad results of these simulations may be due to another reason in addition to the disruption caused by the addition or deletion of connections with their weight value. One would expect that the network architecture that eventually evolves is the architecture that Rueckl *et al.* [13] have found is the best architecture, that is, an architecture with more hidden units dedicated to the more complex What task than to the simpler Where task. This is not the case in our simulations. Although the genetic algorithm does evolve modular architectures (on the average only about 2 hidden units are connected to both the What output units and the Where output units), the network architecture which tends to evolve has more hidden units assigned to the Where task than to the What task. It is not surprising then that the networks' performance on the total task is not good.

The reason for the failure of the genetic algorithm to evolve the appropriate modular architecture seems to be that in the initial generations the algorithm concentrates on the easier task, the Where task, and dedicates many computational resources (hidden units) to this task. When the performance on this task is almost perfect, however, the algorithm is unable to shift computational resources from the Where task to the more difficult What task. More specifically, in the earlier generations the individuals that are selected for reproduction are those that are good at the Where task even if they are not very good at the What task. These individuals tend to have network architectures with more hidden units assigned to the Where task (which decides if they reproduce or not) than to the What task. When in the later generations competition becomes harsher and selection would reward individuals that are good both at the Where task and at the What task, the random genetic mutations are unable to modify a situation in which most hidden units are already assigned to the Where task and evolution is unable to produce individuals that are good at both tasks.

	Hidden units			Error		
Runs	Where	What	Both	Where	What	Total
1	15.8	0.7	1.5	5.5	39.1	44.7
2	6.0	5.1	6.9	4.8	28.7	33.4
3	**3.2**	**11.3**	**3.5**	**9.9**	**27.3**	**37.1**
4	**7.2**	**9.4**	**1.4**	**8.9**	**27.6**	**36.6**
5	16.4	0.3	1.3	5.8	42.2	48.0
6	8.0	5.4	4.5	6.4	31.1	37.4
7	**6.9**	**10.8**	**0.3**	**7.5**	**25.6**	**33.0**
8	8.0	6.5	3.5	8.4	28.6	37.0
9	14.8	2.1	1.1	5.9	41.0	46.9
10	14.8	2.4	0.7	8.2	44.3	52.6

Table 1: Number of hidden units allocated to the Where task, to the What task, and to both tasks, and error on the Where task, the What task, and total error for each of 10 replications of the simulation (average of 100 individuals for each replication). In 3 replications (bold face) more hidden units are allocated to the What task than to the Where task and still the performance is not good.

However, even this may not be the entire story. If we look at Table 1, we see that at least in some replications of the simulation (3 out of 10) more hidden units are correctly dedicated to the What task rather than to the Where task. But even in these replications of the simulation the error on the What task, and therefore also the total error, remains quite high. Hence, the failure of the genetic algorithm to produce efficient networks for the What and Where task appears to be due to its inability to select the appropriate connection weights even for networks which have the appropriate modular architecture.

This may reveal a general inability of genetic algorithms of the type we used in our simulations to evolve the appropriate connection weights for modular networks. We have run an additional set of simulations (not reported here) in which the genetic algorithm tries to find the appropriate connection weights given a fixed modular architecture of the appropriate type, with little success. The reason seems to be that, since the connection weights of different modules are separately encoded in the genotype, a favourable mutation of the connection weights of one module can be accompanied by a nonfavourable mutation of the connection weights of another module, with little total advantage. This seems to be a form of genetic linkage. Either the individual in which the two mutations occur is selected for reproduction - and in this case the nonfavourable mutation in the second module becomes part of the population's pool -, or the individual is not selected for reproduction - and in this case the favorable mutation in the first module is lost.

3.2 Using the Genetic Algorithm to Evolve the Network Architecture and the Backpropagation Procedure for Learning the Connection Weights

Perhaps, then, the solution to the various problems we have seen in the simulations described so far is to use the genetic algorithm to evolve the appropriate network architecture at the population level and the backpropagation procedure to have each individual network learn the connection weights for its inherited network architecture during life. We have run a second set of simulations in which the inherited genotypes encode a variety of possible network architectures but the genotype does not encode the connection weights. The connection weights are not genetically inherited but they are learned during life. At birth each individual is assigned a random set of weights for the particular network architecture it inherits and then the individual learns to do the What and Where task exactly as in the Rueckl *et al.*'s simulations. At the end of learning the terminal error of each individual determines the individual's reproductive chances.

Hidden units			Error		
Where	What	Both	Where	What	Total
4.7	12.2	1.1	0.0	1.6	1.6

Table 2: Number of hidden units allocated to the Where task, to the What task, and to both tasks, and error on the Where task, the What task, and total error for the average individual in the last generation (average of 10 replications of the simulation).

The results are that, first, at the end of the simulation the terminal error is near zero for both the What and the Where tasks - even if it is still somewhat larger for the What task - and, second, the evolved network architectures tend to be the optimal architectures with more hidden units dedicated to the What task than to the Where task (Table 2).

4. Discussion

A neural network that must acquire a capacity to do more than one tasks is better able to acquire this capacity if the network architecture is modular because neural modules prevent the occurrence of neural interference, defined as the arrival of contradictory messages during learning for changing the value of connection weights involved in more than one task. Neural modules contain connection weights involved in only one task and therefore they avoid neural interference.

Neural modules can be hardwired by the researcher or they can spontaneously emerge during development or evolution. Using the genetic algorithm as a model of biological evolution we have simulated the evolution of network architectures that are appropriate for recognizing both the identity and the spatial location of perceived objects. Modular and nonmodular architectures compete in the successive generations of a population of neural networks and modular architectures should emerge as the winning ones.

We have compared two conditions, one in which both the network architecture and the connection weights evolve and are genetically inherited and another one in which only the architecture evolves and is inherited while the connection weights are learned during an individual's life. Only the second condition produces satisfying results, that is, the appropriate modular architecture and high levels of performance in the task. If both the architecture and the connection weights are encoded in the genotype, a change in the network architecture with the addition or deletion of even a single connection can suddenly make a set of weights that has evolved with the preceding architecture inappropriate. Furthermore, evolution may not be the best method for evolving the connection weights for modular networks because a favorable genetic mutation in one module may be accompanied by an unfavorable mutation in another module, although sexual recombination or genetic duplication [3] might help solve this problem.

As suggested by various authors (see, for example, [2]), cooperation between evolution and learning can be the best solution to the problem of acquiring complex capacities, compared with having either evolution or learning completely take care of the problem. However, it is not only that evolution and learning must cooperate and both have a role in the acquisition process but the best solution might be to have evolution take care of the network architecture and learning of the connection weights. Hence, the network architecture is genetically inherited but the connection weights are not. They are learned during life. (This solution has been proposed on the basis of more general considerations by Elman *et al.* [5]).

One should not, however, overemphasize this particular type of division of labor between evolution and learning. A number of other arrangements may exist that maintain the general scheme of entrusting the network architecture to evolution and the connection weights to learning but distribute the details of this scheme differently. For example, the initial connection weights can be encoded in the genotype and then learning modifies these initial weights. It has been shown that the initial weights influence learning [9] and that evolution may find out what are the best initial weights for learning some particular task (see, for example, [1]). Other schemes may involve the genetic encoding not of the actual connection weights themselves but only of various constraints on the connection weights. For example, whether some particular connection weight is positive (excitatory) or negative (inhibitory) may be encoded in the genotype but it is learning that finds out what is the most appropriate absolute value for the weight. Or the range of variation of the value of some weight may be encoded in the genotype, but the actual value within this range is identified by learning. Or, again, evolution can find the appropriate learning parameters and learning the actual weight values [6]. On the other side, learning can change an inherited network architecture by adding and/or deleting connections (cf. the pruning and tiling algorithms [12]). However, it could still be the general case that evolution identifies the general layout of a species' brain and learning refines what is inherited by adjusting the weights on the brain's connections.

References

1. Belew, R. K., McInerney, J., & Schraudolph, N. (1991). Evolving networks: using the genetic algorithm with connectionist learning. In C. G. Langton, C. Taylor, J. D. Farmer, & S. Rasmussen (eds), *Artificial Life II*. Addison-Wesley, Reading, MA.
2. Belew, R. K. & Mitchell, M. (1996). *Adaptive Individuals in Evolving Populations*. Addison-Wesley, Reading, MA.
3. Calabretta, R., Nolfi, S., Parisi, D. & Wagner, G. P. (2000). Duplication of modules facilitates the evolution of functional specialization. *Artificial Life* 6:69-84.
4. Cangelosi A., Parisi D. & Nolfi S. (1994). Cell division and migration in a 'genotype' for neural networks. *Network* 5:497-515.
5. Elman, J. L., Bates, E. A., Johnson, M. H., Karmiloff-Smith, A., Parisi, D. & Plunkett, K. (1996). *Rethinking innateness. A connectionist perspective on development*. The MIT Press, Cambridge, MA.
6. Floreano, D. & Urzelai, J. (2000). Evolutionary robots with on-line self-organization and behavioral fitness. *Neural Networks* 13:431-443.
7. Jacobs, R. A., Jordan, M. I. & Barto, A. G. (1991). Task decomposition through competition in a modular connectionist architecture: The what and where vision tasks. *Cognitive Science* 15:219-250.
8. Jacobs, R. A. & Jordan, M. I. (1992). Computational consequences of a bias toward short connections. *Journal of Cognitive Neuroscience* 4:323-335.
9. Kolen J. F. & Pollack, J. B. (1990). Back-propagation is sensitive to initial conditions. *Complex Systems* 4:269-280.
10. Murre, J. M. J. (1992). *Learning and categorization in modular neural networks*. Harvester, New York, NY.
11. Plaut D. C. & Hinton, G. E. (1987). Learning sets of filters using back-propagation. *Computer Speech and Language* 2:35-61.
12. Reed, R. D. & Marks II, R. J. (1999). *Neural Smithing. Supervised Learning in Feedforward Artificial Neural Networks*. The MIT Press, Cambridge, MA.
13. Rueckl, J. G., Cave, K. R. & Kosslyn, S. M. (1989). Why are "what" and "where" processed by separate cortical visual systems? A computational investigation. *Journal of Cognitive Neuroscience* 1:171-186.
14. Ungerleider, L. G. & Mishkin, M. (1982). Two cortical visual systems. In D. J. Ingle, M. A. Goodale & R. J. W. Mansfield (Eds.), *The Analysis of Visual Behavior*. The MIT Press, Cambridge, MA.

Evolution, Development and Learning - a Nested Hierarchy?

T.E. Dickins & J.P. Levy

Abstract

The Dynamical Hypothesis [22] is gathering force within cognitive science and within biology. Evolutionary, developmental and learning processes can all be characterised by the DH and any models should try to account for this property. The processes differ in terms of their operational time-scale and the resources each has to hand. Evolution sets the parameters for the dynamical interactions in development and learning. Could all three processes possibly be regarded as a nested hierarchy sharing the same dynamical properties? We ask this question and argue that a DH understanding of the potential evolution of cognitive systems could inform subsequent modelling.

1. Introduction

The aim of this paper is to provide a brief survey of the role of the "Dynamical Hypothesis" (DH) [22] within cognitive science at a number of different levels of explanation. We specifically want to make the following main points:

- The DH potentially unifies explanations of cognition at the evolutionary, developmental, and learning levels;
- Evolutionary constraints on cognition need to be seen as "running through" development and learning, but effects at the other levels can influence evolution;
- We need to model and simulate the above in order to support theory.

First, we shall make some general epistemological points to focus our argument.

Science is about discovering the order of the universe and explaining the causes and functions of that order. We observe order in our everyday dealings with the universe and such observations form the foundation of our folk theories. Scientific methods enable scientists to uncover fundamental natural kinds of the universe and understand how they interact in order to produce the higher level phenomena that interest us.

The behavioural and cognitive sciences are interested in explaining how it is that organisms regularly mediate between input and output. Hendriks-Jansen [10] has recently claimed that artificial intelligence (AI) has failed to truly determine natural kinds that provide illuminating explanations. Instead AI has embarked upon an exercise in mimicking input-output relations under the assumptions of a Universal Turing Machine that can implement input-output relations in any number of ways. Thus AI only provides us with formal task descriptions of what is to be explained. If we really want to understand Nature and the actual middle terms she employs we need a better approach than this. We believe this is true of much cognitive science.

Hendriks-Jansen further argued that we need to take an evolutionary perspective because such approaches will better inform our natural history of the original order of interest. In the next section we analyse recent attempts to apply evolutionary thinking to cognition. This will lead us to advocate a different approach to evolutionary theorising in the cognitive sciences, one that is firmly grounded within the DH.

2. Evolutionary Psychology (EP)

Recently some psychologists [2] have examined our complex psychological phenotype in terms of natural selection [6]. For a trait to be naturally selected it needs to provide a solution to a contingent adaptive problem, which needs to be fairly long term and stable. Cosmides and Tooby [5] suggest that the Pleistocene epoch was such an Environment of Evolutionary Adaptedness (EEA). During the EEA various discretely organised cognitive mechanisms were selected that solved these putative problems. For example, Cosmides [4] has demonstrated that people perform much better on the Wason Selection Task when the task is about seeking out violations of social contracts than in its original abstract form. Cosmides argues this is because finding cheats within a social system would have been a contingent ancestral problem and the species has evolved a specific module to deal with this. The fact that people find it hard to transfer their reasoning ability to the abstract task with the same underlying logic is indicative of a cognitive specialism. This approach led to a Massively Modular Hypothesis [20] about cognitive organisation whereby there are specific computations and stored representations for specific problems.

What EP is attempting to do is to show how evolutionary change has led to stable modern order in *Homo sapiens*. The central argument, that specific computational devices were selected for, focuses upon adult end-state and has little to say about ontogeny other than that ontogeny sees the coming on-line of these various mechanisms over time [21]. Thus there are hardwired computational systems whose successive staging is also hardwired. This makes EP a nativist approach that fits well with much contemporary developmentalism [3], but not with all.

3. Evolution and Development

It *may be* the case that evolution has selected for adult-end states that are modularly organised, and that it has also selected for specific successive staging. *But* this theoretical approach does not inform us about how such order emerges over ontogenetic and phylogenetic time, nor does it inform us about how such progression was selected for or why.

Recently a neo-constructivist view of development has emerged that takes a more parsimonious view [8, 19]. The hypothesis is that the infant has a number of low-level in-built systems. From interactions between such systems themselves, and with the environment, new behavioural capacities emerge in a principled manner. Where EP argued for a stable EEA, neo-constructivism says that it is the very stability of the environment, and low level cognitive architecture that allows the emergence of higher order phenomena regularly during development. Such neo-constructivist development is likely to be adaptive and therefore selected for.

Karmiloff-Smith [11] terms this approach as emergent modularisation, arguing that specific computational mechanisms for specific tasks will emerge as a part of this process, but that they are not in-built. The brain functionally organises itself as a response to the activity it is involved in during development. The brain is regarded as sufficiently plastic to afford a number of ontogenetic trajectories. Karmiloff-Smith suggests that "it is plausible that a fairly limited amount of innately specified, domain-specific predispositions (which are not strictly modular) would be sufficient to constrain the classes of inputs that the infant mind computes".

Karmiloff-Smith [12] has found support for this position in work on developmental disorders. Under the "traditional" view of development, disorders can be viewed in terms of the failure of certain modules to either come on line, or to operate properly once on line. There is no notion that this might affect the developmental trajectories of the other extant modules. Karmiloff-Smith has looked at Williams Syndrome (WS). People with WS have low IQ, deficits in spatio-constructive skills, numerical cognition, and in problem solving. The evidence that Karmiloff-Smith has presented suggests that deficits in certain cognitive domains will affect the outcome of the rest of development. For instance, WS people scored in the same range as normal controls for face processing tasks. However, the WS participants solved the problems in a very different manner (componentially rather than holistically). Karmiloff-Smith argues that this is indicative of different cognitive processes leading to the same behavioural outcomes. This might be the result of a process of modularisation affording the WS people a face processing module, but it emerges from an atypical developmental trajectory. New order emerges that satisfies certain functional demands in common with normal development but this order is reached differently.

Underlying neo-constructivism is the older notion that the brain is a dynamic system, consisting of certain variables that can interact in interestingly limited ways. From this interaction order can emerge. Theorists have begun to model just such development with some success [7]. For example, Mataric's robot exhibits wall following equipped only with some sensors and basic pre-set movements that are tripped by specific inputs [10]. It is the interaction of these simpler systems that leads to the emergence of wall following and such systems are arguably coherent natural kinds.

4. The Dynamical Hypothesis

What is the nature of the DH that underlies neoconstructivism? Van Gelder [22] has recently claimed that the DH "is the unifying essence of dynamical approaches to cognition. It is encapsulated in the simple slogan, *cognitive agents are dynamical systems"*. He further splits this into the *knowledge hypothesis* and the *nature hypothesis*. The latter is the claim that we are dynamical systems at the cognitive level, consisting of a number of variables that interact with one another over time in such a way as to exhibit self-organisation. The knowledge hypothesis is the claim that we can understand cognitive agents in dynamical terms.

The key characteristic of DH that is of interest to us is the move away from input-output relations to a conception of ongoing and appropriate change. Appropriate change must be understood in relation to the adaptiveness of the system to its ecological niche. More generally, the idea that the system changes over time and is the result of limited interactions between all of its variables and the environment - that might consist of other systems - marks an important principle. *The question for EP is now "how did such a dynamical system evolve in such a way as to ensure the specific emergence of order that we see in ontogeny?"*

5. What Changes During Evolution?

The preceding question does not assume that evolution has selected a "genetic program" that generates a dynamic cognitive phenotype. Traditional views of evolution have it that order is generated through natural selection weeding out adaptive variance. Kauffman [13] has noted that our notion of the genetic program contains the idea of a sequential processor gradually unfolding standard developmental trajectories by reading them off the genome. He has proposed that this is not the case and that much of the order we see in biological systems has emerged from the self-organising properties of "the genome". It is only when self-organised order has emerged that natural selection then has the opportunity to operate within larger gene space. Biological development can be seen in these dynamical terms too. Such a perspective makes for a better explanation of both the precision of development and the ability of organisms to withstand (some) minor alterations within the genome, as we see in the WS case which is caused by a chromosomal micro-deletion.

Kauffman envisages the genome as a space where each gene can potentially interact (epistasis) with other genes and in so doing affect the expression of the phenotype "controlled" by that gene and the overall "shape" of the organism. Genes are connected to other genes and become active (or inactive) when they receive specified inputs. The system shifts through various stages of activation. The number of genes in this space is N and the number of potential interactions is K. NK modelling allows for the emergence of stable patterns of activity, or state cycles, through the application of local rules within the network. Kauffman's early models were based on the observation that the human genome consists of around 100,000 genes but only 250 types of cell. Kauffman's intuition was that this observed pattern, and a similar one in his simplified models, was indicative of state cycles. It is the mathematical properties of the genome that have determined the number and kinds of cell types that there are - not natural selection. Of course, this poses interesting questions about the origin of these mathematical properties.

6. Where Are We?

Phenotypic stability can emerge from gene space and also from the dynamics of development. The phenotypic stability of evolutionary time scales is played out in a large gene space and produces distinct species that stabilise in a particular region of that space. Within a given species, when we look up close in ontogenetic time, we can see the emergence of phenotypic stability in terms of development of some specific low level systems, and the emergence of subsequent systems as a result of their interaction. We now face the challenge of finding a methodology to satisfactorily isolate and model the appropriate variables involved in such dynamics.

There is a further problem. Development might be regarded as nested within larger evolutionary processes but to some extent the opportunity for variance at the developmental level in terms of structural morphology and from learnt behaviour gives this system sufficient independence to potentially interact with and affect evolutionary processes. This form of interaction, in which two systems simultaneously alter each other's direction of change, is referred to as coupling [22].

Such coupling has arguably been demonstrated in discussion and simulations of the Baldwin Effect [1]. Thus far in the discussion we have been conflating gene space and phenotype space [16, 17, 18]. These are distinct spaces as natural selection operates over phenotypes and genotypes simply code for phenotypes. The Baldwin Effect is about the relationship between these two spaces - if individuals have sufficient phenotypic plasticity of the sort afforded by learning it is possible for them to acquire new adaptive traits. The individuals who are capable of doing this at low cost are likely to be selected for. Such learning can be characterised dynamically. The initial consequence of such selection is that individuals capable of this form of learning will dominate the population. It is also argued that the corresponding genotype for learners will be indirectly selected for. It is possible that this trait may eventually become genetically assimilated perhaps through the selection of faster and faster learners [16, 17]. It is also possible that individuals will be selected for that learn the trait so fast that it appears instinctive to external observers [9]. This will still affect the available genotype.

With the Baldwin Effect we see an effect of learning upon phenotype and then genotype. This is undeniably an effect from learning through to evolution. Potentially the alterations in genotype space will have generated other effects too as the NK parameters are altered, this in turn might lead to changes in the nature of relevant selection pressures that lead to selection of some ordered forms over others. This has been recently discussed in relation to niche construction [14]. The ability to build such things as dams, nests and burrows, for instance, are often learnt but might enter genotype space at some point. However, once the environment is substantially altered then the selection pressure changes too. Darwin's own example is of earthworms whose burrowing activities have radically altered the substructure of much land [14]. This has led to changes in the amount of mucus produced and the epidermal structure of earthworms. It is possible that the emergence of human culture might have wrought similar if not larger scale effects upon the evolution of our cognition [14].

Evolution, development and learning are to some extent hierarchically ordered if only in terms of time scales. What they are not is a nested hierarchy with a straightforward linear relationship. Instead it would appear that there is good argument for interactions between the systems. This evidence does not force us to abandon the role of natural selection, unfolding development or learning in our models of behavioural science but it does suggest a more complex picture than that which the input-output psychologists have painted.

7. Conclusion

The first two aims of this paper have been clearly met. The unification afforded by the DH is a result of taking seriously the self-organising aspects of the genome and cognitive systems. The potential for interaction between levels of explanation is a consequence of each level being regarded as a system in its own right and potentially open to coupling.

If we take the DH the job of cognitive science becomes one of finding out what low-level systems have been selected for. We have to try to understand how these "natural kinds" might interact with one another within a relatively stable environment to produce order over evolutionary time. We also have to understand how this order is able to develop over ontogenetic time and the levels and kinds of perturbation such systems can tolerate. Equally, we have to be aware of how learning can be dynamically understood and emerge from described developmental trajectories. Finally, we need to be aware of the potential role of Baldwin Effects in evolution as well as the possibility of feedback from niche construction and other activities.

Presently, these DH inspired ideas provide an interesting framework to reconceive data about development (and evolution) but the fine detail of such systems is unknown. It is at this point that modelling and simulation work must surely come into their own. Modellers must think about modelling the transition through evolutionary, developmental and learning time. In this way it might be possible to construct parsimonious models of cognition and behaviour constituted by the order of natural kind that Hendriks-Jansen finds so appealing.

We might recognise that cognitive developmental outcomes are dynamical and influenced by evolution but to actually model the order of cognition discussed is not easy. The problem of Leibniz's Law potentially raises its head. On the one hand, in reducing input-output explanations or descriptions to a dynamical model we have to be sure that we do not lose the point of initial reference, we have to be sure the dynamical language refers to the same initial phenomena as the traditional language did. On the other, it might be that the DH forces us to radically reconceive our list of natural cognitive kinds and to stop seeing them in the intentional and computational manner that we currently do. Thus computational accounts of input-output relations might have set the initial question and the answer might deny that antecedent. Indeed, the whole notion of input-output relations is questionable within the DH.

Despite the enormous difficulties of such modelling we feel that there is good reason to conceive of organisms dynamically. As the cognitive sciences are part of the biological sciences they have an epistemological duty of coherence to conform to biological theory. If this is at the expense of traditional or Classicist [15], folk-theory inspired conceptions of cognition, then so be it.

References

1. Baldwin, J. M. (1896) A new factor in evolution. *The American Naturalist*, 30 (June), 441 - 451.
2. Barkow, J. H., Cosmides L. and Tooby J. (eds.). (1992) *The Adapted Mind: Evolutionary Psychology and the Generation of Culture*. Oxford: Oxford University Press.
3. Carey, S. and Spelke, E. (1994) Domain-specific knowledge and conceptual change. In: *Mapping the Mind: Domain Specificity in Cognition and Culture*. L. A. Hirschfeld & S. A. Gelman (eds.) Cambridge: Cambridge University Press.
4. Cosmides, L. (1989). The logic of social exchange: Has natural selection shaped how humans reason? Studies with the Wason selection task. *Cognition* 31, 187 - 276.
5. Cosmides, L. and Tooby, J. (1992) Cognitive Adaptations for Social Exchange. In *The Adapted Mind: Evolutionary Psychology and the Generation of Culture* J. H. Barkow, L. Cosmides, and J. Tooby, (eds.) Oxford: Oxford University Press.
6. Darwin, C. (1859/1985) *The Origin of Species By Means of Natural Selection or the Preservation of Favoured Races in the Struggle for Life*. London: Penguin Books.
7. Elman, J. L. (1993) Learning and Development in Neural Networks: The Importance of Starting Small. *Cognition*, 48, 71 - 99.
8. Elman, J. L., Bates, E. A., Johnson, M. H., Karmiloff-Smith, A., Parisi, D., and Plunkett, K. (1996) *Rethinking Innateness: A Connectionist Perspective on Development*. London: MIT Press.
9. Gould, J. L. and Marler, P. (1987) Learning by Instinct. *Scientific American* (January), 74 - 85.
10. Hendriks-Jansen, H. (1996) *Catching Ourselves in the Act: Situated Activity, Interactive Emergence, Evolution, and Human Thought*. London: MIT Press.
11. Karmiloff-Smith, A. (1992) *Beyond Modularity: A Developmental Perspective on Cognitive Science*. London: MIT Press.
12. Karmiloff-Smith, A. (1998) Development itself is the key to understanding developmental disorders. *Trends in Cognitive Sciences*, 2 (10), 389 - 398
13. Kauffman, S. A. (1993) *The Origins of Order: Self-Organization and Selection in Evolution*. Oxford: Oxford University Press.
14. Laland, K. N., Odling-Smee, J. and Feldman, M. W. (2000) Niche construction, biological evolution, and cultural change. *Behavioural and Brain Sciences*, 23, 131 - 175
15. MacDonald, C. (1995) Classicism versus Connectionism. In: *Connectionism: Debates on Psychological Explanation*. C. MacDonald and G. MacDonald (eds.) London: Blackwell.
16. Mayley, G. (1996a) Landscapes, learning costs and genetic assimilation. *Evolutionary Computation*, 4 (3), 213 - 234.
17. Mayley, G. (1996b) The evolutionary cost of learning. In: *From Animals to Animats: Proceedings of the Fourth International Conference on Simulation of Adaptive Behaviour*. Maes, P., Mataric, M., Meyer, J-A., Pollack, J., and Wilson, S. (Eds.) London: MIT Press.

18. Mayley, G. (1997) Guiding or hiding: Explorations into the effects of learning on the rate of evolution. In: *The Proceedings of the Fourth European Conference on Artificial Life (ecal97).* P. Husbands and I. Harvey (eds.) <http://ai.iit.nrc.ca/baldwin/bibliography.html>
19. Quartz, S. R. and Sejnowski, T. J. (1997) The neural basis of development: A constructivist manifesto. *Behavioural and Brain Sciences*, 20, 537 - 596
20. Samuels, R. (1998) Evolutionary Psychology and the Massive Modularity Hypothesis. *British Journal of Philosophy of Science*, 49, 575 – 602.
21. Tooby, J. & Cosmides, L. (1992) The Psychological Foundations of Culture. In *The Adapted Mind: Evolutionary Psychology and the Generation of Culture* J.H. Barkow, L. Cosmides, and J. Tooby, (eds.) Oxford: Oxford University Press.
22. Van Gelder, T. (1998) The dynamical hypothesis in cognitive science. *Behavioural and Brain Sciences*, 21, 615 - 665

Semantics

Learning Lexical Properties from Word Usage Patterns: Which Context Words Should be Used?

Joseph P. Levy & John A. Bullinaria

Abstract

Several recent papers have described how lexical properties of words can be captured by simple measurements of which other words tend to occur close to them. At a practical level, word co-occurrence statistics are used to generate high dimensional vector space representations and appropriate distance metrics are defined on those spaces. The resulting co-occurrence vectors have been used to account for phenomena ranging from semantic priming to vocabulary acquisition. We have developed a simple and highly efficient system for computing useful word co-occurrence statistics, along with a number of criteria for optimizing and validating the resulting representations. Other workers have advocated various methods for reducing the number of dimensions in the co-occurrence vectors. Lund & Burgess [10] have suggested using only the most variant components; Landauer & Dumais [5] stress that to be of explanatory value the dimensionality of the co-occurrence vectors must be reduced to around 300 using singular value decomposition, a procedure related to principal components analysis; and Lowe & McDonald [8] have used a statistical reliability criterion. We have used a simpler framework that orders and truncates the dimensions according to their word frequency. Here we compare how the different methods perform for two evaluation criteria and briefly discuss the consequences of the different methodologies for work within cognitive or neural computation.

1. Introduction

Distributional statistics are measurements of simple patterns within data, usually based on patterns of co-occurrence. These methods have been used frequently in psychological models of language phenomena including phonology, morphology and word meaning [13]. In many of these models, neural network learning algorithms are employed to measure the co-occurrence statistics. In this paper we report on a study in which the statistics are measured directly. However, the results remain relevant to connectionist and other intelligent systems techniques that involve the use of distributional patterns.

Recently, several groups have made claims that aspects of lexical semantics can be captured by looking at patterns of lexical co-occurrence [9, 10, 5, 8]. We have been looking carefully at the techniques used to collect and make use of these statistics and have shown that varying the precise details of the computational procedures employed can make large differences to the way that the measurements can account for language based data [12, 6].

These techniques have numerous different applications within cognitive science and language technology. Within psychology, the fact that information is there to be extracted counters some "poverty of the stimulus" arguments against the possibility of learning language structure. The representations can also be used in cognitive models and in theories of learning. The techniques have already been successfully applied in language technology to achieve semantic disambiguation and document retrieval [3, 14]. It will be interesting to see whether the interdisciplinary study of exactly how these techniques work will reap benefits for both cognitive science and practical information technologies.

The idea of using distributional statistics to examine aspects of lexical semantics comes from the intuition that some aspects of word meaning may be deduced from the way in which the given word is used in relation to other words. Word usage can be measured by looking at the patterns of co-occurrence between the target word of interest and large numbers of other words in a corpus of real language use. Of course, this technique leaves out several important aspects of word meaning and language use, e.g. reference to objects and events in the world; the influence of general knowledge; speech cues such as stress and prosody. However, recent work has shown that differences in simple patterns of word usage do appear to reflect differences in word meanings.

There have been two broad classes of methods used to measure lexical semantic similarity as distances between vectors of co-occurrence counts (e.g. see [11] p 286): *document space* and *word space*. A notable example of work using document space is due to Landauer & Dumais [5] who used a technique that was developed for a practical use – information retrieval. They call their method "Latent Semantic Analysis" or LSA and stress the importance of dimensionality reduction. Using an encyclopaedia designed for children with 4.6 million words and 30,473 articles (with long articles truncated to the first 2,000 characters) they generated vectors for each word with components corresponding to the number of occurrences of the word in each article. They then transformed their vectors using an entropic measure and extracted the 300 most important dimensions using "singular value decomposition" (SVD), a procedure related to principal component analysis. They showed that the learning rate of their model mirrors the pattern of vocabulary acquisition of children and how a child can induce the rough meaning of a previously unseen word from its present context and a knowledge of past word co-occurrences.

Methods that employ word space (e.g. [14, 4, 10, 6]) rely on a more fine-grained counting scheme whereby each component of the co-occurrence vector for each particular (target) word corresponds to a another particular (context) word. The value of each component is an appropriately transformed count of how often the corresponding context word appears close to (i.e. within a particular sized window around) the target word of interest.

2. Our Vector Generation Approach

We [12, 6, 7] use a similar word space method to that used by Lund & Burgess [10, 9]. The set of word co-occurrence vectors forms a matrix consisting of columns for each target word type and rows which represent the counts of how often each

	lorry	apples	bananas
sweet	1	1	2
trees	0	2	2
shop	0	0	1
eat	0	0	0
peel	0	2	2
driver	1	0	0
road	1	0	0
diesel	2	0	0
pollution	1	0	0
wheels	2	0	0

The <u>lorry</u> *driver* swerved on the road. As well as causing *pollution*, a <u>lorry</u> also has large *wheels*. A <u>lorry</u> requires *diesel* to work. A <u>lorry</u> might carry *sweet* <u>apples</u> and <u>bananas</u>. <u>Bananas</u> are easier to *peel* than <u>apples</u> but <u>apples</u> have nicer *trees*. <u>Bananas</u> are cheaper than <u>apples</u> in a *shop*.

Table 1: Simple example of collecting the word co-occurrence statistics.

context word type occurs within a small window around the target word. Our counts are usually derived from the textual component of the *British National Corpus* (BNC) [1] which consists of 90 million words from a wide variety of sources, though we have also explored using vectors from an 168 million word corpus of USENET newsgroup text. Whatever the source, we create what is effectively a simple probability distribution by normalising each raw count by word frequency and window size to get the conditional probability of each context word type appearing around each target word.

The manner in which we measure the word co-occurrence statistics can be seen more clearly in the simple example shown in Table 1. The block of text is our corpus, the <u>underlined</u> words are our target words, and the *italicised* words are the context words. We count the number of times each context word appears in a window of plus or minus five words around the target words. This gives frequency counts that describe the typical context of each of the target words in terms of the context words they co-occur with. In this way we derive ten dimensional vectors for the three target words from the corpus of 50 words. This example serves to show how the raw statistics are counted before being normalised. The co-occurrence vectors (columns in the table) become probability distributions when the raw counts are divided by the overall frequency of occurrence of the target word with an adjustment made for window size. In practice, of course, we would use a very much larger number of target and context word types and a considerably larger corpus.

If the vector components are ordered according to the total number of occurrences in the whole corpus of the corresponding context words, we then have a straightforward procedure for reducing the dimensionality of the vectors by removing the lowest frequency components.

Various claims have been made in the literature about better ways of restricting or reducing the numbers of dimensions used. Some of these methods are merely practical considerations for computational convenience, and some may reflect the statistical nature of the language data, but others may have important implications for how such statistics are used cognitively or neurally. This paper will concentrate on comparing which context word dimensions really are most useful for this kind of work.

3. Other Previous Work

3.1 Finch & Chater

Finch & Chater [4] explored how co-occurrence vectors might serve as a basis for inducing syntactic categories. They took a 40 million word corpus of USENET newsgroup text and used the 150 most common words in the corpus as context words with a window of two words either side of each target word. They analysed the data from the 1000 most frequent words and found that a simple vector correlation technique revealed a considerable amount of information about syntactic categories. They found that cluster analysis dendrograms could be interpreted as a hierarchy of syntactic categories that is remarkably close to a standard linguistic taxonomy and included structure right up to phrasal categories.

Although this work concentrated on inducing syntactic regularities, they also found that some of their clusters exhibited semantic regularities. The most common 150 words in a corpus of English are likely to be mostly closed class or grammatical function words such as determiners (e.g. "the", "a"), conjunction, prepositions etc. It is not terribly surprising that the syntactic category of a word is found to be related to its pattern of co-occurrence with function words. What is more surprising is that function word co-occurrence gave some flavour of semantic properties. The use of closed class word co-occurrence patterns to induce measures of semantic similarity will be examined further below.

3.2 Lund & Burgess

Lund & Burgess have published several recent papers using co-occurrence statistics within a framework that they call "Hyperspace Approximation to Language" or HAL. In one particular study [10], they used a 160 million word corpus of USENET newsgroup text which they claim to be a source that gives them natural conversational language. They used a weighted window of size 10 to produce their word co-occurrence counts. The Euclidean distances between words in the high dimensional space were then used to predict the degree of priming of one word with the other in a lexical decision task.

They found that their results were unchanged by including more than the first 200 most variant context words. Their use of only the most variant words appears to be made on the grounds of computational convenience rather than a claim for limiting the scope of context words that might be used in the brain.

3.3 Lowe & McDonald

Lowe & McDonald [8] have described the use of co-occurrence vectors to model mediated priming. They collected their word co-occurrence statistics using a window size of 10 and took a positive log-odds ratio as a measurement of lexical association. They chose the context word dimensions to use in their model by selecting only those that were most "reliable". These were chosen conservatively using an ANOVA to judge how consistent the co-occurrence patterns of the context words were across different sub-corpora. Using a rather conservative criterion, this method yielded 536 context words. Before they measured the reliability they ruled out a "stop-list" of 571 words including closed class words and other mostly very common words which are usually seen as uninformative in the information retrieval literature.

4. Comparing Different Context Word Dimensions

In this paper we shall concentrate on studying the choice of context word sets that form the basis for the acquisition of the co-occurrence vectors. We begin by comparing how the different strategies for choosing the context word dimensions performed on an evaluation measure based on the synonym portion of the "Test of English as a Foreign Language (TOEFL)" task used by Landauer & Dumais [5]. This consisted of a set of 80 test words and the task was to choose the word most closely related in meaning to each test word from a set of four alternatives. Tom Landauer kindly provided us with the materials for this task. Although this test was originally used to demonstrate the utility of their LSA framework, we shall compare the different context word choices within our (word space) framework.

The LSA program scored around 64% by using a strategy of choosing the word with the largest cosine (i.e. smallest angle) between it and the target. This score is comparable to the average score by applicants to U.S. colleges from non-English speaking countries and is apparently high enough to fulfil that part of the admission requirements for many U.S. universities.

Before we can compare the various different context word sets, we need to make various other design choices, such window type and size, distance measure, corpus size, and so on. We have already considered these decisions in some detail elsewhere [12, 6, 7], and from these studies we choose what appears to be the best overall set-up for this task. We consequently used a window size of two words to the left and two to the right to collect co-occurrence statistics from the 90 million words of the written component of the BNC. We compared word vectors p and q using the Hellinger distance measure

$$H(p,q) = \sum_i (\sqrt{p_i} - \sqrt{q_i})^2$$

which is one of a sequence of information theoretic measures (including the Kullback-Leibler divergence) that is appropriate for comparing probability distributions [15].

4.1 Choosing the Context Word Sets

Figure 1 compares the performance on the TOEFL task within the methodological framework outlined above for four different methods of choosing and ordering the context word dimensions:

Frequency – ordering the context words by their frequency of occurrence in the BNC. We stop after 8192 words by which point, for most of the evaluation measures we have used in the past, performance has levelled off.

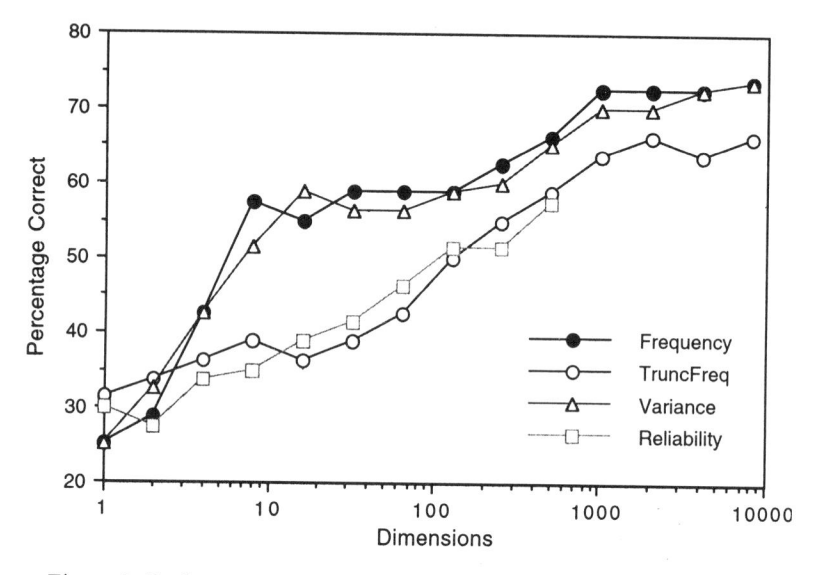

Figure 1: Performance on TOEFL task against number of dimensions.

TruncFreq – again ordering by frequency but with the most frequent 147 words (i.e. words with counts over 50,000) removed. This is to evaluate claims about the utility of the most frequent (usually closed class) context words.

Variance – ordering the context words by the variance of their components across all the target words in the corpus. This is the method used by Lund & Burgess although they used a large (weighted) window and a different corpus.

Reliability – the most reliable 536 words (ordered by frequency) kindly provided by Will Lowe and Scott McDonald from their study [8].

We can see from Figure 1 that there is a clear superiority for the ***Frequency*** and ***Variance*** approaches over the other two, particularly for dimensions less than a few hundred. The ***Reliability*** method does about as well as the ***TruncFreq*** method up to its limit of 536 dimensions where it achieves about 9% less than the simple ***Frequency*** method.

We repeated the above experiment for 19 equal sized non-overlapping subcorpora of 4.6 million words within the written component of the BNC. These are roughly the same size as the corpus used by Landauer & Dumais [5] and may be closer to the actual number of words actually read by a person during their school education. The general pattern of results was similar but uniformly lower than those for the full 90 million word corpus.

To check the extent to which this pattern of results was specific to the TOEFL task, we repeated the experiment using a second evaluation measure that we have developed and tested in the past, namely a simple semantic categoriser with a large number of candidate categories [6]. For this task, we first chose 10 of the highest frequency members from each of 53 of the Battig & Montague [2] semantic categories. Then for each category we computed the centroid (i.e. the geometric

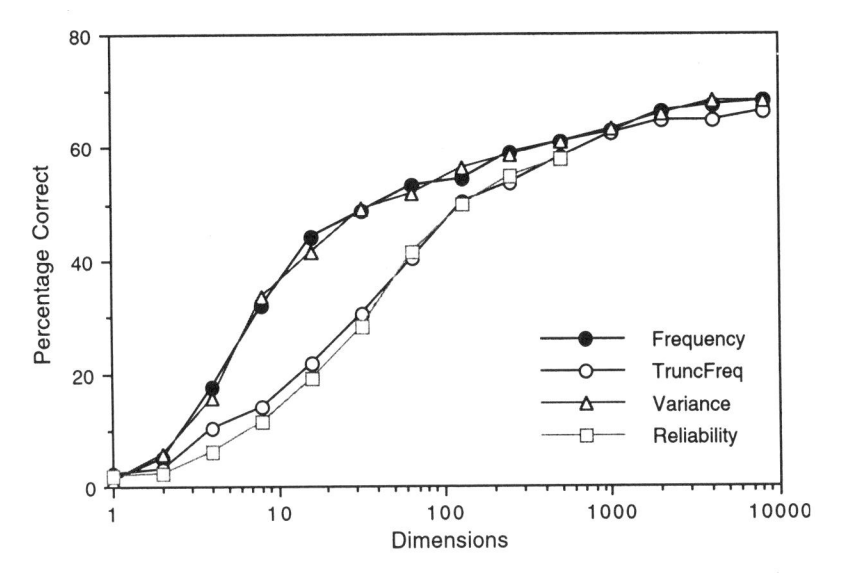

Figure 2: Semantic categorisation scores against number of dimensions.

mean) of the co-occurrence vectors of its 10 members. These are the points in the vector space about which we would expect the category members to cluster. The performance on this task is the percentage of the 530 vectors that are correctly classified in the sense of being closest to the correct one of the 53 centroids. (In practice we need to avoid the bias caused by having each vector used in the definition of its own centroid by excluding it, leaving in each case the 53 centroids defined by the remaining 529 vectors.)

Figure 2 shows the outcome of this second study. We again determined vector closeness with the Hellinger distance measure, but used a larger window size of four words to each side, as we have previously found these to be optimal for this task [6]. We see that the results are broadly in line those for the TOEFL task. *Frequency* and *Variance* again produced the best scores, but their advantage over *Reliability* and *TruncFreq* is reduced above around 500 components.

4.2 Frequency versus Variance

We can understand why the *Frequency* and *Variance* results are so similar from Figure 3. We see that the mean values of each component across the targets words (which are clearly proportional to component word frequencies) are highly correlated with the variances. Truncating the co-occurrence vectors according to one will clearly have a similar effect to truncating according to the other.

4.3 Open and Closed Class Words

In both Figures 1 and 2 it is notable that the results are good but not optimal for the first 100 or 200 dimensions. *Frequency* and *Variance* do best over this range despite the fact that for both these methods (and almost exclusively for *Frequency*)

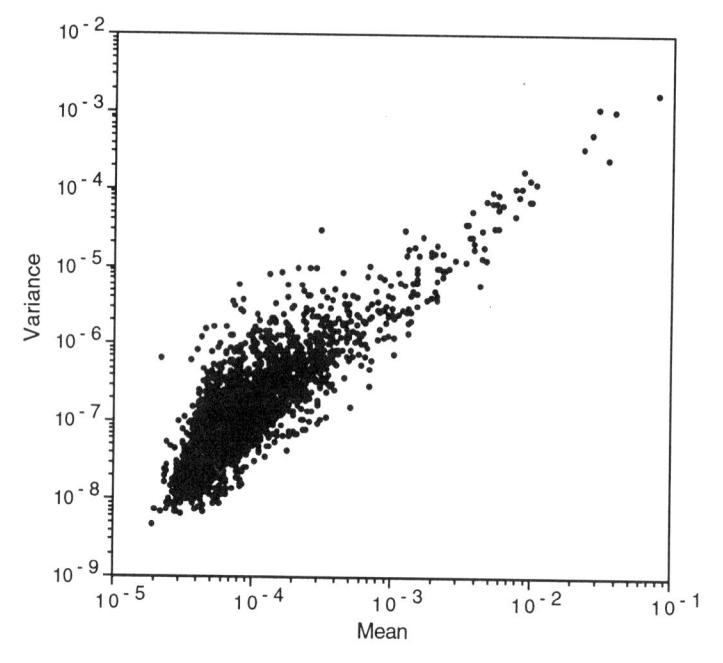

Figure 3: Mean and variance of vector components across target words.

these context word dimensions correspond to closed class or function words. It is sometimes assumed that these words are so frequent and unvarying in their patterns of co-occurrence that they would not give much information about the semantic usage of a word. Lowe & McDonald [8] follow the practice used in the information retrieval literature and exclude a "stop list" of closed class and other presumed unimportant words from consideration as context dimensions. It appears from the results presented here that these words *are* useful for the tasks we have examined and that their method for choosing context word dimensions suffers by excluding them from consideration. In fact, we have found that if the most frequent 147 words from the corpus (which are mostly closed class) are added to the 536 words of the *Reliable* set, then results are boosted significantly.

4.4 Dimensionality reduction

Using their document space based approach, Landauer & Dumais [5] found that as long as they chose the first 300 principal components of their dimensionally reduced matrix they achieved good results on the TOEFL test (around 64%). However, the performance on the TOEFL test was poor if too few or too many dimensions were used. They went on to describe a model of vocabulary acquisition in children which accounts for their extraordinarily high rate of word learning. Their paper stresses the importance of using an optimal dimensionality of data derived from the co-occurrence statistics.

We have shown here and elsewhere [6] how, using a larger corpus and our word space based method, we can achieve very good results on the TOEFL test without any form of dimensionality reduction, apart from ignoring the very low frequency words as a matter of convenience. We have also presented [7] some preliminary results on using singular value decomposition on our co-occurrence vectors. There seemed to be no improvement demonstrated over the original vectors, even when we used our small corpora of 4.6 million words. Clearly, dimensionality reduction is useful in some cases, but only for certain ways of using the co-occurrence statistics. For our word space based method using a small window size and an information theoretic distance measure, dimensionality reduction appears to be unnecessary. Clearly this feature warrants further investigation in the future.

5. Discussion

It is clear that all the methods we have discussed and compared in this paper extract useful and interesting statistical regularities from corpus-based co-occurrence counts. We believe that the methodological claims that are made in this field should always be tested using different data and different parameters to determine how general they really are. Our approach has been to start with as simple a procedure as possible and to tune the various parameters (such as window type and size) by observing empirically how well the different parameter combinations perform under various different evaluation tasks (such as the TOEFL test and our semantic categorisation task). We have demonstrated here that simply ordering the context word dimensions in order of their frequency and using as many of them as possible or practical produces good results. Closed class or function words appear to provide useful contributions to the co-occurrence statistics. Ordering the context words by variance or reliability appears to confer no advantages within our framework and the computationally expensive procedure of singular value decompositions seems not to increase performance on our evaluation tasks.

Choosing between methodological variations is often a matter of computational convenience, or corresponds to a claim or assumption about the statistical language data, rather than a statement about how cognitive and neural systems might realise and utilise such data. However, the use of closed class words probably does have implications for cognitive models because there are claims in the psycholinguistic literature that open and closed class words are processed differently in the brain. We have found no evidence that this has to be true because including closed class words does seem to improve performance in the corpus based studies discussed in this paper. Moreover, there are ways in which connectionist and neural systems might instantiate dimensionality reduction, but again we have found no evidence that this has to take place to make good use of word co-occurrence statistics.

References

1. Aston, G. & Burnard, L. (1998). *The BNC Handbook: Exploring the British National Corpus with SARA*. Edinburgh University Press.
2. Battig, W.F. & Montague, W.E. (1969). Category norms for verbal items in 56 categories: A replication and extension of the Connecticut category norms. *Journal of Experimental Psychology Monograph*, 80, 1-45.

3. Dagan, I., Marcus, S. & Markovitch, S. (1993). Contextual word similarity and estimation from sparse data. in *Proceedings of the 31st Annual Meeting of the ACL*, 164-171.
4. Finch, S. & Chater, N. (1992). Bootstrapping syntactic categories. In *Proceedings of the 14th Annual Meeting of the Cognitive Science Society*, 820-825.
5. Landauer, T. & Dumais, S. (1997). A solution to Plato's problem: The latent semantic analysis theory of acquisition, induction, and representation of knowledge. *Psychological Review, 104(2),* 211-240.
6. Levy, J.P., Bullinaria, J.A. & Patel, M. (1998). Explorations in the derivation of word co-occurrence statistics. *South Pacific Journal of Psychology, 10(1),* 99-111.
7. Levy, J.P. & Bullinaria, J.A. (1999). The emergence of semantic representations from language usage. Paper given at the *EPSRC Workshop on Self-Organising Systems - Future Prospects for Computing*, UMIST, October 1999.
8. Lowe, W. & McDonald, S. (2000). The direct route: Mediated priming in semantic space. *In Proceedings of the 22nd Annual Meeting of the Cognitive Science Society.*
9. Lund, K., Burgess, C. & Atchley, R.A. (1995). Semantic and associative priming in high-dimensional semantic space. In *Proceedings* of the 17th Annual Meeting *of the Cognitive Science Society*, 660-665.
10. Lund, K. & Burgess, C. (1996). Producing high-dimensional semantic spaces from lexical co-occurrence, *Behavior Research Methods, Instruments, & Computers, 28(2),* 203-208.
11. Manning, C.D. & Schütze, H. (1999). *Foundations of Statistical Natural Language Processing.* Cambridge, MA: MIT Press.
12. Patel, M., Bullinaria, J.A. & Levy, J.P. (1998). Extracting Semantic Representations from Large Text Corpora. In Bullinaria, J.A., Glasspool, D.W. & Houghton, G. (eds), *4th Neural Computation and Psychology Workshop, London, 9-11 April 1997: Connectionist Representations*, 199-212. London: Springer-Verlag.
13. Redington, M. & Chater, N. (1997). Probabilistic and distributional approaches to language acquisition. *Trends in Cognitive Sciences, 1 (7),* 273-281.
14. Schütze, H. (1993). Word Space. In S.J. Hanson, J.D. Cowan & C.L. Giles (Eds.) *Advances in Neural Information Processing Systems 5,* 895-902. San Mateo, CA: Morgan Kauffmann.
15. Zhu, H. (1997). Bayesian Geometric Theory of Learning Algorithms. In: *Proceedings of the International Conference on Neural Networks (ICNN'97),* Vol. 2, 1041-1044.

Associative Computation and Associative Prediction

Andrzej Wichert

Abstract

We introduce a biologically and psychologically plausible neuronal model which could explain how pictorial reasoning and learning is carried out by the human brains. This biologically inspired model sheds some light on how some problem solving abilities might actually be performed by the human brain using neural cell assemblies by forming a chain of associations. The model uses picture representation rather than symbolic representation to perform problem solving. The computational task concerning problem solving corresponds to the manipulation of pictures. A computation is performed with the aid of associations by the transformation from an initial state represented as a picture to a desired state represented as a picture. Picture representation allows for the presence of noise and also enables learning from examples. The solved problems are reused to speed up the search for related or similar problems. Either an observer chooses relevant examples or the model learns by experience of failures and successes. The learning from examples is demonstrated by empirical experiments in block world and on a robot in a labyrinth. It is shown that learning improves the behaviour of the model in a statistically significant manner.

1. Problem Solving and Associative Prediction

Human problem-solving can be modelled by a production system which implements a search algorithm [10]. Production systems theory describes how to form a sequence of actions which lead to a goal, and offers a computational theory of how humans solve problems [2]. Production systems are composed of if-then rules which are also called productions. The complete set of productions constitute the long term memory. Productions are triggered by specific combinations of symbols which describe items. These items represent a state and are stored in short term memory. A computation is performed with the aid of productions by the transformation from an initial state in the short term memory to a desired state.

In a pure production system which was proposed as a formal theory of computation [10] the system halts if no production can fire in a state. In systems which model human behavior backtracking to a previous state of working memory is allowed [10]. By allowing backtracking and the exclusion of loops, a search from the initial state to the desired state is executed. The search defines a state space, problems are solved by searching in this space whose states includes the initial situation and the desired situation [10]. Production systems are often also the atomic part of more effective planning algorithms, like for example the SOAR model [12], which will be not examined here.

A production system by itself does not adequately explain learning from experience, where the solved problems are reused to speed up the search for related or similar problems. The definition of similarity between different problems is very difficult. The difficulties arise, because the states are represented by symbols which are used to denote or refer to something other than themselves, namely to other things in the world. In this context symbols do not in themselves, represent any utilizable knowledge which could be used for a definition of similarity criteria between themselves. One solution to this problem would be the representation of states by pictures.

To ease comprehension for the human reader, additional pictures like two dimensional binary sketches are often used beside the symbolical representation. (see Figure 1). Pictures correspond to vectors. Vectors define a space in which the similarity between themselves can be computed.

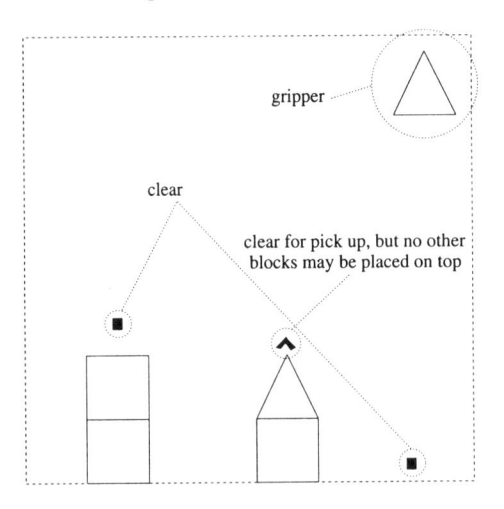

Figure 1: A state in the blocks world.

A sequence of pictorial states describing a plan can be stored in an associative memory. The sequence commences with the initial state and ends with the penultimaste state before the desired state. After the storage of some plans, a succeeding state of one stored plan can be recalled, given the current and the desired state. The current state together with the desired state represent the question, the next state the answer. A question vector is composed by concatenation of the pattern representation of the current state and the desired state. The answer vector corresponds to the pattern representation of the next state. Both vectors can be stored in the associative memory.

From each current state together with the desired state a question vector is formed and stored together with the answer vector corresponding to the next state. After "learning" the sequence can be recalled. By posing the question vector which is composed of the initial state, holding the desired state part, and by feedback, the answer vector to the question vector, the state sequence is determined (see Figure 2 and [13]).

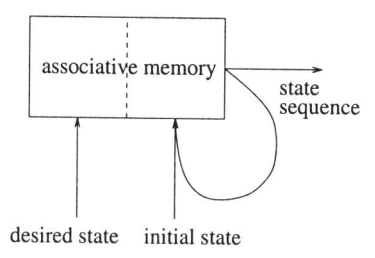

Figure 2: The sequence begins with the question vector which is form by the concatenation of the vectors representing initial and desired state. Via the feedback connections a new question vector is formed out of the concatenation of the answer vector and the vector representing the desired state. The corresponding answer vector is the next state of the sequence.

Because of the ability of the associative memory to determine the most similar stored pattern to a currently not stored pattern, similarity between different problems can be defined by the pictorial state representation. An associative memory in which a sequence of states describing a plan is stored is called the prediction associative memory.

1.1 Neural Assembly Theory

Neural assembly theory, which was first suggested by Donald Hebb (1949) [6] is related to the production systems theory. Neural assembly theory describes the bridge between the structures found in the nervous system and in a high level cognition such as problem solving. The process of problem solving is described as the transformation of thoughts by a group of assemblies [3, 13]. It was shown that Donald Hebb's hypothesis of cell assemblies as a biological model of internal representation of events and situations in the cerebral cortex corresponds to the formal associative memory model [13]. However, traditional symbolic forms of representation which are used in production systems appear to be inappropriate for use with cell assemblies because they lack a concept of similarity. In this case, distributed representation of knowledge seems more suitable. Pictorial reasoning is a kind of known human reasoning [9] which uses distributed representation. Visual mental images are one form of representation of short-term memory [9].

1.2 Associative Memory

Binary represented states can be represented by the associative memory. The associative memory [15, 7] is composed of a cluster of units which represent a simple model of a real biological neuron. The unit is composed of weights which correspond to the synapses and dendrides in the real neuron. They are described by w_{ij} in fig. 3. T is the threshold of the unit. Two pairs of binary vectors are associated, this process of association is called learning. The first of the two vectors is called

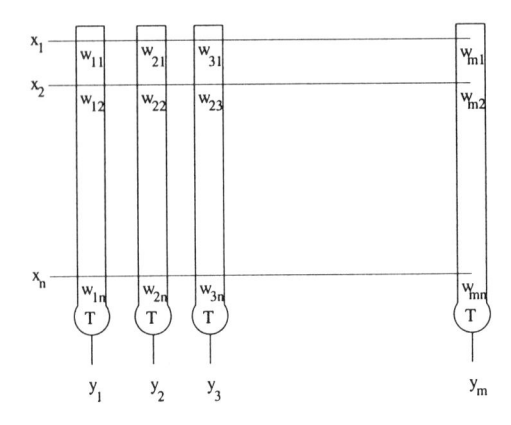

Figure 3: The associative memory is composed of a cluster of units.

the question vector and the second, the answer vector. After learning, the question vector is presented to the associative memory and the answer vector difference is determined.

Reliability of the Answer Once an answer vector is determined, it would be useful to know how reliable it is. Let $\vec{x}$ be the question vector and $\vec{y}$ the answer vector that was determined by the associative memory. First, the vector $\vec{x^l}$ which belongs to the vector $\vec{y}$ is determined. The vector $\vec{x^l}$ is determined by a backward projection of the vector $\vec{y}$. The synaptic matrix used in the backward projection is a transpose of the matrix W which is used for the forward projection. In the second step, the similarity of the stored question vector $\vec{x^l}$ to the actually presented vector $\vec{x}$ is determined. The greater the similarity of the vector $\vec{x^l}$ to the vector $\vec{x}$, the more reliable the answer vector $\vec{y}$.

2. Associative Computation

The computational task associated with problem solving corresponds to the manipulation of pictures, but how can this be done? A structured state representation by pictures is needed so that objects in the picture can be manipulated. Gross and Mishkkin(1977) [5, 9] suggest that the brain includes two mechanisms for visual categorization: one for the representation of the object and the other for the representation of the localization [9, 14]. The first mechanism is called the "what" pathway and is located in the temporal lobe. The second mechanism is called the "where" pathway and is located in the parietal lobe [9, 14]. According to this division, the identity of a visual object can be coded apart from the location and the size of the object. A visual state represented by a vector can be also represented by meaningful pieces of the vector [8, 1]. Pieces which represent objects of the scene are called cognitive entities. Each cognitive entity represents the identity of the object and its position by the coordinates. The identity of an object is represented by a binary pattern which is normalised for size and orientation. Its location corresponding to the abscissa is represented by a binary vector of the dimension of the abscissa of the picture representing the state. The location corresponding to the ordinate is likewise represented by a binary vector of the dimension of the ordinate of the picture representing the state. A binary bar of the size and position of the object in the picture of the state represents in each of those vectors the location and size (see Figure 4, Figure 1). Cognitive entities can represent associations which represent transitions

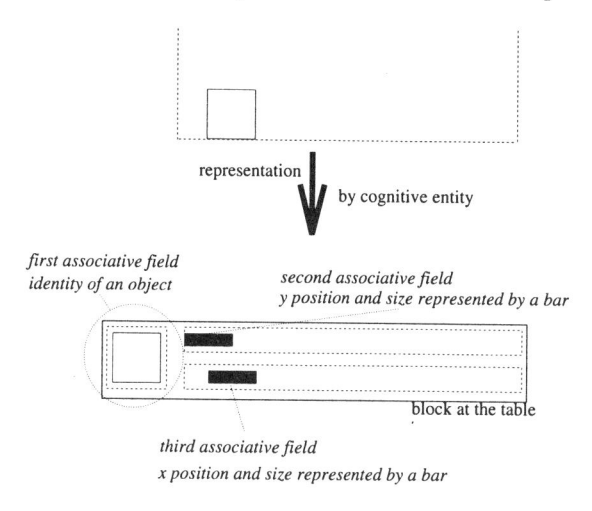

Figure 4: A cognitive entity.

between states. The first pattern represented by the cognitive entities describes the state which should be present before the transition (the premise). The second pattern describes the world state after the transition (the conclusion). In order to preserve the equality of cognitive entities in the premise and in the conclusion pattern, a notation for an empty cognitive entity is used (see fig 5). In Figure 5 an example from the block world is shown. Both, an empty robot arm, which is represented by the right corner, or a "clear" position are represented by a dot.

In the robot-in-a-maze task, a robot can move from one position to another [4, 11] in a maze (Figure 8). The robot itself is represented by one cognitive entity. Its shape is specified by the first associative field. Its position is specified by bars in the associative fields two and three. The maze is represented by the other cognitive entities and it is decomposed into objects corresponding to walls, the intersections of walls, and passages. An object is specified by the first associative field, its position by the two remaining associative fields. For the robot the associations describe the possible moves. The premise represents the position of the robot and the nearest passage way present by two cognitive entities, so that the robot can move through it. The conclusion describes the new position after the robot moved through the passage (see Figure 6).

The cognitive entities of the premise pattern are replaced by the conclusion pat-

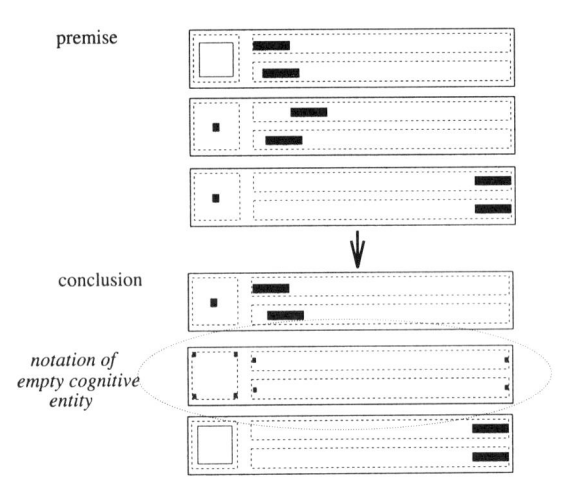

Figure 5: Representation of the association: If a block is at a certain position and above it, it is clear and the gripper is empty then the block is grasped by the gripper. The old position of the block is marked as clear, avoiding the frame problem [10]. One cognitive entity of the conclusion pattern is not used. In the inverse association, the premise pattern is interchanged with the conclusion pattern.

tern in case the similarity between the condition pattern and the corresponding part of the state picture is sufficient.

2.1 Permutations

A state is represented by η cognitive entities. Associations represent transitions between the states representing pictures. The premise of an association is represented by δ cognitive entities which describe a correlation of objects which should be present (see Figure 7). If present, they are replaced by δ cognitive entities of the conclusion. Generally the premise is describe by fewer cognitive entities then the state, $\delta \leq \eta$. In the recognition phase, all possible δ-permutation of η cognitive

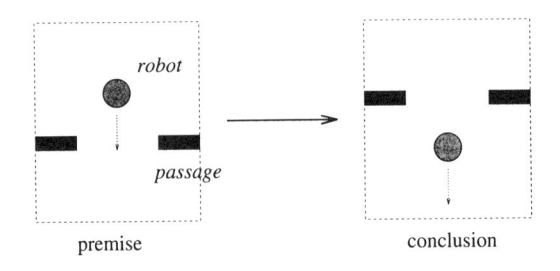

Figure 6: One association describes a possible move of the robot.

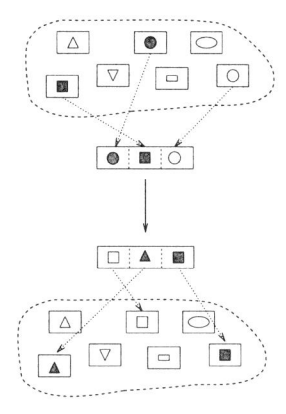

Figure 7: A copy of the state representation is formed and the corresponding cognitive entities are replaced by the conclusion pattern.

entities should be composed to test if the premise of an association is valid.

$$\Xi := P(\eta, \delta) = \frac{\eta!}{(\eta - \delta)!}$$

This is done because the premise can describe any correlation between the cognitive entities. In the retrieval phase after learning Ξ permutations are formed. Each permutation represents a question vector $\vec{x_i}$, $i \in \{1, \ldots, \Xi\}$. To each question vector $\vec{x_i}$ an answer vector $\vec{y_i}$ with the quality criterion is determined. If the reliability value of this answer vector is above a certain threshold, the association can be executed. A copy of the state representation is formed and the corresponding cognitive entities are replaced by the conclusion pattern (see Figure 7). Associative memory which perform this task is called the permutation associative memory. Given a state represented by a unit, the permutation associative memory recognizes s question vectors, s copies of state representation are formed. The cognitive entities which form the question vectors (premise) are replaced by the cognitive entities of the answer vectors (see fig. 7). The resulting s states are represented by s units.

2.2 Search and Associative Prediction

The state space is represented by a chain of units in which values are propagated by local spreading activation. After the temporary parallel execution of a chosen state, s new states emerge. From the s^t states, one state is chosen and the new s^{t+1} states are determined. A state can cause an impasse when no valid transition to a succeeding state exists. In this case, backtracking to the previous state is performed. Another state can be chosen, if possible, or backtracking is repeated. The resulting search strategy is the "deep search" strategy.

After learning sequences of states describing plans associative prediction can be performed. The permutation associative memory determines $b(i)$ states of a chosen state. At the same time, the prediction associative memory determines to this chosen state together with the desired state the answer pattern $Pred$. The state $b(i)$, whose pattern is most similar to $Pred$ is chosen and the search is continued.

3. Learning Strategy

Either the learning phase and the retrieval phase of the prediction associative memory are separated or they are not. If the phases are separated, a kind of "teacher" exists in the learning phase. An observer chooses relevant examples and acts as a supervisor who guides the search. He shows how to solve a problem. This solution is stored in the prediction associative memory. This kind of learning is called supervised learning. If there is no separation, then the prediction associative memory learns by experience of failures and successes. After a solution of a problem was found using the prediction heuristic which guides the search, it is stored in the prediction associative memory. The resulting search strategy is the hill climbing search strategy [16]. At the beginning of learning session the prediction memory is empty, the prediction heuristic has no information. However the prediction heuristic is improved during learning. This kind of learning is called "unsupervised" learning.

Three tasks were examined. The results are shown in Table 1.

- In geometric blocks, blocks can be placed in three different positions and picked up and set down by a robot arm. There are two different classes of blocks: squares and triangles. No other block may be placed on top of a triangle, while either type of block may be placed on top of a square.
- In ABC blocks, three blocks differ by attributes, but not by form. In our representation the A, B, C marks correspond to the marks at the corner of the counter representing the blocks.
- Figure 8 is an example of the robot-in-a-maze task, in which a robot can move from one position to another [4].

task	examples	steps	us	sup
geometric blocks	9	14.89	13.43%	47.75%
ABC blocks	34	62	40.52%	81.21%
robot	72	7.19	20.86%	-

Table 1: For a certain task for which a number of examples was presented, shown in column *task* and *examples*; the blind search strategy needed on average the number of steps shown in column *steps*. The prediction heuristic which was formed by unsupervised learning *(us)* brought a significant improvement in % which is shown in column; the prediction heuristic which was formed by supervised learning *(sup)* brought a significant improvement in % which is shown in column *sup*.

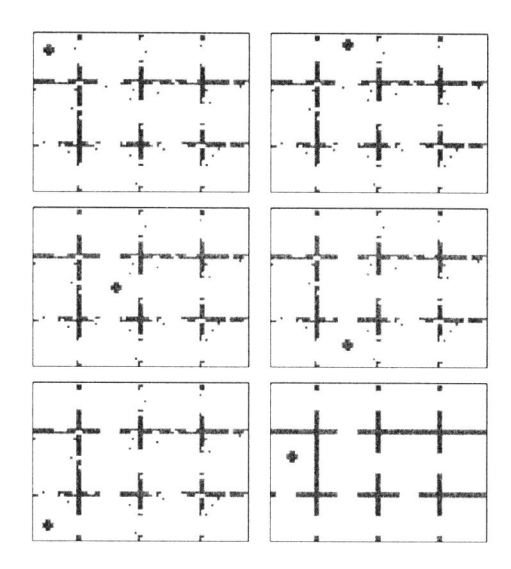

Figure 8: In this example the robot has to go from upper left corner room of the labyrinth to the room under it. The shortest path is chosen because the robot had previously explored and already learned the labyrinth before by unsupervised learning. The search sequence is represented by noisy pictures, the initial state and desired state are included. The desired state is represented without noise.

4. Conclusion

The computational task concerning problem solving in our model corresponds to the manipulation of pictures. A problem is described by the associations in the long term memory which is represented by permutation associative memory, by the initial state, and by the desired state. The solution to the problem is represented by a chain of the associations which successively change the state from the initial state to the desired state. The basic behaviour of our model corresponds to the behaviour of a symbolic production system which performs a depth-first search strategy. However the representation of states through pictures enables the access of knowledge which was formed by learned experience during problem solving. The learned knowledge speeds up the search for related problems significantly.

One general conclusion from the experiments is the claim that it is possible to use systematically associative structures to perform reasoning by forming chains of associations. In addition, beside symbolical problem solving, pictorial problem solving is possible.

References

1. Anderson, J. A. (1995). *An Introduction to Neural Networks.* MIT Press,.
2. Anderson, J. R. (1995). *Cognitive Psyhology and its Implications.* W. H. Freeman and Company, fourth edition,.
3. Braitenberg, V. (1978). Cell assemblies in the cerebral cortex. In R. Heim, & G. Palm, (Eds.) *Theoretical Approaches to Complex Systems,* pp. 171–188. Springer-Verlag,.
4. Ferber, J. (1995). *Les Systèmes Multi-Agents: Versus une intelligence collective.* Paris: InterEditions.
5. Gross, C., & Mishkin. (1977). The neural basis of stimulus equivalence across retinal translation. In S. Harnad, R. Dorty, J. Jaynes, L. Goldstein, & Krauthamer, (Eds.), *Lateralization in the nervous system.* New York: Academic Press.
6. Hebb, D. (1949). *The organization of behaviour.* New York: John Wiley.
7. Hecht-Nielsen, R. (1989). *Neurocomputing.* Addison-Wesley,.
8. James, W. (1985). *Psychology, the Briefer Course.* University of Notre Dame Press, Notre Dame, Indiana,. (Orginally published 1892).
9. Kosslyn, S.M. (1994). *Image and Brain, The Resolution of the Imagery Debate.* MIT Press,.
10. Luger, G. F., & Stubblefield, W. A.. (1998). *Artificial Intelligence, Structures and Strategies for Complex Problem Solving.* Addison-Wesley, third edition,.
11. McCarthy, J., & Hayes, P. (1969). Some philosophical problems from the standpoint of artificial intelligence. In B. Meltzer & D. Michie (eds.) Machine Intelligence 4. Edinburgh, Scotland: Edinburgh University Press.
12. Newell, A. (1990). *Unified Theories of Cognition.* Harvard University Press,.
13. Palm, G. (1982). *Neural Assemblies, an Alternative Approach to Artificial Intelligence.* Springer-Verlag,.
14. Posner, M. I., & Raichle, M. E. (1994). *Images of Mind.* New York: Scientific American Library.
15. Steinbuch, K. (1961). Die Lernmatrix. *Kybernetik, 1,* 36–45,
16. Winston, P. H.(1992). *Artificial Intelligence.* Addison-Wesley, third edition.

The Development of Small-world Semantic Networks

Eric Postma, Alard Roebroeck, and Joyca Lacroix

Abstract

Cognitive processes rely on knowledge structures that can be represented by networks of interconnected concepts, i.e., semantic networks. We study the developmental dynamics of generic semantic-network models. We focus on two measures: (i) the *characteristic path length* between an arbitrary pair of concepts, and (ii) the *clustering coefficient* of groups of concepts. Short path lengths facilitate the dynamics of mental processes through spreading activation and a large clustering coefficient reflects the existence of structured representations. We analysed semantic-network models generated from behavioural data. We measure the characteristic path length and clustering coefficient at various stages of development. The results reveal semantic networks to be small-world networks, i.e., networks that combine short path lengths with high clustering. In addition, developing semantic networks are characterised by an increase in small-worldliness by maintaining a constant path length and clustering coefficient despite the increase in the number of concepts.

1. Background

The ability to associate concepts lies at the heart of many mental processes. The association structure underlying mental processes allows for the rapid association of arbitrary concepts. The process of association is often expressed in terms of spreading activation in semantic networks [4]. Activated concepts spread their activation to associated concepts, which in turn spread their activation to their associates, and so forth. Since the activation decays with distance [2] (i.e., the number of associative links), the average separation among two concepts should not be too large. At the same time, however, the network should accommodate the formation of semantic domains (i.e., tightly clustered groups of concepts), to reflect the associative structure of categories. In this study these two requirements are quantified in terms of *characteristic path length* and *clustering coefficient*, respectively. Using these two measures, we analyse artificial semantic networks generated from age-dependent word-association data to characterise the underlying association structure. More specifically, our analysis reveals the degree to which semantic networks are *small-world networks* [15] which combine the maximisation of clustering with the minimisation of characteristic path length.

The outline of the paper is as follows. Section 2 reviews Watts and Strogatz's [15] work on small-world networks and discusses the suitability of four network types for describing the human associative network. Section 3 outlines the behavioural data and their translation into a sequence of age-dependent semantic networks. In section 4 the artificial semantic networks are analysed and the results presented. Section 5 discusses our findings and concludes on the small-worldliness of developing semantic networks.

2. Small-World Semantic Networks

In what follows, we discuss Watts and Strogatz's network analysis in terms of semantic networks. In characterising network (or graph) structures, Watts and Strogatz [15] defined two measures: (1) the characteristic path length L and (2) the clustering coefficient C.

L equals the average number of associative links separating a pair of concepts along a shortest path. A high average association length indicates large distances among concepts from different semantic domains, such as, for instance, *apple* (as part of the semantic domain *fruit*) and *Newton* (as part of the semantic domain *physicists*) in figure 1.

C expresses the degree to which neighbouring concepts, i.e., the concepts associated with a given central concept, are associated with each other. A high value of C signifies a densely interconnected domain of concepts. Consider, for instance, the hypothetical semantic network shown in figure 1. The central concept *apple* is associated to the neighbouring concepts *pear, orange,* and *lemon*. The maximum number of associative links among k neighbouring concepts equals $k(k-1)/2$. The maximum number of associative links among the $k=3$ neighbours of *apple* equals 3. Since only two out of the three associations are present (i.e., *lemon-orange* and *pear-orange*), the concept *apple* contributes 2/3 to the clustering coefficient. The concept *Newton* contributes a value of 1 to the characteristic path length, because all its neighbours are interconnected.

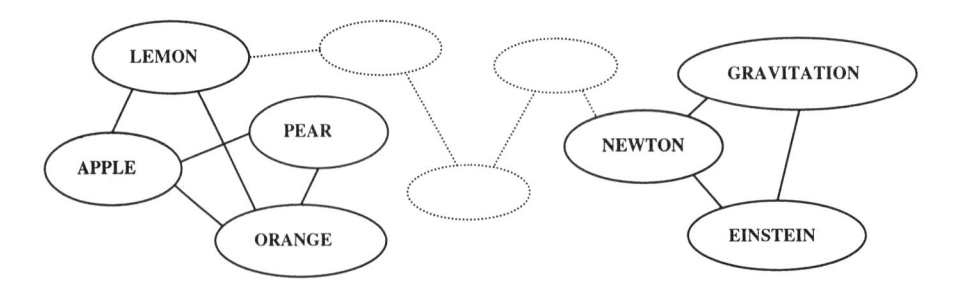

Figure 1: Example of part of a hypothetical semantic network showing two semantic domains (i.e., *fruit* and *physicists*). Concepts are represented by ellipses, their associations by lines.

2.2 Four Network Types

We discuss four types of network structures in terms of their characteristic path length L and clustering coefficient C. First, we consider a fully connected semantic network (average connectivity $k = N-1$; see figure 2a). Full connectivity has the advantage that the semantic network has a small characteristic path length, i.e., $L=1$. However, full connectivity leads to a highly unstructured knowledge representation (i.e., everything is associated with everything else). The second type of network is a sparse random semantic network ($k << N$; see figure 2b) in which each concept is associated with k randomly-selected other concepts. The average association distance of a random semantic network is small (see, e.g., [8]). Unfortunately, random associations lead to highly unstructured knowledge representations. The third type of network is a regular semantic network ($k << N$; see figure 2c). The network is made up of small domains of densely interconnected concepts. Each domain is connected to a few neighbouring domains. The regular semantic network allows for structured knowledge representations. However, the average association distance is large. Finally, the fourth type of network is a *small-world semantic network* [12].

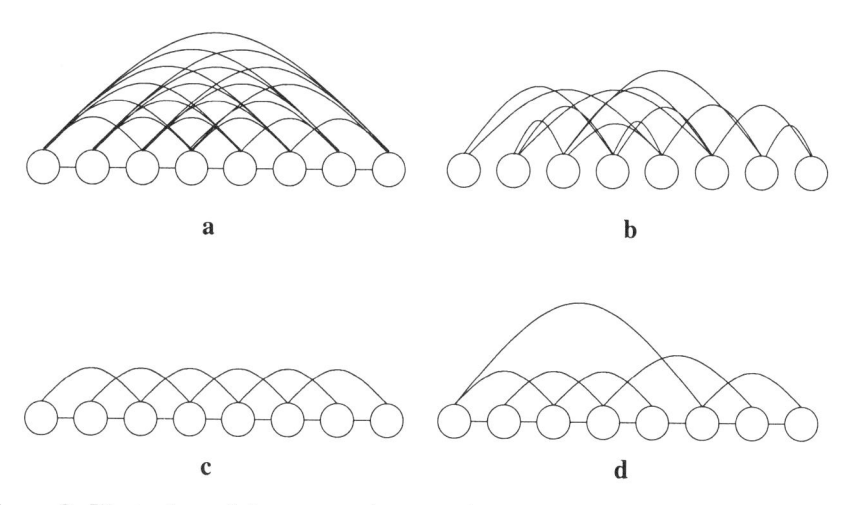

Figure 2: Illustration of four semantic-network types ($N = 8$). (a) fully-connected network ($k=N-1$), (b) random network ($k=4$), (c) regular network ($k=4$), (d) small-world network ($k=4$). The disks represent concepts, the lines associations among pairs of concepts. In the networks shown in (b), (c) and (d), boundary conditions are ignored for clarity of presentation (i.e., $k=2$ and 3 for the outer two nodes).

As illustrated in figure 2d, small world semantic networks consist of domains of densely interconnected concepts (as in regular semantic networks), but have a moderate number of associations with randomly-selected other domains (as in random semantic networks).

Watts and Stogatz [15] showed small-world networks to combine the best of both worlds: a large clustering coefficient and a small characteristic path length. The interested reader is referred to [11] for a gentle introduction to small-world networks. Table 1 summarises the properties of the four networks in terms of their connectivity, clustering coefficient, and characteristic path length. Solely the small-world network combines sparse connectivity with large clustering coefficient and small characteristic path length. We therefore expect semantic networks to be small-world networks [12].

Type of network	K	C	L
Fully-connected	$N-1$	Large	Small
Random	$<<N$	Small	Small
Regular	$<<N$	Large	Large
Small-world	$<<N$	Large	Small

Table 1. Overview of the characteristics of four types of semantic network. The first column specifies the type of connectivity. The second column gives the associated average value of k. The third and fourth columns indicate the presence of a large or small clustering coefficient (C) and characteristic path length (L), respectively.

2.3 Small-World Networks

Watts and Strogatz studied the continuum ranging from regular networks (figure 2c) to random networks (figure 2b). Starting from a regular network, they rewired each link with probability p, without changing the (average) number of links per node k. The two extremes of the continuum correspond to $p=0$ (regular network) and $p=1$ (random network). Watts and Strogatz determined the normalised values of C and L as a function of p, i.e., $C(p)/C(0)$ and $L(p)/L(0)$, expressing the clustering coefficient relative to a regular network and the characteristic path length relative to a regular network, respectively. Their results from computational analyses of artificial graphs are illustrated in figure 3. The figure shows the normalised values of C and L as a function of p on a logarithmic scale. Strikingly, only a small amount of randomness (i.e., a small rewiring probability p indicated by the arrow in figure 3) is needed to reduce the characteristic path length to a small value. Small-world networks are associated with a degree of randomness, that combines a small normalised L with a large normalised C. It is important to note that at the position indicated by the arrow in figure 3, the clustering coefficient remains virtually unchanged. Consequently, the transition from a large world to a small world is hardly detected at the local scale of concepts and the density of their interconnections.

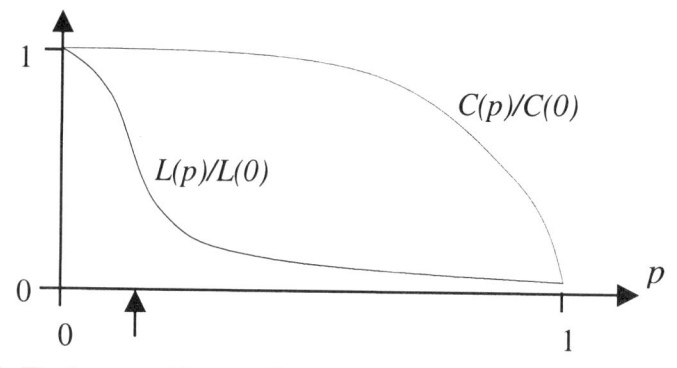

Figure 3: The large-world to small-world transition as evident from the normalised clustering coefficient C(p)/C(0) (dotted curve) and the normalised path length L(p)/L(0) (solid curve), as a function of randomness p, drawn on a logarithmic scale (redrawn from [15]). At a small value of p (arrow), the path length decreases drastically whereas the clustering coefficient remains virtually unchanged.

Framed in terms of the large-to-small world transition, our analysis aims at characterising the development of semantic networks in terms of a large or a small world. Before turning to our analyses, the following section discusses the creation of artificial semantic networks from behavioural data.

3. Artificial Semantic Networks

To assess the presence of associative links in the human semantic network, we considered various publicly-available dictionaries and thesauri. The MRC psycholinguistic database containing word-association and age-of-acquisition data published by Michael Wilson[1] on the Internet, turned out to be most suitable for our purposes. The word-association data were obtained in an experimental study [9] where subjects were asked to associate freely on target words. As the aim of the study was to "to obtain a reasonably large complete mapping of the associative network for a large set of words", the appropriateness for our study is evident. Table 2 shows the associates of the concepts "apple" and "Newton". The age-of-acquisition data are expressed in age times hundred and range from 200 to 700 [6].

concept	associated concepts (associative strength)
APPLE	PIE (20) PEAR (17) ORANGE (13) TREE (8) CORE (7) FRUIT (4)
NEWTON	APPLE (22) ISAAC (15) LAW (8) ABBOT (6) PHYSICS (4) SCIENCE (3)

Table 2. Two examples of concepts and their associates. The associative strength (specified between brackets) is given on a scale from 1 to 100.

---

[1] http://www.ling.ed.ac.uk/help/mrc/

The association data were used in two ways. First, in combination with age-of-acquisition data they were translated into age-dependent semantic networks. Second, to simulate semantic development they were used in isolation (i.e., without the age-of-acquisition data) for training a hetero-associative memory model of semantic development. Both models are discussed in more detail below.

3.1 Age-dependent Semantic Networks

The age-dependent association matrices, i.e., semantic networks, are created as follows. The presence of an associative relation between any two concepts is represented by the $N \times N$ association matrix A. Our model consists of a age-dependent sequence of binary association matrices $A(t)$ for $t \in \{200,300,...,700\}$ (the values of t represent age in years × 100). Each element $A(t,i,j) = A(t,j,i)$ of the matrix $A(t)$ specifies the presence ($A(t,i,j) = 1$), or absence ($A(t,i,j) = 0$), of an association between the concepts i and j at age t. The number of concepts N corresponds to the number of concepts acquired at age $t = 700$. For $t < 700$, a concept h that is not acquired yet, is represented by $A(t,h,j) = A(t,j,h) = 0$, $1 \leq j \leq N$. For each target word w, a response word r was generated by a proportion $\lambda(w,r)$ of the subjects. The entries of $A(700)$ are defined as $A(700,w,r) = s[\lambda(w,r)-\theta]$, ($\theta = 0$) with the function $s[\,]$ defined as

$$s[x] = \begin{cases} 1 & x > 0 \\ 0 & x \leq 0 \end{cases}$$

The entries of $A(t)$ for $t < 700$ are defined as

$$A(t,w,r) = s[aoa(w)-t] \times s[aoa(r)-t] \times A(800,w,r),$$

where $aoa(w)$ is the age of acquisition of word w. For each age t, the largest connected sub-graph was identified and used for further analysis. The number of concepts in the largest sub-graph is defined as the number of concepts acquired at age t. Table 3 lists the sizes of the artificial semantic networks so obtained.

t	200	300	400	500	600	700	adult
N_{graph}	30	322	728	1066	1185	1205	8210
k	3.2	10.5	14.7	17.0	17.7	17.7	58.7

Table 3. Size (N_{graph}) and average number of associations per concepts (k) for seven different ages (t). The last column (designated "adult") specifies N and k for the entire set of concepts for which association data are available.

3.2 Hetero-Associative Memory

To simulate semantic-network development, we presented the concepts acquired at age $t = 700$ to a hetero-associative memory (HAM) consisting of M input nodes, M output nodes, and M^2 adaptive weights. Concepts are represented by binary vectors **p**, with $|p| = m$. Associations among concepts were encoded by means of clipped Hebbian learning, i.e., at each presentation of a pattern-pair all weights are set to $w(i,j)=\min[1.0, w(i,j)'+\lambda p(i)q(j)]$. (Initially, all weights are set to zero.) To simulate the developmental stages, the network is trained 7 times, with $\lambda=0.2$. At each stage kN pairs of concepts are randomly selected with a probability proportional to their association strength. For a sufficiently large network ($M = 1024$), the value of $m = 4$ yields a storage capacity of the network that is large enough to store the kN associations [3]. Standard HAMs associate a single input vector to a single output vector. To accommodate the association of a single input vector to multiple output vectors, as is required to store semantic networks with $k > 1$, we used stochastic synapses [7]. During recall, given an input vector p, the elements of the integer output vector q are defined as

$$q(i) = \sum_{j=1}^{M} f(w_{i,j})p(j),$$

with the function f defined as

$$f(x) = \begin{cases} 1 & \text{if } rand < x \\ 0 & \text{otherwise} \end{cases},$$

where *rand* is a random value taken from a uniform distribution on the unit interval. Of the output values $q(i)$ so obtained, the m largest values are set to 1 and the others to 0. The resulting binary vector is matched to all concept vectors. The best-matching vector is defined as the response. Each input pattern is presented 100 times. All responses to the input are defined as concepts associated with the input concept. Table 4 lists the k values of the resulting semantic networks.

t	100	200	300	400	500	600	700
N_{graph}	20	35	266	677	999	1011	1205
k	1.2	3.0	9.4	12.1	16.4	17.0	18.0

Table 4. Average number of associations per concepts (k) for 7 different ages (t) of the HAM-based semantic networks.

4. Semantic Network Analysis

We measure the clustering coefficient and path length (using the Floyd-Warshall algorithm [5]) of the semantic networks (for which the rewiring probability p is unknown) and compare their values with the case $p=1$ obtained by randomly rewiring the network while preserving the average number of associations per concept k. In case $L \approx L(1)$ and $C >> C(1)$ the human semantic network exhibits the small-world phenomenon. The small-worldliness, defined as $\mu = (C/L)/(C_{random}/L_{random})$ is a quantitative measure of "small-worldliness" [14], quantifies the combined maximisation of clustering and minimisation of path length. The graphs in figure 4 present the results of our small-world analysis. The graph in figure 4a shows the characteristic path length as a function of age for the semantic network (solid curve) and the random network (dashed curve). Figure 4b displays the clustering coefficient as a function of age for the semantic (solid curve) and random (dashed curve) networks with equal values of k and N. The triangles represent the values for the HAM network. In figure 4c, the results shown in figures 4a and b are combined yielding the small-worldliness as a function of age.

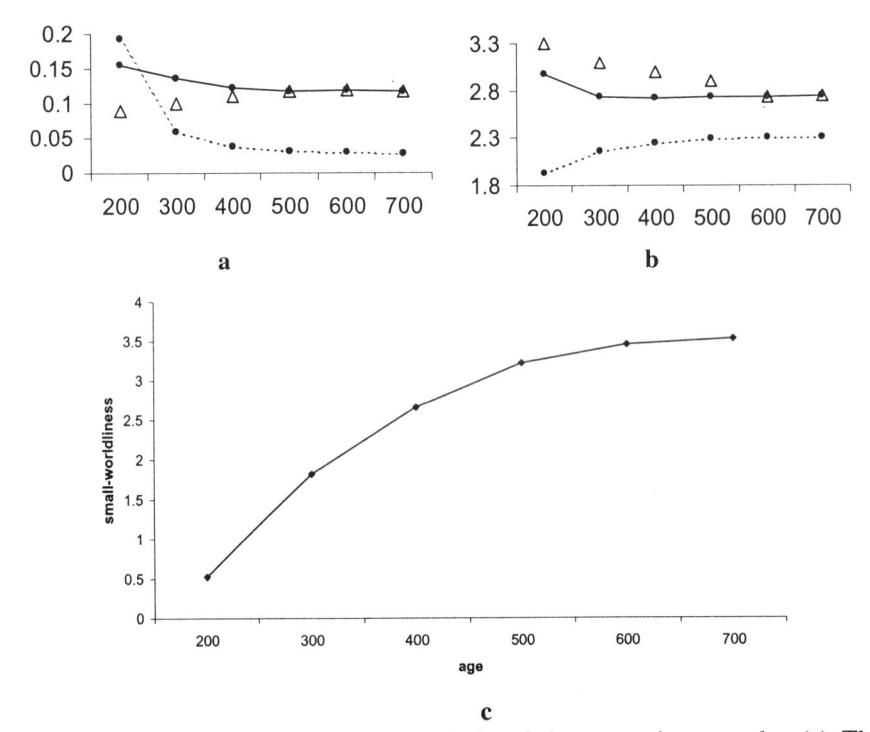

Figure 4: Results of the small-world analysis of the semantic networks. (a) The clustering coefficient C and (b) characteristic path length L as a function of age. (c) Small-worldliness as a function of age. Data points connected by solid and dashed curves in (a) and (b) represent values for semantic and random networks, respectively. The triangles represent the values obtained for the HAM network.

Evidently, during development the clustering coefficient stabilises at a value well above the "random" clustering coefficient, indicating the emergence and preservation of structure with the growth of the semantic network. Strikingly, the path length does not change with the growth of the network. The path length of the adult network, which is almost seven times larger than the network at age = 700, is L ≈ 2.7[2]. The small-worldliness increases with age and saturates at a maximum value of $\mu \approx 3.5$. Interestingly, this value falls within the range of the small-worldliness values of the neural network of *C.elegans* ($\mu = 4.7$) and the functional connectivity of macaque cortex ($\mu = 2.4$) [13][15]. Apparently, the small-worldliness of anatomical and functional networks match fairly well.

The development of the path length and clustering coefficient for the HAM-based semantic networks agrees reasonably well with these results. At early ages (i.e., 200-400) there is a difference with the results of the standard semantic network. Undoubtedly, this difference is due to the fact that no age-of-acquisition data were used to guide the development. At later ages, however, the values of L and C agree very well, again indicating that small-worldliness (being directly related to L and C) is insensitive to the underlying type of connectivity, i.e., physical (as in the standard semantic network) vs. functional (as in the HAM network).

5. Discussion and Conclusions

The main result obtained is that the characteristic path length remains virtually constant despite the expansion of the semantic network. The average distance of 2-3 links between arbitrary pairs of concepts which is maintained during development, is consistent with results of mediate-priming experiments which reveal priming effects for prime and target concepts separated by two associative links [10]. Although our binary-association semantic networks are only crude approximations of the human association strengths, preliminary experiments using real-valued associative strengths revealed a similar pattern of results as reported here, i.e., a path length of $L \approx 2.7$ for late stages of development. In our future work we intend to focus on the growth dynamics of semantic networks. Of particular interest are the implications of the preferential attachment of novel concepts to central concepts such as MONEY, SEX, and HOME with respect to the robustness and tolerance to damage of semantic networks [1].

We conclude by stating that human semantic networks tend to optimise small-worldliness by maintaining a constant path length and clustering coefficient, despite the increase in the number of concepts and that small-worldliness appears to be independent of the underlying type of connectivity.

---

[2] Estimated from the path lengths of 100 randomly selected concepts to all other concepts using Dijkstra's algorithm.

References

1. Albert, R., Jeong, H., & Barabási, A-L. (2000). Error and attack tolerance of complex networks. Nature, 406, 378-382.
2. Anderson, J.R. (1983). *The Architecture of Cognition.* Harvard University Press.
3. Buckingham, J. & Willshaw, D. (1992). Performance characteristics of the associative net. *Network: Computation in Neural Systems, 3*, 407-414.
4. Collins, A.M. & Loftus, E.F. (1975) A spreading-activation theory of semantic processing. *Psychological Review*, 82: 407-429.
5. Floyd, R.W. (1962). Algorithm 97: Shortest path. *Communications of the ACM*, 5: 345.
6. Gilhooly, K.J. & Logie, R.H. (1980). Age of acquisition, imagery, concreteness, familiarity and ambiguity measures for 1944 words. *Behaviour Research Methods and Instrumentation*, 12, 395-427.
7. Graham, B. & Willshaw, D. (1999). Probabilistic synaptic transmission in the associative net. *Neural Computation*, 11, 117-137.
8. Herzel, H. (1998). How to quantify 'small-world networks'? *Fractals*, 6(4): 301-303.
9. Kiss, G.R., Armstrong, C., Milroy, R., & Piper, J. (1973). An associative thesaurus of English and its computer analysis. In A.J. Aitken, R.W. Bailey, and N. Hamilton-Smith (Eds), *The Computer and Literary Studies,* University Press, Edinburgh.
10. McKoon, G. & Ratcliff, R. (1992). Spreading activation versus compound cue accounts of priming: Mediate priming revisited. *Journal of Experimental Psychology: Learning, Memory and Cognition*, 18, 1155-1172.
11. Newman, M.E.J. (1999). Small worlds: the structure of social networks. Working paper 99-12-080. Santa Fe Institute. (http://www.santafe.edu/sfi/publications/99wplist.html)
12. Postma, E.O., Wiesman, F., & Herik, H.J. van den (2000, in press) Small-World Semantic Networks. *Proceedings of the Twelfth Belgian-Dutch Conference on Artificial Intelligence.*
13. Stephan, K.E., Hilgetag, C-C, Burns, G.A.P.C., O'Neill, M.A., Young, M.P., Kötter, R. (2000). Computational analysis of functional connectivity between areas of primate cerebral cortex. *Philosophical Transactions of the Royal Society of London B*, 355, 111-126.
14. Walsh, T. (1999). Search in a small world. In T. Dean, editor, *Proceedings of the Sixteenth International Joint Conference on Artificial Intelligence (IJCAI'99), Volume 2*, pp. 1172-1177, Morgan Kaufmann Publishers, San Francisco.
15. Watts, D.J. & Strogatz, S.H. (1998). Collective dynamics of 'small-world' networks. *Nature*, 393: 440-442.

What is the Dimensionality of Human Semantic Space?

Will Lowe

Abstract

McDonald and Lowe [15] showed that cosines in a semantic space of several hundred dimensions reflect human priming results for a wide range of semantic and associatively related words [16, Exp.2]. Previously, Lowe [11, 10] argued that the intrinsic dimensionality of semantic space is much lower, and that high-dimensional structure can be effectively captured in just two dimensions as the surface of a neural map. This paper provides a replication of McDonald and Lowe's results in two dimensions using the Generative Topographic Mapping [2], a statistically motivated neural network architecture for topographic maps.

1. Semantic Space

Semantic space model have proved very successful models of semantic memory [9, 3]. A semantic space operationalizes the idea, initially introduced in distributional linguistics, that words are semantically similar to the extent that they behave in the same way in text; verbs that subcategorize for the same sorts of arguments and nouns that can be modified by the same kind of adjectives are to that extent semantically similar. Semantic space representations use vectors of surrounding word counts as a substitute for knowing the distributional profile and argument structures in advance [12]. Words with similar vector representations share more similar linguistic contexts and are thus more semantically similar.

The success of semantic space models in modeling psycholinguistic phenomena to a large extent vindicates the approach to meaning underlying distributional linguistics, but it also raises a number of technical questions relating to the 'non-parametric' nature of the approximations made when constructing a space. If argument structures are known in advance then it is obvious how to measure similarity: for any two words look in each argument slot and compare the sorts of words found there. Thinking of all words as having subcategorization preferences may not be immediately intuitive but is explicit in Link and Dependency grammar [8], and helps connect semantic space work to syntactic perspectives. When argument frames are not known we count all surrounding words up to a maximum window size. The question is then: how many words do we need to count to get a good approximation? Semantic spaces are vector spaces of typically high dimensionality, so the the generalization of this question addressed in this paper is: what is the appropriate dimensionality of human semantic space?

2. Previous Work

Landauer and Dumais [9] have argued that there is an optimal number of dimensions for psychological modeling, and that data of appropriate dimension should be generated by taking large numbers of word counts and subjecting them to linear dimensionality reduction. Their claim is that human semantic space is of fairly low dimension compared to the dimensionality of vector data from a semantic space model. They estimate [9, Fig.3] that the 300 directions of principal variance should be retained from the thousands generated by their model to optimally predict human behaviour.

The idea that the intrinsic dimensionality of data is typically lower than its observed dimensionality motivates Multidimensional Scaling (MDS) and Factor Analytic approaches in psychology. MDS performs a similar function to neural models of topographic map formation in computational neuroscience [5, 7]. With the development of the Generative Topographic Mapping (GTM; [2]), a non-linear extension of Factor Analysis, these models are now not usefully distinguished. Ritter and Kohonen [17] used a self-organizing map to project simple vector representations of word co-occurrence counts onto a two dimensional map surface. Similar approaches have been taken by Scholtes [18] and Lowe [11]. Implicit in this work is the assumption that co-occurrence data is inherently very low-dimensional.

3. Interpreting Semantic Space Models

There are two distinct ways to interpret semantic space models. A space may be a description of the lexical semantic structure of a language. In this sense, constructing a semantic space is a methodology for finding semantic structure in English using a distributional similarity measure. Alternatively a semantic space may be a theory of semantic representation in people. On the first interpretation when distances in a space correlate reliably with human performance on some psychologically interesting measure we can infer that there is sufficient statistical regularity in the linguistic environment to be able to perform the psychological task. However, for a computational approach to psychology this is only half the story; there needs to be another theory of how that information is represented in the mind/brain. Semantic spaces *can* be psychological models: e.g. we might assert that each person has vectors of lexical associations and performs similarity computations on them to determine semantic similarity. However, this interpretation is not the one being tested when semantic distances are correlated with a human experimental performance. When the Hyperspace Analogue to Language (HAL; [14]) or Latent Semantic Analysis [9] is compared to human data there is no analysis by subjects, only by items. This is true of most previous work in semantic space. There are no subjects; we are testing a theory about items.

The work reported below treats neural network models as subjects and thus doubles as a theory of semantic representation, as well as a theory of the intrinsic dimensionality of semantic space itself.

The next section briefly reviews earlier work modeling associative and multiple types of semantic priming using a semantic space of high-dimension. The next section shows how substantially the same results are obtained if the dimensionality of the data

305

is reduced dramatically. Finally we consider the implications of this work for estimating the dimensionality of human semantic space.

3.1 Experiment 1: Priming in High-dimensional Space

Moss and colleagues [16] showed that semantic priming occurs for a wide range of semantic relations, both with and without association. Stimulus words named members of the same taxonomic category (category coordinates), either natural objects or artifacts, or they were related functionally (functional items), through script or instrument relations. Moss and colleagues showed separate semantic and associative priming effects for all categories. They also showed that the semantic priming effect was greater in the presence of association (the associative boost).

McDonald and Lowe [15] demonstrated that Moss *et al.*'s results can be modeled in a high dimensional space. We briefly review the details of model construction and results using the latest version of the model for comparison to the low-dimensional results described below.

We constructed a semantic space from 100 million words of the British National Corpus (BNC), a balanced corpus of British English [4]. Word vectors were generated by passing a moving window through the corpus and collecting co-occurrence frequencies for 536 of the most reliable context words within a 10 word window either side of each stimulus item. Context words were the same as those used in previous work modeling graded and mediated priming [13]. The method for choosing reliable context words is described elsewhere [12, 13]. We used positive log odds-ratios to measure the amount of lexical association between each context word and each of the experimental stimuli. The odds ratio is well-known to be a measure of association that takes chance co-occurrence into account [1].

We also created vectors for 1000 filler words of frequency ranks 1000 to 2000 in the BNC (114.55 to 49.15 occurrences per million). Stimulus frequencies ranged from 0.02 to 1639.23 per million, with a median frequency of 33.95 per million. 114 unrelated primes were chosen randomly from the set of filler words

As in the original experiment we varied three factors: association (associated, non-associated), semantic type (category coordinate, functional relation) and relatedness (related, unrelated). Semantic subtypes were nested under Semantic Type.

For the purposes of modeling priming, the cosine between a prime and target should be inversely proportional to the corresponding reaction time. The size of a priming effect is calculated by subtracting the cosine between the unrelated prime and target from the cosine between the related prime and target. Cosines for the unrelated prime-target pairs was taken to be the cosine of the target with another prime in the same condition. Cosines are entered directly into analyses of variance.

3.2 Results

Cosines in the semantic space are shown in Table 1. There was a main effect of relatedness, $F(1, 108) = 314.922$, $p < .001$, and of association, $F(1, 108) = 16.433$, $p < .001$, replicating associative priming. There was also an interaction between association and relatedness, $F(1, 108) = 9.939$, $p < .01$. This replicates the associative boost.

	Associated			Non-associated		
	Related	Unrelated	U-R	Related	Unrelated	U-R
Cat. Coord.	0.553	0.173	**0.379**	0.458	0.163	**0.295**
Functional	0.547	0.181	**0.366**	0.394	0.168	**0.226**

Table 1: Cosines from the high-dimensional semantic space with unrelated primes chosen randomly from an alternative source. Bold face numbers are priming effects for each semantic type.

We then considered the associated and non-associated items. Semantic priming occurred in the associated condition, $F(1, 54) = 205.972, p < .001$ and in the non-associated condition, $F(1, 54) = 113.309, p < .001$. The priming effect for category coordinates appeared slightly larger than for functional items which is also consistent with the human results, but this difference was not significant.

Among the category coordinates semantically related pairs were more similar than unrelated pairs, $F(1, 52) = 165.567, p < 0.001$. There was also associative priming, $F(1, 52) = 5.607, p < 0.05$. There was no associative boost, $F(1, 52) = 2.623, p = .111$, and no other significant interactions. The associative boost did not occur due to a low level of similarity between the associated artifact targets and their related primes. The delicacy of the boost also follows human results.

Functional pairs also showed a semantic priming effect, $F(1, 52) = 154.771, p < .001$, a main effect of association, $F(1, 52) = 11.555, p < .01$, and a reliable associative boost, $F(1, 52) = 8.661, p < .01$. There was also a main effect of subtype, $F(1, 52) = 4.58, p < .05$, due to steadily decreasing amounts of similarity across subtypes relative to a stable baseline (associated related script > associated related instrument > non-associated related script > non-associated related instrument).

Detailed analyses for each semantic subtype are reported elsewhere [12].

3.3 Discussion

The space replicates Moss *et al.*'s finding that semantic priming occurs for a wide range of semantic categories, with and without association. We also see an associative boost.

3.4 Experiment 2: Priming in Low-dimensional Space

In this experiment we use 20 GTM networks as subjects. 20 GTM models [2] were trained on 1689 transformed semantic space vectors. The large number of irrelevant word vectors were intended give each network a better idea of the overall shape of semantic space, rather than just a small set of words of interest. 1000 words were the filler from Experiment 1, 224 were from Moss's materials, and the rest were experimental stimuli from 5 other priming experiments. Results from the latter are reported elsewhere [13, 12].

The entire augmented set of 534-dimensional semantic space vectors was then transformed linearly into 50 dimensions by 20 independently generated stochastic matrices [6], one for each GTM model. Each GTM was initialized with random parameters and saw a distinct random mapping of the semantic space vectors.

Ideally each network would have been trained on vectors generated by sampling from a much larger corpus. However, this is computationally extremely demanding, even were such a corpus available. Using newsgroups is a possible next step in this research.

3.5 Random Mapping

Neural networks are often criticized for relying crucially on intelligent prior tranformations of the data. Consequently although principal component analysis of the vectors would also reduce dimensionality to tractable levels, it would also represent a substantial modeling assumption that is not obviously motivated from a neural perspective. Random mapping reduces the dimensionality of the data to a level that is tractable for reasonable network training times while making the fewest possible assumptions about the nature of the input, save that it derives from vectors of lexical associations.

Random mapping also introduces variability into the input data that ensures that no net trains on the same data set. The psychological interpretation of this process is that networks are subjects that have been exposed to roughly the same language data but with significant amounts of noise. We then test the claim that representing this information using topographic maps generates accurate predictions about priming.

Any specific random mapping for the semantic space vectors into d-dimensional space is a $534 \times d$ matrix, $\mathbf{R}$, with i.i.d. zero mean Normally distributed elements. Each column of $\mathbf{R}$ is normalized to unit length to create a non-orthogonal basis. To give an idea of how much structure is preserved in a random mapping, Kaski [6] has shown that the inner product between two low dimensional projections $\mathbf{a}_1 = \mathbf{R}\mathbf{h}_1$ and $\mathbf{a}_2 = \mathbf{R}\mathbf{h}_2$ of high-dimensional unit length vectors $\mathbf{h}_1$ and $\mathbf{h}_2$ is

$$\mathbf{a}_1^T \mathbf{a}_2 = \mathbf{h}_1^T \mathbf{h}_2 + \delta \qquad (1)$$

where δ is approximately $\mathcal{N}(0, 2/d)$. This result essentially defines an error bar on similarity estimates in the low-dimensional space relative to their 'real' values in high-dimensions[1]. It is intuitively surprising that similarities are (on average) so well preserved by a completely random mapping; this phenomena alone deserves further attention.

3.6 Generative Topographic Mapping

The GTM is a non-linear extension of Factor Analysis with strong similarities to the Self-organizing map. It attempts to build a generative model of the variance structure in

---
[1] See Kaski's paper for error estimates for non-normalized high-dimensional vectors, and a detailed derivation.

data points **a** on the assumption that they are generated by a smooth non-linear mapping from a two-dimensional manifold **x**. The GTM thus explicitly assumes that the intrinsic dimensionality of the data is two-dimensional, and that off-manifold structure is simply noise. Clearly this is an extremely strong and falsifiable assumption to make about even 50-dimensional data. Since the GTM defines a mapping from a low-dimensional latent space into the data space, it is straightforward to invert this mapping for any data point to obtain $p(\mathbf{x} \mid \mathbf{a})$. The mean of this distribution is a point estimate of the point in latent space that is most likely to have generated **a**. In this respect the model is used similarly to Factor Analysis.

To make specific predictions about priming effects, we compute posterior means as described above for each related prime, unrelated prime and target vector. Each mean is a two element vector that describes a point in two-dimensional space. We take cosine measures in this reduced space, just as in the high dimensional model.

3.7 Results

	Associated			Non-associated		
	Related	Unrelated	U-R	Related	Unrelated	U-R
Cat. Coord.	0.689	-0.151	**0.839**	0.485	-0.250	**0.735**
Functional	0.690	-0.113	**0.803**	0.440	-0.134	**0.574**

Table 2: Mean cosine similarity measures from the networks on Moss *et al.*'s data with independently chosen unrelated baseline. Bold numbers are priming effects for each semantic category, with and without association.

Mean similarity measures are shown in Table 2. There was a main effect of relatedness, $F_1(1, 19) = 2391.276, p < .001$, $F_2(1, 108) = 195.478, p < .001$. There was also a reliable effect of association, $F_1(1, 19) = 94.703, p < .001$, $F_2(1, 108) = 7.703, p < .01$, replicating the associative priming effect. The associative boost was significant by subjects, $F_1(1, 19) = 21.316, p < .001$, $F_2(1, 108) = 1.949, p = .166$.

There were main effects of semantic relatedness in both associated and non-associated conditions, $F_1(1, 19) = 2358.761, p < .001$, $F_2(1, 54) = 103.68, p < .001$ and $F_1(1, 19) = 643.53, p < .001$, $F_2(1, 54) = 92.124, p < .001$.

The category coordinates showed a semantic priming effect, $F_1(1, 19) = 1754.725, p < .001$, $F_2(1, 52) = 104.371, p < 0.001$, and an associative priming effect, $F_1(1, 19) = 44.774, p < 0.001$, $F_2(1, 52) = 4.647, p < .05$. No associative boost appeared in either analysis due to the surprisingly large priming effect for non-associated artifacts.

Semantic priming was present in the functional relations, $F_1(1, 19) = 892.731$, $p < .001$, $F_2(1, 52) = 96.693$, $p < .001$. Associative priming was significant across subjects, $F_1(1, 19) = 47.997$, $p < .001$, and marginally significant in the items analysis, $F_2(1, 52) = 3.403$, $p = .07$, The associative boost was significant for subjects, $F_1(1, 19) = 51.055$, $p < .001$, and approached significance in the items analysis, $F_2(1, 52) = 3.168$, $p = 0.081$.

Separate analyses for each semantic subtype are reported elsewhere [12].

3.8 Discussion

Table 2 shows that low-dimensional simulation gave results very similar to those found in the original experiment, and in its replication in high-dimensions. As is typically the case in dimensionality reduction, related items become more similar. This can be seen in Table 2 where priming effects are much larger, whereas the unrelated baseline is essentially unchanged. The reduction also brings out previously weak trends in the data, e.g. the fact that the non-associated semantic priming effects is much stronger for category coordinates, and that instrument relations do not require association to prime strongly.

4. Conclusion

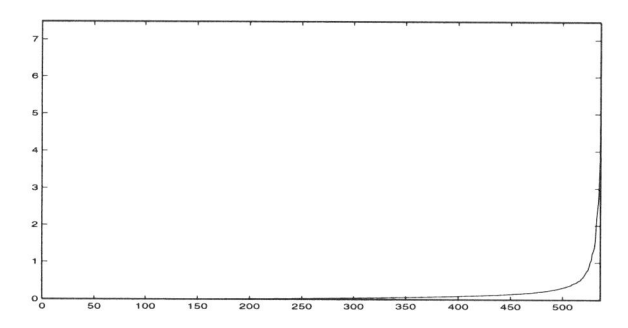

Figure 1: Eigenvalues of the covariance matrix for the high-dimensional semantic space vectors, ordered by size.

Experiment 2 also suggests that the the intrinsic dimensionality of the semantic space data is quite low. Another complementary way to see this is to look at linear measures of the variance structure. Figure 1 shows the eigenvalues of the covariance matrix for the high-dimensional data, sorted by size. It is clear from the figure that the majority of the data variance extends in only a few directions. The first handful of values contain more than 80% of the total variance. Another way to understand this is to consider linear reconstructions of the data: only a handful of real numbers representing data projections onto the principal eigenvectors would be necessary to reconstruct this data to 80% accuracy.

Looking at orthogonal directions of variance is a useful baseline for understanding the success of the GTM because the intrinsic dimensionality of the data can only be smaller than a linear estimate would suggest. On the other hand the procedure is only approximate since the interpretation of eigenvalue structure in terms of variance component holds only for jointly Normally distributed data. This assumption is unlikely to hold exactly for semantic space vector elements.

In any case it is interesting to compare this to Landauer and Dumais' claim that several hundred dimensions are necessary for semantic space. It is possible that the tasks used in that work require significantly different dimensionality spaces than for even fairly detailed priming studies. Replicating the Landauer and Dumais' tasks in the current framework is current work. But this is clearly not the case for this data. We have also shown elsewhere that many other priming results can be captured in very low-dimensional models [12]. These studies support the claim that the dimensionality of human semantic space may be very low indeed.

Acknowledgements

Thanks to Daniel Dennett for supporting this work, and to two anonymous reviewers for helpful comments.

References

1. Agresti, A. (1990). *Categorical Data Analysis*. John Wiley and Sons.
2. Bishop, C. M., Svensén, M., and Williams, C. K. I. (1998). GTM: The generative topographic mapping. *Neural Computation*, 10(1):215–235.
3. Burgess, C., Livesay, K., and Lund, K. (1998). Explorations in context space: Words, sentences, discourse. *Discourse Processes*, (25):211–257.
4. Burnage, G. and Dunlop, D. (1992). Encoding the British National Corpus. In *Papers from the Thirteenth International Conference on English Language Research on Computerized Corpora*.
5. Goodhill, G. J. and Willshaw, D. J. (1994). Elastic net model of ocular dominance: Overall stripe pattern and monocular deprivation. *Neural Computation*, 6:615–621.
6. Kaski, S. (1989). Dimensionality reduction by random mapping: Fast similarity computation for clustering. In *Proceedings of the International Joint Conference on Neural Networks*, pages 413–418.
7. Kohonen, T. (1995). *Self-organizing maps*. Springer, Berlin.
8. Lafferty, J., Sleator, D., and Temperley, D. (1992). Grammatical trigrams: A probabilistic model of link grammar. Technical report, CMU School of Computer Science.
9. Landauer, T. K. and Dumais, S. T. (1997). A solution to Plato's problem: the latent semantic analysis theory of induction and representation of knowledge. *Psychological Review*, (104):211–240.

10. Lowe, W. (1997a). Meaning and the mental lexicon. In *Proceedings of the 15th International Joint Conference on Artificial Intelligence*, pages 1092–1097, San Francisco. Morgan Kaufmann.
11. Lowe, W. (1997b). Semantic representation and priming in a self-organizing lexicon. In Bullinaria, J. A., Glasspool, D. W., and Houghton, G., editors, *Proceedings of the Fourth Neural Computation and Psychology Workshop: Connectionist Representations*, pages 227–239, London. Springer-Verlag.
12. Lowe, W. (2000). *Topographic Maps of Semantic Space*. PhD thesis, Institute for Adaptive and Neural Computation, Division of Informatics, Edinburgh University.
13. Lowe, W. and McDonald, S. (2000). The direct route: Mediated priming in semantic space. In Gernsbacher, M. A. and Derry, S. D., editors, *Proceedings of the 22nd Annual Meeting of the Cognitive Science Society*, pages 675–680, New Jersey. Lawrence Erlbaum Associates.
14. Lund, K., Burgess, C., and Atchley, R. A. (1995). Semantic and associative priming in high-dimensional semantic space. In *Proceedings of the 17th Annual Conference of the Cognitive Science Society*, pages 660–665. Mahwah, NJ: Lawrence Erlbaum Associates.
15. McDonald, S. and Lowe, W. (1998). Modelling functional priming and the associative boost. In Gernsbacher, M. A. and Derry, S. D., editors, *Proceedings of the 20th Annual Meeting of the Cognitive Science Society*, pages 675–680, New Jersey. Lawrence Erlbaum Associates.
16. Moss, H. E., Ostrin, R. K., Tyler, L. K., and Marslen-Wilson, W. D. (1995). Accessing different types of lexical semantic information: Evidence from priming. *Journal of Experimental Psychology: Learning, Memory and Cognition*, (21):863–883.
17. Ritter, H. and Kohonen, T. (1989). Self-organizing semantic maps. *Biological Cybernetics*, (61):241–254.
18. Scholtes, J. C. (1991). Neural nets and their relevance for information retrieval. Technical Report CL-91-02, University of Amsterdam, Institute for Language, Logic and Information, Department of Computational Linguistics.

Author Index

Ans, B.	13	Lőrincz, A.		73
Barone, R.	43	Lowe, W.		303
Bartos, P.	143	Molenaar, P. C. M.		83, 197
Bernarda-Ludermir, T.	63	Monaghan, P.		3
Bharucha, J. J.	175	Page, M.		105
Bigand, E.	175	Parisi, D.		113, 253
Bryson, J.	53	Postma, E.		293
Bullinaria, J. A.	231, 273	Raijmakers, M. E. J.		83, 197
Calabretta, R.	253	Roebroeck, A.		293
Cleeremans, A.	185	Rousset, S.		13
Cooper, R.	133	Sanchez, E.		153
De Mazière, P. A.	33	Schlesinger, M.		113
Di Ferdinando, A.	253	Shapiro, J. L.		43
Dickins, T. E.	263	Shillcock, R.		3
Diniz-Filho, J.	63	Sougné, J.		23
Done, J.	163	Stein, L. A.		53
Van Overwalle, F.	219	Szatmáry, B.		73
Frank, R. J.	163	Szirtes, G.		73
French, R. M.	13, 209	Takács, B.		73
Gale, T. M.	163	Teuscher, C.		153
Glasspool, D.	133	Tillmann, B.		175
Hartley, S. J.	95	Timmermans, B.		185, 219
Hüning, H.	243	Van Hulle, M.M.		33
Labiouse, C. L.	209	Visser, I.		197
Lacroix, J.	293	Wearden, J.		43
Le Voi, M.	143	Westermann, G.		123
Levy, J. P.	263, 273	Wichert, A.		283

Subject Index

Action selection problem	56
Action selection	56-60, 133, 134, 137, 139, 141
ACT-R	55
Adrenergic	63
Affordance	133, 134, 141
Agent coordination problem	56
Agrammatic aphasia	124, 129, 130
ALCOVE	107, 145
Amnesia	13, 14, 16
Amygdala	58, 63, 64
Anomia	16
Aphasia	123, 129
Architectural modules	54, 55, 57
ART	83, 84, 88, 89
Artificial grammar learning	186-188, 193
Associative memory	284-286, 289-291
Attractor network	217
Augmentation	219-221, 224-227
Autoassociator	15, 185, 192, 194, 219, 222, 224, 227
Autocatalytic set	244-246
Autoencoder	209-211, 216, 217
Backpropagation	7, 15, 16, 101, 106, 145, 189, 190, 197, 199, 212, 243, 254-256, 258, 260
Backtracking	283, 289
Backward competition	219-222, 226, 227
Backward revaluation hypothesis	221, 223, 226
Baldwin effect	231-242, 267, 268
Basal ganglia	58
Bayesian maximum-likelihood decision	108, 111
Behaviour arbitration problem	56, 58
Behaviour Based Artificial Intelligence (BBAI)	54-56, 58
Behaviour Oriented Design (BOD)	59
Biconditional Grammars	185, 188, 189, 194
Bifurcation	83-90
Binding	141
Blocking of conditioning	150

Catalytic networks	243-251
Catastrophic forgetting	15, 84, 88, 216
Catastrophic interference	13-15, 17, 18, 30, 31, 216
Categorisation	95, 99, 100, 163, 165, 167, 168, 279, 281
Category learning	143-145
Category-specific deficits	13, 14, 16, 163
Causal attribution	209
Causal induction (reasoning)	219, 221
Cell assemblies	283, 285
Chaining rule problem	27, 28
Chaos theory	34
Cholinergic	63, 64
Clean-up units	17, 165
Coarse-/fine-coding	3, 10
Coarse-coding	5, 6, 9, 10, 168-170
COGENT	135
Cognitive development	84, 90, 123, 124, 268
Cognitive dissonance	209
Competitive learning rule	246
Complete agents	53-60
Conditioned inhibition	227
Conditioned stimuli	46, 65-67, 70
Configural cue model	143, 145, 148
Constructivism	123-131, 265
Constructivist neural network	123-131
Content Addressable Memory (CAM)	83, 90
Contention Scheduling	134, 141
Control systems	115, 231, 234, 242
Crossover	117

DALR	145
Delay learning	23-31
Delayed rewards	43
Delta learning rule	106, 136, 150, 209, 221, 224
Dendritic tree	79
Depth-from-motion processing	33, 34
Development	83, 84, 90, 95-97, 100-102, 113-116, 121, 123-126, 131, 263-268, 293, 297-301
Discounting	219-221, 224-227
Discrimination learning	164, 167
Dissociations	10, 111, 123, 129-131, 163

Dream sleep	15
Dual-network	13-18, 21
Dynamical hypothesis	263, 265

Econets	115-120
Eigenvalue	86, 105, 109, 245, 309, 310
Eigenvector	216, 310
English past tense	124, 128-130
Entorhinal cortex	79
Entropy	78, 202-205
Epigenetic theories	83, 84, 90
Episodic memory	17, 21, 57, 186
Euclidean distances	18, 101, 276
Event-related potentials	181
Evolutionary psychology	264
Exact ART	83, 84, 88-90
Excitatory Post Synaptic Potential (EPSP)	25, 26
Exemplar-based inference	210, 211, 213, 217
Explicit learning	186, 241

Face expressions	105-111
Fahlman correction	212
Feature based representations	95-97
Feature correlations	102
Feedforward networks	74, 79, 192, 211, 221
Finite state grammar	175, 186, 194, 198, 202
Finite-state automaton	190, 197-199
Firing rate	23, 65, 67, 68
Floyd-Warshall algorithm	300
fMRI	33-37, 39, 40
Fold-bifurcation	86-88, 90
Forward competition	219-222
Fovea	4

Gaussian activation functions (See also RBF)	126
General Linear Model (GLM)	33, 38, 39
Generative network	73, 74, 78, 79
Generative Topographic Mapping (GTM)	303, 304, 306-308, 310
Genetic algorithm	113, 116, 153, 159, 255, 257-261
Gtree algorithm	101

Hand-eye coordination	113
Harmonic priming	177, 181
Hebbian learning	24, 26, 28, 31, 67, 75, 299
Hemisphere	3-6, 9, 129
Hetero-Associative Memory (HAM)	298, 299
Hidden Markov models (HMM)	197, 200, 201
Hill climbing	290
Hippocampus	13, 14, 16-18, 55, 78
Homunculus fallacy	73, 78
Hopf-bifurcation	86, 88
Hopfield network	15, 216
Hypothesis testing	186, 188
Illusory correlation	209
Implicit learning	175, 177-179, 182, 185-195, 199, 204
Impression formation	209, 217
Independent Component Analyses (ICA)	73-79
Induction	90, 219, 221
INFERNET	24-26, 30
Inferotemporal cortex	164
Inhibitory Post Synaptic Potential (IPSP)	25
Input Variable Selection (IVS)	33-35, 40
Interactive activation model	133, 134, 136, 178, 251
Jacobian matrix	86
K-means clustering	198, 199, 205
Latent semantic analysis	274, 304
Lateral connections	71, 83, 85, 86, 90
Lateralisation	4
Laws of effect	133, 134
Laws of exercise	133, 134
Leaky integrator	232
Learned inattention	150, 151
Learning by trial-and-error	115, 116
Leave-one-out crossvalidation regime	108, 110
Left Hemisphere (LH)	3-10, 129
Leibniz's Law	268
Lempel-Ziv entropy	203-205
Lexical semantics	273, 274, 304

Limbic system	58
Line-bisection	3, 7-9
Localist representation	105, 106, 111, 165, 168, 243, 251
Long Term Depression (LTD)	26, 29, 30
Long term memory	16, 18, 59, 60, 73, 78, 291
Long Term Potentiation (LTP)	26, 28, 29
Lorenz-attractor	34
Luce choice function	149
Luce ratio	190, 191, 193
Maximum likelihood	108-111, 204, 205
Memory	13, 14, 16-18, 21, 46, 53, 57, 59, 60, 63, 73, 78, 79, 95-97, 102, 181, 182, 186, 211, 283-286, 289-291, 303
Mixtures of experts	148
Modular neural networks	63, 71, 111, 165, 253-261
Modularity	13, 53-58, 63, 78, 253, 254
Molecular evolution	243-244
Multi-agent systems	56
Multidimensional scaling	304
MUSACT	178, 180
Music cognition	175-182
Mutation	116, 231-233, 235, 241, 242, 253, 257-261
NAND gate	155, 158
Natural kind category	96
Nature-nurture debate	231, 234, 242
Nearest-neighbour	106, 107, 111
Neglect dyslexia	3, 4, 9-11
Neo-constructivist	265
Neocortex	13, 14, 16, 21, 78
Neural clock	43
Neuro-imaging	33
Neurotrophins	124
Noradrenergic	64
Object recognition	163-167, 169
OR gates	158
Orbitofrontal cortex	63
Parietal cortex	5, 6, 8
Parietal lobule	4

Pattern completion		216, 219
Pattern recognition		13
Perceptron		165
Person perception		209-217
PET		33
Piaget's stage theory		84, 90
Pictorial reasoning		283, 285
Pitch perception		178
Plan		56-59, 284, 285, 289
Planning		55, 57-59, 161, 283
Post Synaptic Potential (PSP)		24-26
Post-natal maturation		83, 86
Prereaches		114, 115
Primary visual cortex		75, 78, 125, 164
Priming		64, 177, 181, 273, 276, 301, 303-310
Principal Component Analysis (PCA)		33, 34, 37, 38, 75, 105, 109-111, 216, 274, 307
Prismatic adaptation		120
Prismatic displacement		119, 120
Problem solving		265, 283-291
Procaccia & Grassberger theorem		35
Production system		283-285, 291
Proprioception		113-121
Pseudopatterns		13-20

Quickprop		126

Radial Basis Functions (RBFs)		5, 6, 106, 109
Random boolean networks		153, 155, 161
Reaching		113-121
Reactive planning		57-59
Reconstruction networks		74
Recurrent connections		155, 199, 200
Recurrent network		79, 185, 189, 197, 199, 209-211, 222-224, 227
Refractory period		25, 26
Reinforcement learning		43, 45, 46, 50, 121, 133, 134, 142
REM sleep		14, 21
Right Hemisphere (RH)		3-10
Rule-based knowledge		186
Rule-based vs.memory-based learning		194

Scalar expectancy theory		45

Selective attention	144, 145
Self Organizing Map (SOM)	163-169, 178-180, 304, 307
Self-organisation	243, 244, 266
Self-organizing network	109, 111, 175, 176, 181
Semantic memory	95, 96, 303
Semantic networks	293-301
Semantic priming	273, 305, 306, 308, 309
Semantic similarity	274, 276, 304
Semantic space	303-310
Sensory functional theory	163
Sexual reproduction	117
Short-term memory	60, 182, 223, 283, 285
Short-term memory span	182
Shunting Neural Network (SNN)	83-90
Sigma-pi units	243
Similarity judgments	181, 182
Simple Recurrent Network (SRN)	185, 189-191, 194, 197, 199-205
Skill modules	54-59
Small-world semantic network	293-301
SOAR	55, 283
Sparse coding	18, 30, 31, 78, 165, 167, 179, 180, 216, 296
Sparseness of connections	30, 31
Speech recognition	178, 200
Spiking neurons	23, 24, 43, 46, 47, 50
Stability-plasticity dilemma	216
Statistical Parametric Mapping (SPM)	33, 34, 38-41
Stereotypes	209-217
Supervisory attentional system	141
Synchrony	23
Synfire chains	23, 24, 29-31

Tactile perception	113
Talairarch-Tournoux space	39
Temporal independent components analyses	75-79
Time intervals	43, 44, 48, 49
Tonal music	175, 176
Topographic map	303, 307
Turing neural networks	159
Two-armed bandit problem	19, 20

Unconditioned stimulus	46, 65-67, 70
Unorganized machines of Turing	153-161
U-shaped learning	123, 127, 131

Ventral visual pathway		164, 169
Vision		113-121
Visual agnosia		163
Visual Object Recognition (VOR)		163, 164, 166, 167, 169
Visuospatial neglect		3-5, 8, 9

Wason selection task		264
What and where task		253-262
What pathway where pathway		286
Winner-takes-all		85, 87-90, 246
Word frequency		273, 275
Word recognition		3, 4, 178, 251
Working memory		59, 60, 283

Zero mean martingale increment processes		200